THE ULTIMATE SITE GUIDE TO SCARCER BRITISH BIRDS

LEE G R EVANS

To
Chris,

With best wishes,

Lee G R Evans

Illustrated by Steven Brown, Mike Langman, Phil Jones, Richard Millington, Paul Semmens, James Hanlon, Andrew Mackay, Antony Disley and Pete Dennis.

This publication is a culmination of knowledge gleaned during my 28 years of involvement with British birds. It is dedicated to my family and friends, whose loyalty and time are more important to me than ever before.

Love as always

Lee

THE ULTIMATE SITE GUIDE TO SCARCER BRITISH BIRDS

LEE G R EVANS

LGRE PRODUCTIONS INC.

LGRE PRODUCTIONS INC.

First published in 1996 by LGRE Productions Inc., a subsidiary of Lee Evans Enterprises Ltd, 8 Sandycroft Road, Little Chalfont, Amersham, Bucks, HP6 6QL.
Printed by Lake - Shore Graphics Tel: 0115 956 1178

ISBN 1 898918 15 5

Evans, Lee G R
The Ultimate Site Guide to British Birds

CONTENTS

Acknowledgements	6
Introduction	7
Birdline Telephone Numbers	8
Code of conduct for birdwatchers	9
Species Accounts	10
Migration and Rarity Hotspots	256
Checklist of the Birds of Britain	282
Schedule 1 species	289
List of Illustrations	290
Index to Species	291
Index to Sites	292

ACKNOWLEDGEMENTS

The inspiration for this guide was provided by the many hundreds of enthusiastic, keen and often young novice birdwatchers who have sought my advice over the last twenty or so years on where, when and how to see the more difficult to find species of British birds. I hope that the information that I provided them enhanced their knowledge and encouraged their interest, and I am rewarded by the fact that many of these observers are now at the forefront of British ornithology and are adding immensely to the knowledge of the avifauna in their local areas.

I am particularly indebted to Ted Abraham, Richard Allison, Simon Cox, Nigel Driver, Richard Fairbank, Paul Harris, John Hewitt, Dave Holman, Gerald Jobson, Adrian Kettle, John McLoughlin, Lee Marshall, Angus Murray, Simon Nicholls, Barry Nightingale, Dave Odell, Bob Proctor, Stuart Read, Brett Richards, Graham Speight, Keith Vinicombe, Steve Whitehouse, Barry Wright and Steve Young, for providing essential information on local sites of interest.

I am also extremely grateful to Dave Holman and Richard Millington for commenting on sections of the draft manuscript and for making valuable contributions to my knowledge.

I particularly wish to thank the various artists that provided the fine vignettes that adorn this book, including Phil Jones, Steven Brown, Mike Langman, Andrew Mackay, Pete Dennis, James Hanlon and Paul Semmens.

Additional thanks are due to Phil Jones for his excellent portrayal of the Scottish White-tailed Eagles on the front cover.

Lastly, Derek Turner expertly exercised his considerable editorial skills on the entire text and a special debt of gratitude extends to him.

Lee G R Evans April 1996

INTRODUCTION

The aim of this publication is to guide birdwatchers to the most reliable sites in the British Isles in which to see the more sought-after species of birds and is the culmination of 28 years of involvement by the author in finding and seeing these birds. It includes details on how, when and where to see 105 of the country's most difficult species but does not attempt to address the whereabouts of the most common or, in fact, to detail sites of general interest - this type of information has previously been published by Nigel Redman and Simon Harrap in their excellent *Birdwatching in Britain: a site by site guide*.

Species Accounts

Specific instructions on how to see 105 species of bird are detailed in this publication, with each account concentrating on a number of best locations in which to find them. Following a brief summary on distribution, status, pattern of occurrence and habitat preference, a list of prime sites is offered with detailed instructions on how to get there. Furthermore, additional information is often applied to help one find and identify the species once one has arrived at the site. It is not, however, intended to be a guide for prospective egg-collectors or those intent in causing harm or destruction in any form. Although the treatment of each species is comprehensive, it is not necessarily exhaustive, some sites being omitted due to security precautions. The aim of this guide is to detail sites in which there is a high degree of opportunity of finding that particular species, but, with some scarcer migrants, it is not possible to offer any firm sites. In the case of some rare breeding birds, particularly grebes and raptors, it is not possible to divulge information which could cause disturbance or jeopardise their nesting success. Because birdwatching, and particularly the interest in rare birds is a growing phenomenon in Britain, it is necessary to establish some strict guidelines in the publication of sites and the following principles have been generally adhered to:

Is increased disturbance to the site going to have detrimental effects to the species or its habitat?

Is there public access to the site?

Is the site wardened so that visitors can be adequately controlled and monitored?

Of course, publication of sites of any kind will not please some individuals or organisations, but it is my belief that ordinary birdwatchers should be given the opportunity to see such birds, and their actual presence at a site may well deter less well-meaning individuals. Obviously, the information offered here involves some element of risk to the birds and their habitat, but in my opinion this is outweighed by the benefits that a wider audience may bring. My only hope is that each of you behaves responsibly at each site and adheres to the Code of Conduct for Birdwatchers (see page 9) Remember, if you do see anything suspicious or witness acts of irresponsibility, do not hesitate to contact the nearest police station or RSPB regional office.

Whilst every effort has been made to ensure that the site accounts are as accurate and up-to-date as possible, there will inevitably be errors or misinformation. I would therefore be grateful for any corrections to be bought to my attention. Furthermore, additional information on sites or status updates at particular locations will be particularly welcomed. Please write to: Lee G R Evans, 8 Sandycroft Road, Little Chalfont, Amersham, Bucks, HP6 6QL.

Apart from the Landranger series of Ordnance Survey 1: 50 000 maps, an essential companion to this publication is the 1½ miles to 1 inch *Philip's Great Road Atlas of Britain*, published in June 1996.

The following site guides contain essential information on general birdwatching sites in Britain and should be used in conjunction with this book:

Birdwatching in Britain - A site by site guide by Nigel Redman and Simon Harrap.
Where to Watch Birds in Scotland by Mike Madders and Julia Welstead
Where to Watch Birds in Wales by David Saunders.
Where to Watch Birds in Cumbria, Lancashire & Cheshire by Guest & Hutcheson.
Where to Watch Birds in Northeast England by Dave Britton and John Day.
Where to Watch Birds in Yorkshire and North Humberside by John Mather.
Where to Watch Birds in East Anglia by Peter & Margaret Clarke.
Where to Watch Birds in the West Midlands by Graham Harrison and Jack Sankey.
Where to Watch Birds in Bedfordshire, Berkshire, Buckinghamshire, Hertfordshire & Oxfordshire by Brian Clews, Andrew Heryet and Paul Trodd.
Where to Watch Birds in Kent, Surrey and Sussex by Don Taylor *et al.*
Where to Watch Birds in Somerset, Avon, Gloucestershire & Wiltshire by Ken Hall and John Govett.
Where to Watch Birds in Dorset, Hampshire & the Isle of Wight by George Green and Martin Cade.
Where to Watch Birds in Devon & Cornwall by David Norman and Vic Tucker.

Birdline Telephone Numbers

The following telephone information services offer up-to-the-minute details of rare bird sightings in the UK :-

0891 700 222 *Birdline*
0891 700 234 *Birdline Scotland*
0891 700 249 *Birdline Northwest*
0891 700 246 *Birdline Northeast*
0891 700 248 *Birdline Wales*
0891 700 247 *Birdline Midlands*
0891 700 245 *Birdline East Anglia*
0891 700 240 *Birdline Southeast*
0891 700 241 *Birdline Southwest*
0891 700 243 *Birdline Scilly*

These premium rate lines are charged at 49p per minute (peak) and 39p per minute at all other times. Prices include VAT.

Code of Conduct for Birdwatchers

Today's birdwatchers are a powerful force for nature conservation. The number of us interested in birds rises annually and it is vital that we take seriously our responsibility to avoid any harm to birds. We must also present a responsible image to non-birdwatchers who may be affected by our activities and particularly those on whose sympathy and support the future of birds may rest. There are ten points to bear in mind:

1.The welfare of birds must always come first, whether your particular interest is photography, ringing, sound recording, scientific study or simply watching.

2.The habitat must be protected - it is vital that activities do not cause damage to a site.

3.Keep disturbance to a minimum. Never disturb nesting birds and avoid stress to tired migrants or victims of severe weather conditions.

4.When you find a rare breeding bird think carefully about whom you tell. If you feel that protection is necessary, consult the appropriate RSPB Regional Office or Species Protection Department at The Lodge, Sandy (Beds). Remember that disturbance at or near the nest of Schedule 1 species is a criminal offence.

5.Do not harass tired migrants or vagrants. Act responsibly in a crowd and discourage flushing.

6.Abide by the bird protection laws at all times.

7.Respect the rights of landowners and do not enter land without permission. Comply with permit schemes and if you are leading a group, always give advance notice of your visit. Always obey the Country Code.

8.Always respect the rights of other people in the countryside. Have proper consideration for other birdwatchers and do not interfere with the activities of members of the public enjoying the countryside.

9.If possible, make your records available to the local bird recorder so that our knowledge of birds and their occurrences can continue to advance.

10.Always behave abroad as you would in Britain as well behaved birdwatchers can be essential ambassadors for bird protection.

The Site Accounts

Each site for a particular species of scarce bird is graded in its reliability and this is indicated by an asterisk varying in size depending upon the chances of success.

✱ - a site that under suitable conditions and at the correct time of year will guarantee success.

❶ - a site that offers a 60% chance or more of success.

❷ - a site that offers a 40% chance of connection.

A site with no symbol is one which can offer sightings but with only the minimum chance of success.

Black-throated Diver *Gavia arctica*

Breeding Black-throated Divers are restricted to lochans in northern Scotland, especially in Sutherland, Wester Ross and Lewis, with a population estimate of at least 151 pairs. The vast majority winter off Scottish coasts and in south Cornwall, and occur rarely inland, with 20 or less records annually.

The Black-throated Diver can be encountered off virtually any coastal site in winter but in general is difficult to locate. Between mid September and late November, small numbers of adults and moderate numbers of juveniles congregate in the southern North Sea and are best observed off the north Norfolk coast and in Yorkshire. The best views are obtained however, as with all species of diver and grebe, when individuals wander or are storm-blown inland. The regional birdlines will be most helpful in this case.

Prime sites

The following sites are the most reliable in Britain for getting unequivocal and good views of this species in winter. Although small numbers are present offshore in most coastal waters throughout the winter, sightings are sporadic and by no means guaranteed. The sites offered below, if visited under suitable conditions, will provide virtually guaranteed telescopic views between November and April.

✱Gerrans and Veryan Bays - Cornwall

Up to eight per cent of the UK wintering population can be found off the south coast of Cornwall during November to May, predominantly in the Dodman Point to Carrick Roads area. The best site is the south-facing Gerrans Bay, running from Portscatho in the west to Nare Head in the east, and often holding up to 70 Black-throated Divers.

The best vantage points are at Pendower Beach (park and watch from viewpoints beside the hotel) and from along the coastal public footpath to the south of Portscatho village or from the National Trust car park above the village. Closer views can sometimes be had by watching from the concrete bunker at Padnavadan, accessed from the public footpath leading from the minor road to Rosevine. The divers are best observed at high tide and on a day with slack or northerly winds creating calm sea conditions. Smaller numbers can be found in the neighbouring Veryan Bay to the east and in Falmouth Bay. This site is also very good for Great Northern Diver and up to twelve Slavonian Grebes in winter.

Elsewhere in **Cornwall**, much smaller numbers of Black-throated Divers can be found in Mounts Bay, Penzance, from January to April (2-5 birds) and in St Ives and Carbis Bay (1-4). The former can be viewed from various points along the main promenade between Newlyn and Penzance and the latter from St Ives Island or from Carbis Bay.

❷Portland Harbour - Dorset

Up to twelve Black-throated Divers regularly winter in the sheltered waters of Portland Harbour and can be best observed from the relatively high point of Sandsfoot Castle. Birds can be well scattered or in tight groups but are generally found towards the centre of the harbour or near the groynes. Portland Harbour is best approached from the A354 between Weymouth and Portland, and can be generally overviewed or accessed from Ferrybridge car park or from beside the Salvation Army building in Castletown, Portland. Sandsfoot Castle, which is the best vantage point in which to view the northern part of the harbour, can be reached by taking Old Castle Road south-westwards from the sharp bend of the A354 in Rodwell, a district of Weymouth.

Portland Harbour in winter can also be very good for large numbers of wintering Red-breasted Mergansers, up to ten Eider, up to four Great Northern Divers, 8-10 Black-necked Grebes, 4-5 Slavonian Grebes and 1-2 Red-necked Grebes. Black Guillemot and Little Auk are also occasionally seen.

Occasionally, 1-2 Black-throated Divers may be present in Studland or Shell Bay, Poole Harbour (**Dorset**), and can be viewed from the Knoll car park, at the south end of Studland Bay.

❷Lower Largo Bay - Fife

Although Black-throated Divers can be encountered off virtually any coastal point in Scotland between October and April, the most reliable site is Largo Bay, where up to ten birds are regularly present. The bay is best observed from Ruddon's Point, which can be reached via the A917 eastwards from Lower Largo. Turn right about one mile north of Elie into Shell Bay caravan park and follow the minor road through the caravan site to the Point (please note that between October and March the entrance gate is locked from 4.00pm to 9.00am daily and one must be prepared to walk the last mile to the Point). The divers can be seen by looking westwards from the point, but beware of large numbers of Red-throated Divers, an abundant species here. Also, Slavonian and Red-necked Grebe should be present in small numbers. This site is also extremely good for large numbers of seaduck, including up to five Surf Scoters, 200+ Velvet Scoter, 300+ Long-tailed Ducks and 200+ Scaup.

Great Northern Diver *Gavia immer*

The Great Northern Diver is a winter visitor to the waters around Shetland and Orkney, with smaller numbers reaching the North Sea as far south as Humberside, and in the Irish Sea, off Anglesey and west Wales and off Cornwall and around Scilly. Furthermore, it is the most frequently recorded diver inland, with between 22-40 records annually.

The following is a county-by-county guide where Great Northern Divers can almost certainly be observed between November and May.

Scotland

The Great Northern Diver is a relatively common species in Scotland, wintering off all coasts but in larger numbers off the western side. The following sites are recommended:

Highland: Dunnet Bay, in **Caithness** (ND 20 76, Sheet 12), is an important winter resort for this species with up to 30 present between November and May. It can be reached by taking the B855 turning off the main A836 Thurso-John O'Groats road and is to the west of Dunnet village. Much further west towards Durness, Loch Eriboll (NC 43 59, Sheet 9) is also very good and can be viewed from the A838.

Furthermore, large numbers of Great Northern Divers gather in late spring in the seas around the low-lying offshore island of Eilean Hoan (NC 440670, Sheet 9) and can be 'scoped from the A838 Durness-Tongue road.

Further south and in the Ullapool area, several large and very deep sea lochs hold large numbers of wintering Great Northerns, including Loch Broom (NG 12 93, Sheet 19) and Loch Ewe (NC, Sheet 19)(viewable from A832). However, the best area, holding between 35 and 70 birds until at least May is in Gruinard Bay. Birds here are best observed from Mellon Udrigle village (accessed north from Laide on the A832) or from Rubha Beag headland (accessed by walking north across the coastal moorland from Mellon Udrigle village). A long-staying White-billed Diver was present here from December 1994 until at least March 1996.

Loch Gairloch and Applecross are also good vantage points in which to see Great Northern Divers, especially in spring.

Ardnamurchan Point (NM 670 410, Sheet 40) is also very good and can be accessed via the B8007, 17 miles west of Kilchoan. Observe from the lighthouse, this spot being also productive for various species of cetaceans.

Much further south, Turnberry Point, Ayrshire (**Strathclyde**) is good for this species, with 5-10 being present throughout the winter. Drive west on the minor track across the golf course to the point by the lighthouse.

On the east coast, birds are scarce and rather unpredictable in their haunts. However three main sites are used: Burghead Bay, NW of Elgin (**Grampian**) (view from Findhorn village or Burghead); Largo Bay (**Fife**) (view from Lower Largo village or Ruddons Point) and Gullane Bay (**Lothian**) (NT 470830). Generally speaking, 1-5 birds will be present at each of these sites between October and April.

Northern England and Wales

Northumberland: the species is rather uncommon and only occasionally seen off Holy Island, Bamburgh and Druridge Bay.

North Yorkshire: Relatively infrequent but present in small numbers in winter in Robin Hoods Bay, Scarborough North and South Bays and Filey Bay.

Gwynedd: One of the most reliable sites in Wales is Holyhead Harbour, Anglesey. There is normally 1-3 birds present at this site, which can be viewed from the coast road running between Holyhead ferry terminal and Soldier's Point.

Dyfed: Surprisingly scarce, although 1-2 remain in Carmarthen Bay each winter. Best vantage points are from Pendine, Telpyn Point and Cefn Sidan. Further north, Fishguard Harbour regularly attracts up to three birds.

Glamorgan: 1-2 birds are known to winter off the Gower coast and are best observed from Port Eynon Point.

East Anglia

Norfolk: Favoured haunts by 1-2 wintering birds are offshore of Holme Dunes and Hunstanton Cliffs and Golf Course and between Cley Coastguards and Sheringham.

Suffolk: in recent years, at least one bird has been recorded in the vicinity of the Orwell Estuary, and has been normally viewable from Woolverstone. During severe weather, Ipswich Docks and Alton Water become favoured haunts.

Essex: By no means regular, but occasionally present in the Blackwater Estuary or off Mersea Island.

South-east England and the South Coast

Kent: Up to four birds are regularly recorded, including 1-2 in the Medway Estuary, one offshore of Reculver Towers and one in the complex of gravel pits at Dungeness.

Hampshire: A minimum of two birds overwinter in Hampshire on average, with sightings in Langstone Harbour and off Needs Ore Point being most regular.

Dorset: Up to three birds regularly winter in Portland Harbour and are best observed from Sandsfoot Castle (see Black-throated Diver). Further east, 2-4 birds can be found in Poole Bay, normally between Branksome Chine car park and Knoll car park, Studland.

Devon: a scarce visitor occasionally observed from Dawlish Warren, Slapton Beach or Goodrington Beach, Paignton, in winter.

Cornwall: The most widespread species of diver in Cornwall with good numbers wintering in St Austell Bay, Looe Bay, Veryan and Gerrans Bay, Falmouth Bay, Mounts Bay and St Ives Bay. Prime sites for observation include Charlestown and Par Beach (St Austell Bay), Dodman Point and Portloe (Veryan Bay), Portscatho (Gerrans Bay), Swanpool Beach (Falmouth Bay), Marazion Beach car park, Penzance Promenade and Newlyn Beach (Mounts Bay) and St Ives Island and Carbis Bay (St Ives Bay). The majority of birds are solitary but form 'loose rafts' offshore.

Scilly: Up to five birds remain about the islands from late October until early May and are best observed from Bar Point/Watermill Cove, St Mary's, or from the Garrison, St Mary's, looking towards Samson.

Red-necked Grebe *Podiceps grisegena*

A scarce winter visitor with small numbers appearing along the south and east coasts, and up to 60 regularly in the Firth of Forth from mid August to November. Up to ten individuals remain at potential breeding sites, with nesting occasionally attempted. Between September and May, birds are regularly recorded inland, with between 48-90 birds annually. It has a preference for sheltered coastal localities or inland waters in winter. On autumn passage, Red-necked Grebes can be encountered at numerous sites along the east coast, and in particular, off the north Norfolk coast between Cley Coastguards and Weybourne.

Coastal Sites

✱Gosford Bay - Lothian
This is probably the best site in Britain for this species with up to 60 adults present at the site from late August until March. The site can be viewed from the B1348, east of Port Seton, and the A198, which it joins immediately adjacent to Ferny Ness. There is a parking area at Ferny Ness, which makes a good vantage point, whilst a footpath leaves the A198 at the north end of the bay and follows the headland round into Aberlady Bay. Up to 160 Slavonian Grebe also winter in this area.

❶Largo Bay - Fife
A regular wintering group of up to eight birds can be found in the bay and are generally best viewed from Lower Largo village or from Ruddon's Point (accessed from Shell Bay Caravan Park, one mile north of Elie).

✱Dungeness RSPB Reserve - Kent
At least one individual (occasionally up to four) winters in the complex of pits in the Dungeness area. The best site is the main pit complex on the reserve and viewable from a number of hides overlooking the area - please remember to report to reception on arrival. Otherwise, the nearby New Diggings and ARC Pit are worth checking (view from Lydd to Dungeness road) or Lade GP, to the west of Greatstone-on-Sea.

❶Portland Harbour - Dorset
Up to four birds regularly winter in Portland Harbour between November and February and are best observed from Ferrybridge or Sandsfoot Castle (see page 11).

❷Poole Harbour/Studland Bay - Dorset
Up to three birds remain in Studland Bay from November to March and are best observed from Shell Bay or the Knoll car park, Studland. It is essential to have calm sea conditions at this site.

✱St Austell Bay - Cornwall
Up to twenty Red-necked Grebes winter in west Cornwall and are mainly concentrated on the stretch of coastline between St Austell Bay and St Ives Bay. The best localities for potential sightings are off Carlyon Bay and Porthpean in St Austell Bay, off Porthmellon, off Caerhays and Portholland in Veryan Bay, off Pendower and Portscatho in Gerrans Bay and to the north and north west of St Anthony Head in the Carrick Roads. Swanpool Beach in Falmouth Bay and Carbis Bay in St Ives Bay are also traditional sites. This information was meticulously gathered by Roger Lane and published in *Birds in Cornwall 1987*, pages 122-125.

Inland Sites

A large proportion of wintering Red-necked Grebes in Britain choose to winter at inland sites and are very faithful to these sites. In harsh winters, numbers are swollen by large influxes from the western Baltic or Continental North Sea coast, with such movements occurring in January 1937 and February 1979.

❶Staines Reservoirs - Surrey
In most winters, a bird takes up residence in the Staines Reservoirs/Wraysbury GP complex. The south basin is favoured, but visits are often made to nearby Heron Lake or the British Airways Pit. Occasionally, Datchet Reservoir will be preferred, but this has much disturbance at weekends from recreational facilities.

❶Rutland Water - Leicestershire
A regular haunt for 1-2 birds from September to May inclusive. Favoured sites are at the dam end and from Normanton Church or the sailing club. Visitors should report to the Birdwatching Centre at the Egleton Reserve, signposted off the A6003, one mile south of Oakham.

✱Pitsford Reservoir - Northamptonshire
In recent years, an adult has been present at Pitsford Reservoir intermittently throughout the year, and has been best observed from the main causeway or in the Scaldwell or Walgrave arms. Hides overlook both of these bays and are open to members of Northamptonshire Bird Club and visitors. The reservoir is situated five miles north of Northampton Town Centre and is accessed from Brixworth, off the A508. The main causeway is on the Brixworth-Holcot road.

❷Hanningfield Reservoir - Essex
This large reservoir situated south of Chelmsford is the regular haunt of at least one Red-necked Grebe throughout the winter, although up to four may be present in September-October. Access is restricted however to permit holders and the only real chance of seeing one is if it is viewable from the causeway that runs across the south arm at TQ 723 972 (sheet 167).

❷Chew Valley Lake - Avon
In the last five winters (to 1995), a Red-necked Grebe has commuted between Chew Valley Lake and nearby Blagdon Lake, and has been present from August to May. It favours the western half of Chew, and can be seen from several of the hides overlooking the western shore or from Herriotts Bridge car park or Herons Green Bay car park.

❷Cheddar Reservoir - Somerset
Cheddar Reservoir is a regular haunt for Red-necked Grebe in winter but its presence may be determined by hard weather. It is likely that birds visiting Chew Valley and Blagdon Lakes (**Avon**) are the same. Cheddar Reservoir is situated due west of Cheddar village and is restricted to permit access only (write to: Bristol Waterworks Company, Fisheries and Recreations Dept., Woodford Lodge, Chew Stoke, Bristol BS18 8XH).

❶Carnsew Basin, Hayle Estuary - Cornwall
A regular site in recent years for a wintering Red-necked Grebe has been Carnsew Pool, adjacent to the Hayle Estuary. It is separated from the main estuary by a raised embankment, but although tidal, always retains a large amount of water. It can be viewed from a public footpath which runs along the embankment on three sides of the pool and also gives good views over the mudflats. Access is from the B3301 west of Hayle or from the town itself - through a small industrial area and boatyard opposite the railway viaduct.

Slavonian Grebe *Podiceps auritus*

The Slavonian Grebe is a scarce breeding species restricted to the Scottish highlands, with at least 45 pairs nesting in 1994. It is also a regular winter visitor to coastal areas of the British Isles and up to 140 may appear inland. It first bred in Scotland in 1908, and after slowly increasing in numbers to a peak of about 80 pairs in 1979, has now been seriously declining to a low of 40-50 pairs in 1992-5. The main reason for this decline is from persecution and human disturbance in the breeding season, but a decrease in invertebrate populations due to overstocking of fish and a change in the nutrient status of the loch water due to commercial operations has also had some effect. Slavonian Grebes breed on large and small freshwater lakes and pools, but primarily those with sheltered and protected bays in which to nest. Because of its precarious state as a nesting bird in Scotland, only one site for this beautiful species is detailed - birds here are protected on a nature reserve owned and managed by the RSPB. However, it is common knowledge that they are present at other sites and birders will continue to visit these sites. Please be vigilant and do not harm or disturb these confiding birds. Please inform the RSPB or the local police immediately if you witness any nest-raiding or suspicious activity.

Slavonian Grebes in Summer

✷Loch Ruthven RSPB Reserve - Inverness-shire
This 85 hectare RSPB reserve is situated to the east of Loch Ness, about 14 miles south of Inverness, and is best accessed from the A9, leaving west on the B851 six miles south of Inverness. Fork right after about eight miles onto a minor road that leads down to a public parking area at the east end of the loch. Access is restricted to the south-east shore only, but this is ample for rewarding views of Slavonian Grebe. Up to ten pairs nest in the emergent vegetation surrounding the loch edge and can be encountered along the entire 1.5 mile-long walk to the hide. The birds return in late April and remain to about early September.

Coastal Winter Sites for Slavonian Grebe

The Slavonian Grebe is the most maritime of all our grebes, but even so, prefers the shelter offered by secluded bays or estuaries. It is an annual winter visitor but has a more widespread and more northerly distribution than Red-necked or Black-necked Grebe. A total of about 400 birds winter in British waters, primarily in Orkney, the Western Isles, the Dornoch Firth, the Firth of Forth, the Blackwater Estuary in Essex, Poole and Portland Harbours in Dorset and in SE Devon. The following sites are those with populations of ten or more Slavonian Grebes in winter and guaranteed to produce sightings.

✷Largo Bay - Fife
Up to 30 Slavonian Grebes winter on the north side of the Firth of Forth and can be viewed either from Lower Largo village or from Ruddon's Point in the east. Calm sea conditions and a telescope are essential.

✷Aberlady Bay - Lothian
Some 100-160 Slavonian Grebes frequent Aberlady Bay between late September and early April, and are best observed in Gosford Bay. The main vantage points are from the B1348, east of Port Seton, or from the Ferny Ness public car park on the A198.

➊ Bridlington Bay - North Humberside
A moderate number of Slavonian Grebes spend the winter in Bridlington Bay and can be identified by 'scope from the North Promenade in Bridlington. This area is also very good for Red-throated Divers, sea-duck and Little Gulls.

✱ Hunstanton-Holkham - North Norfolk
A total of about 15-20 Slavonian Grebes can be found off north Norfolk in winter, primarily between Holkham and Brancaster Bays. The largest gatherings can be located off Holkham Gap (immediately east of the Gap or west up to a mile and a half), Titchwell Beach or Holme Dunes.

✱ Blackwater Estuary - Essex
Up to 30 Slavonian Grebes winter in the Blackwater Estuary and are best observed from the north shore. Access is best obtained from Rolles Farm, south of Tollesbury village, or from one of the two public footpaths running south to the seawall from the east of the village.

✱ Church Norton, Pagham Harbour - West Sussex
A small gathering of Slavonian Grebes is present off Pagham Harbour mouth from November to March and is best observed from the shingle beach at Church Norton. Take the B2145 southwards from Chichester towards Selsey Bill and after passing the Ferry Pool at Sidlesham, turn left to Church Norton after 1.5 miles. Park at the end of the road by the church and then follow the short track to the harbour. Walk right for about 600 yards to the shingle beach. The grebes can be offshore anywhere between the harbour mouth and the Severalls.

✱ Poole/Portland Harbours - Dorset
Like most divers and grebes, one of the best sites on the south coast in which to see them in winter is Poole or Portland Harbours. Small congregations of Slavonian Grebes can be identified on calm days in Shell Bay and Studland Bay and off Sandsfoot Castle in Portland Harbour (see page 11).

✱ Dawlish Warren and Tor Bay - Devon
A large number of Slavonian Grebes and other species of diver, grebe and sea-duck winter off the south-east coast of South Devon and can be relatively easily encountered on days of calm sea conditions. Langstone Rock, on the SW side of the Exe Estuary, is a good vantage point in which to see this species and can be accessed from the A379 Exeter-Dawlish road. Turn south around Cockwood Harbour bridge and then east in Dawlish town and drive under the railway bridge onto the Warren. Follow the seawall to the rock and scan.

Further south, the mainly sheltered and relatively shallow bay at Torbay is excellent for this species and Black-necked Grebe. One of the best spots to watch from is Goodrington Beach, accessed off the Paignton-Brixham road at the mini-roundabout opposite the sports and leisure complex. Alternatively, look from Preston beach promenade or from Paignton Pier, although this is the best site for wintering Black-necked Grebes.

Slapton Beach can also be productive for this species.

Black-necked Grebe *Podiceps nigricollis*

The Black-necked Grebe is a rare breeding bird in Britain, with between 32-45 pairs nesting annually, as well as a scarce winter visitor and passage migrant. After first breeding in Wales in 1904 (at least six pairs discovered on Llyn Llywenan, Anglesey, on 8th June) it subsequently bred in England in 1918 and in Scotland in 1930. It prefers lowland eutrophic meres, ponds, lochs and reservoirs in which to nest, particularly those with thriving and extensive emergent vegetation. Outside the breeding season, birds can be seen on passage in small numbers at many inland sites during July to October, and in spring from March to May. Additionally, breeding-plumaged pairs often linger at suitable sites during the spring before moving further north. The winter distribution of Black-necked Grebe is more southerly and westerly than both Red-necked and Slavonian Grebe and extremely localised. Apart from one site in SW Scotland, the majority are found along the English south coast, in sheltered bays between Hampshire and Devon. The total wintering population is in the region of 120 individuals.

Black-necked Grebes in Summer

As with all species of diver and grebe, Black-necked Grebe is highly susceptible to disturbance and for this reason, the author is only prepared to publicise one protected site.

✱Loch of Kinnordy RSPB Reserve - Tayside

This site, consisting of 81-hectares of open water and freshwater marsh, is situated 1.5 miles west of Kirriemuir to the north of the B951 Glen Isla road. Between April and September, several pairs of grebes are present at the site and nest in the dense areas of Amphibious Bistort. They can be viewed from the main hide, reached by walking 200 yards west from the car park. Early visits are recommended as the thriving Black-headed Gull colony can affect viewing from mid June onwards. The reserve is open from 9.00am until sunset daily between April and August and on Sundays only from September to November. Do not venture beyond the hides.

Winter Coastal Sites

✱Loch Ryan - Dumfries and Galloway

Loch Ryan is a very large and mostly enclosed sealoch located in southwest Galloway. A flock of up to ten birds is present from September to February inclusive wintering along the shallow and relatively sheltered western shore. The favoured area is that known as The Wig at NX 035 678 (Sheet 82), which can be checked from suitable stopping places along the A718. Loch Ryan is also renowned for its impressive winter rafts of sea-duck, particularly Scaup.

❶Conwy Bay - Gwynedd

Up to five birds winter in Conwy Bay, in the somewhat sheltered waters between the Bangor-Llandudno coastline and Anglesey. Birds can be anywhere between Great Orme and Penmon Point but are best observed from Llanfairfechan Promenade or nearby Penmaenmawr. Calm sea conditions and a telescope are essential.

❶Porthmadog - Gwynedd

In recent winters, at least four birds have chosen to winter in Tremadog Bay and have been visible from Borth-y-gest village (just south of the town) or from Harlech beach.

✱Langstone Harbour - Hampshire

Langstone Harbour has the biggest concentration of wintering Black-necked Grebes in Britain with up to thirty birds frequenting the inter-tidal Langstone Channel from October to March. They are best observed from the seawall bordering the oyster-beds in the NW corner of Hayling Island. Park at SU 720040 (Sheet 197) and walk west along seawall to about SU 715033. The birds are generally in the channel looking towards South Binness Island. Alternatively, check from the raised plateau to the south of Brockhampton Sewage Works, Havant. Look south into harbour from SU 708050.

✱Studland Bay - Dorset

The second largest wintering concentration in Britain with up to twenty birds present from November to March. Birds wander between Shell Bay and Studland south beach and are best observed from the beach south of South Haven Point (by Studland Ferry Terminal) or from the Knoll car park or Studland village.

✱Portland Harbour - Dorset

Up to ten birds winter in Portland Harbour between November and February and are essentially present off Sandsfoot Castle, to the north of the harbour. Access is via Old Castle Road, off the A354 in Rodwell.

✱Tor Bay - Devon

A total of up to twelve birds can be found in Tor Bay between November and February. Most winter off Paignton Pier and Preston Beach accessed from the main coast road between Paignton and Torquay. Further south, Goodrington Bay is well worth a check. Leave the Paignton-Brixham road at the mini-roundabout opposite the sports and leisure complex.

Passage Site

✱Staines Reservoirs - Surrey

Black-necked Grebes can regularly be seen on passage at Staines Reservoirs, with up to six birds regularly present in April-May and up to 20 from mid-August to late September. Both basins are favoured. Access is straightforward, either from the A3044 or B378.

Black-browed Albatross *Diomedea melanophris*

Since 1972, a female Black-browed Albatross has visited the Hermaness Gannetry on Shetland on virtually an annual basis, and is present intermittently throughout the summer. The date of arrival fluctuates between the end of March and mid April, and she remains on average until mid July. Most years sees her return to the same spot on Saito, an outcrop of rock just offshore, and if lucky she occupies a small ledge on which she sometimes lays infertile eggs. It is recommended that you visit the site as early as possible after she returns, as this is the best period in which to connect. It is not unusual for her to be present for three or four days at a time in April, but as the weeks progress, mounting pressure from territorial Gannets normally results in premature departure from the colony and the visits become much more erratic. However, in 1994 and 1995, appearances at the colony were almost daily between April and May.

Saito Outcrop (HO 594 157) is on Unst, the most northerly of the Shetland islands, and can be reached by parking at the sharp right hand bend in the road at Burrafirth (HO 607 140) and walking north-westwards across the boggy moorland, keeping the small stream on your left and the highest point of the hill slightly right and ahead of you. Continue walking northwest for about a mile (or for about 20-25 minutes). One will eventually appear at the cliff edge and it is here that you look for an obvious buttress formation protruding out at right angles to the main cliff face, about 200 feet below you. This is Saito and is generally saturated by nesting Gannets. In 1994 and 1995, the albatross was favouring the bottom left corner of the upper colony of nesting Gannets on the outcrop. Please exercise extreme caution at this site as it is very dangerous. Avoid visits during strong wind and rain and remember to keep clear of the edge - there are no safety barriers.

Cory's Shearwater *Calonectris diomedea*

The Cory's Shearwater is an annual visitor to SW Wales and SW England (especially Cornwall) between mid July and late September, with numbers fluctuating from year to year (i.e., 82 in 1991 and 1377 in 1990). It is rare elsewhere, but noted regularly off NE England in August-October. Small numbers are also occasionally recorded in the English Channel and North Sea in May and June.

The best county in which to see this seabird in Britain is without doubt Cornwall, particularly the southwest facing headlands. Ideal conditions are light to moderate south-westerly winds combined with clear weather or occasional squalls. The peak period for sightings is between 17th July and 5th September, with movements often peaking over a couple of days.

The best sites are Porthgwarra, Pendeen Watchpoint, St Ives Island and Lizard Point in Cornwall; Strumble Head in Dyfed; Prawle Point in Devon and Flamborough Head in North Humberside. However, only two of these sites have a sixty per cent or more chance of producing sightings under ideal conditions. Birds at Flamborough are best seen in periods of light easterly winds between 25th August and 25th September.

➊Gwennap Head, Porthgwarra - Cornwall

This southwest-facing headland is situated at the SW tip of west Cornwall and is accessed from Polgigga. Leave Penzance west on the A30 and at Catchall, turn left onto the B3283 to St Buryan. After St Buryan, join the B3315 and continue towards Land's End for two miles. At the sharp right hand corner in Polgigga, turn left to Porthgwarra. Follow this long, winding and very narrow road to the end and park in the pay-and-display car park just beyond the tea-room and telephone box. Gwennap Head is then reached by taking the coastal footpath from near the tea-room westwards for about 300 yards. An obvious promontory of rocks allows for cover from rain and an excellent vantage point in which to view passing seabirds. Most Cory's pass in a line just beyond the Runnel Stone and can be seen in large numbers. A light to moderate south-westerly wind is essential.

➊Pendeen Watchpoint - Cornwall

If the wind is moderate to storm-force west or north-west (Force 5-10), Pendeen Watchpoint is the place to visit. The site lies at the far north-western tip of the exposed Land's End peninsula, about twelve miles west of St Ives. It is best approached from Penzance, leaving the A30 north-westwards on the A3071 towards St Just. After 2.5 miles turn right towards Pendeen and follow to crossroads. Drive directly across and follow minor road to lighthouse, where parking is permitted alongside the road or beside the lighthouse. Once parked, follow the track bordering the right hand side of the lighthouse wall to the lower track and then follow the steps down to the sheltered eastward-facing wall. One of the attributes of this site is the presence of a reef offshore, which helps as a focal point. Under prime conditions, Cory's, Great, Sooty, Mediterranean and large numbers of Manx Shearwaters can be seen flying west just beyond these rocks.

Great Shearwater *Puffinus gravis*

The Great Shearwater is an annual visitor to Britain, mainly to SW England and the SW approaches, with varying numbers of between 20 and 2000 being seen between early August and late September. It is rare elsewhere, but almost annually recorded off Flamborough Head (Humbs) in autumn.

It differs mainly from Cory's in its dark-capped appearance and its white patch at the base of the tail, but also has diagnostically dark-tipped under secondary coverts forming a diagonal band and a dark patch on the belly. The flight is more shearwater-like and the pale nape band can occasionally be seen. The bill is wholly dark.

Cory's Shearwater is larger and longer-winged and has wholly white underparts. The head and hindneck are greyish-brown and contrast well with the upperparts, whilst the underwing coverts do not show any diagonal bars, just a neat brown border to the secondaries. The flight is diagnostic, being much more relaxed and gliding, with the wingtips being held slightly above the horizontal and distinctly angled backwards. There is more shearing than flapping and if viewing conditions are sufficient. the pale lemon-yellow bill may be seen. Beware of confusion with immature Gannet or shearing Fulmar.

As with Cory's Shearwater, the best sites are in west Cornwall, although the winds need to be more west or northwest biased to produce sightings. The optimum period for observations are between 18th August and 16th September and a force 5-7 wind with squally showers is preferred.

➊Pendeen Watchpoint - Cornwall
In a moderate westerly/north-westerly wind in late August/early September, there is a 90% chance of connecting with this species at Pendeen, with up to eighty birds being recorded on some days. These sometimes pass between the reef and the lighthouse rocks, affording observers with excellent views through the 'scope. For access instructions, see under Cory's.

➊St Ives Island - Cornwall
If it is late August or early September and the wind direction veers from south-westerly to north-westerly (preferably overnight) and is above force 6 in strength, St Ives Island is the place to be. Moderate numbers of Great Shearwater (and small numbers of Cory's along with an assortment of other shearwaters, petrels, skuas, Sabine's Gulls, terns and Grey Phalaropes) should be encountered. Large numbers of seabirds are forced by the strength of the wind or poor visibility into the shelter of St Ives Bay and work their way slowly back out past the headland often affording some exceptional views.

The island is situated at the north end of St Ives town and can be reached by following the harbour road north towards the Island car park. A number of very narrow winding streets eventually terminates at the pay-and-display car park. The seawatching vantage point is adjacent to the coastguards lookout at the top of the headland.

Both Cory's and Great Shearwaters can also be seen from the Penzance-Scilly ferry crossing.

Sooty Shearwater *Puffinus griseus*

The Sooty Shearwater is a regular visitor to Britain, being recorded in good numbers off all coasts between early August and mid October. Prime sites include SW Cornwall and NE England, between Spurn Point, Humberside, and the Farne Islands, Northumberland.

The following are the best sites in which to see this species in Britain, although it can be encountered at a number of other sites under suitable conditions.

✱Fife Ness - Fife
Fife Ness is one of the premier seawatching sites in NE Scotland and regularly produces large numbers of Sooty Shearwaters in August. E-NE winds are best, access being achieved from Crail.

Elsewhere in Scotland, Girdleness (**Grampian**) and St Abb's Head (**Borders**) regularly record double figures of Sooty Shearwaters.

✱Seaton Sluice - Northumberland
Under the right conditions, large numbers of Sooty Shearwaters can be seen flying north past Crag Point or the coastguards lookout at Seaton Sluice. Access from the main A193 coast road between Whitley Bay and Blyth.

✱Whitburn - Tyne & Wear
The stretch of coastline between Seaham and South Shields will produce Sooty Shearwater under suitable conditions. The best vantage point is Souter Point, slightly north of Whitburn. A public footpath runs due east from the A183 to the site.

✱Hartlepool Headland - Cleveland
The promontory at Hartlepool Headland is an extremely good site in which to seawatch from. Ideal conditions are fresh to gale force northerly winds, but any element of east will also produce results. Sooty Shearwaters can be seen in small numbers between late July and late September. Leave Hartlepool eastwards on the A1048. Turn right onto the A1049 to Middleton and continue to end. A minor road runs along the promenade from which to view.

Nearby Seaton Carew and South Gare can also be extremely productive for seawatching.

✱Flamborough Head - Humberside/East Yorkshire
Large numbers - up to 300 a day - of Sooty Shearwaters are recorded flying north past Flamborough Head between late July and early September. Although relatively distant, sightings are virtually guaranteed if the wind is between northwest and southeast. The seawatching point at Flamborough Head is accessed from Flamborough village. Drive east on the B1259 to the lighthouse pay-and-display car park and then walk east to the foghorn at the eastern tip of the headland. Carefully negotiate the footpath leading down the slope to the somewhat sheltered seawatching spot.

Elsewhere in Yorkshire, Hornsea and Spurn Point are good vantage points in which to see this species, whilst further south, Gibraltar Point (**Lincs**) and Sheringham (**Norfolk**) may yield positive results in suitable conditions.

✱Cornwall - Pendeen Watchpoint and Porthgwarra
Both Pendeen Watchpoint and Porthgwarra produce large numbers of Sooty Shearwaters under suitable conditions. Any wind (moderate strength upwards) with a westerly element between mid July and early October is likely to produce this species.

Balearic Shearwater *Puffinus mauretanicus*

This seabird breeds on islands in the Western Mediterranean and is considered by some authorities to be a separate species from the closely related Levantine Shearwater *Puffinus yelkouan*, found on islands in the central and eastern Mediterranean. It can be separated from Levantine in its larger size and more extensive and darker brown plumage. Both forms differ from Manx Shearwater in having dusky-brown rather than black upperparts and dusky-brown flanks and underwing coverts. The underparts are highly variable, with some Balearics being dusky-white with variable brown mottling on the sides of breast, flanks and undertail coverts and others being almost wholly brown below. These darkest individuals can easily be confused with the larger, longer-winged, stronger-flying and stiff-winged Sooty Shearwater.

Balearic Shearwater is a late summer visitor to the English Channel, with 500-2,000 birds appearing annually between Portland Bill (**Dorset**) and Porthgwarra (**Cornwall**) from late June to mid October. It is scarcely recorded elsewhere, with occasional sightings off North Norfolk and Humberside.

✱Pendeen Watchpoint/St Ives Island/Porthgwarra - Cornwall
West Cornwall is the premier region in which to witness this Mediterranean nesting seabird, with most of the headlands producing sightings under suitable conditions. From mid July to late September, birds wander north from the Bay of Biscay to British Waters, peaking during the period 26th July to 15th September. Gwennap Head at Porthgwarra is possibly the best site, with birds passing west during south to westerly winds, whilst at St Ives and Pendeen, up to 30 per day will pass during moderate to fresh west/northwest winds.

✱Portland Bill - Dorset
In most years, groups of up to 100 post-breeding Mediterranean Shearwaters disperse into the English Channel and Irish Sea from late June to mid August. For a period of up to three weeks in July, visits during early morning and evening to the obelisk at Portland Bill should result in sightings of this species offshore. Light to moderate S/SW winds are best with clear visibility.

Wilson's Petrel *Oceanites oceanicus*

Wilson's Petrel is a vagrant seabird to British Waters, with just 15 recorded from the British mainland (mainly Cornwall and Dyfed) and at least 224 seen from various pelagic trips (mainly in the SW approaches). At least another 100 have been recorded in Irish waters.

It is apparent from the statistics that variable numbers of Wilson's Petrels are present in the Western Approaches throughout August and early September and may be encountered during pelagic trips between 15 and 90 miles S/SW of Bishop Rock (Scilly). The favoured area is that known as 'Wilson's Triangle', an area of ocean about 50-55 miles SW of the rock, where factory ships or Spanish trawlers congregate in numbers. The Wilson's Petrels can be identified here amongst the large numbers of feeding British Storm Petrels attracted to the ships. The ratio is normally about 1: 800 birds.

The only regular means of attempting to see this seabird is from a professionally-organised pelagic trip and at the time of writing (1996), only the Penzance-based *Isles of Scilly Steamship Company* offering such an option. Almost annually since 1989, the *Scillonian III* has set sail from Penzance Harbour in July or August in pursuit of this species and other interesting seabirds and cetaceans. The trips are generally planned on a Sunday and depart Penzance at 5.00am and return at approximately 9.30pm. Obviously, they are hugely dependent on weather conditions and trips may be cancelled at short notice. Ideally, winds should be light westerly on the day, with light sea-swell and clear visibility, and follow a period of light to moderate W/SW winds. For details and bookings, contact the *Isles of Scilly Steamship Company Ltd.* at Penzance - ☎ 01736 62009.

Wilson's Petrels can otherwise only be reasonably expected to be seen off west Cornwall. A visit to St Ives Island or Pendeen Watchpoint during force 6-9 W/NW winds may provide dividends, particularly if they follow uninterrupted periods of south-westerly gale-force winds. Large numbers of petrels are forced into St Ives Bay to shelter and often congregate around the sewage outfall as the day progresses. All Wilson's Petrels identified at this site have been recorded between mid day and early evening, generally during a Storm Petrel or Leach's Petrel wreck.

British Storm-Petrel *Hydrobates pelagicus*

A nesting seabird restricted to the Northern Isles and islands off the west coast of Britain, with an estimated population of at least 90,000 pairs, some 43% of the World population. They winter in the Atlantic but visit breeding colonies in Britain between late April and October. The largest colonies occur in the Western Isles (mainly St Kilda and the Summer Isles), Shetland, Orkney, Dyfed and the Isles of Scilly. Away from these breeding areas, birds appear regularly off Dorset and Cornish coastlines between late May and late September and numbers are trapped regularly on the NE coast during overnight tape-luring sessions in late summer.

Due to their nocturnal habits, breeding Storm Petrels are very difficult to observe and are best studied at sea. The following sites are those recommended for late summer or autumn viewing.

Storm Petrels in Summer

The following ferry services offer opportunities in which to study Storm Petrels at sea during May to July:

Aberdeen (Grampian) to Lerwick (Shetland) - best observed in the region of Fair Isle.

Uig, Skye (Highland) to Lochmaddy, North Uist (Western Isles) - best observed during last hour of crossing.

❶Penzance (Cornwall) to St Mary's (Scilly) - primarily seen within eight miles of the islands.

✸Ilfracombe/Bideford (Devon) to Lundy Island (Devon) - primarily recorded about midway in the crossing, mostly in May and June.

✸Newquay - Cornwall
Storm Petrels can regularly be seen during summer evenings feeding at the sewage outfall off Newquay headland.

Storm Petrels in Autumn

Cornwall is primarily the best site in which to watch Storm Petrels in autumn, large numbers occurring off Porthgwarra and St Ives Island during the right conditions. The first signs of passage begin in late July, but from early August to mid September, birds can be seen almost daily, especially from Gwennap Head, Porthgwarra. Light to moderate southwest/westerlies are best, but birds are mainly distant and generally equidistant to the Runnel Stone. Strong north-westerly winds during this period will see large numbers congregate in St Ives Bay, especially during the afternoon, and some of the best land-based views in Britain can be achieved from birds feeding at the sewage outfall. These same conditions will see similar numbers passing Pendeen Watchpoint, but these will always be distant.

Small numbers can also be seen on a regular basis off Portland Bill (Dorset) from late July to mid August. Evenings are best.

An alternative method for seeing Storm Petrels is to attend one of the regular ringing sessions held on the east coast throughout July. These are held at Tynemouth (Tyne & Wear), Filey Brigg (North Yorks) and Flamborough Head (Humbs).

Leach's Petrel *Oceanodroma leucorhoa*

The main breeding grounds of Leach's Petrel are on islands off Greenland, southern Labrador, Newfoundland, New Brunswick and Maine, with smaller populations on the Faeroes and Westmann Islands. The majority winter in the tropical Atlantic off the west coast of Africa and the north-eastern coast of Brazil, and the eastern tropical Pacific Ocean.

In Britain, this truly oceanic seabird breeds only on the Northern Isles (Ramna Stacks, Shetland, Foula, Sule Skerry, North Rona, the Flannan Isles and St Kilda), with an estimated population of at least 10,000 pairs nesting annually. Elsewhere, it is a regular autumn migrant off the coasts of West Scotland, NW England, Wales and SW England, and appears in much smaller numbers off NE and SE England and East Anglia. Peak passage is between late September and early November, with small numbers occasionally appearing in winter, following Atlantic storms.

The most reliable areas for observation are in Lancashire, Merseyside, Dyfed and Cornwall, and generally follow north-westerly gales or strong westerly winds in the period between 4th and 22nd September.

➊Heysham Promenade - Lancashire
Heysham Promenade in suitable conditions will provide ample opportunities for viewing Leach's Petrels, with birds at high tides flying within 75 yards of the promenade. A hide along the foreshore also offers a vantage point.

➊Blackpool South Promenade - Lancashire
The stretch of coastline between Blackpool South Pier and Fairhaven Lake can be excellent for observing Leach's Petrels during the right conditions, with many birds often passing at ridiculously close range. Two hours either side of high tide are best.

✱Formby Point - Lancashire
The coastguard lookout at Formby Point is excellent for Leach's Petrel on strong north-westerly winds with birds passing close offshore. High tide visits are essential.

✱Crosby Marina/Seaforth NR - Merseyside
During periods of strong north-westerly winds, large numbers of Leach's Petrels get driven into the Mersey river mouth and occasionally on to Crosby Marina. Watch from the shelter of the coastguard lookout immediately west of the perimeter fence. High tide is best.

✱Leasowe Embankment/New Brighton - Merseyside
This is the best site in Britain for Leach's Petrel, where birds can be watched from the luxury of a warm car. In force 5-9 north-westerly winds and two hours either side of high tide, large numbers of Leach's Petrels pass south-westwards along the shoreline often only yards offshore. The car park can easily be reached by driving north from Junction 1 of the M53 towards Wallasey. After crossing the A551 turn left after 600 yards into the golf links car park and continue on to the dune coastal car park.

✱Red Rocks, Hoylake/Hilbre Island - Cheshire
At high tide and during strong NW winds, large numbers of Leach's Petrels pass this promontory. Access is via Hoylake.

➊Point of Ayr - Clwyd/Great Orme - Gwynedd
As birds move west from the Wirral coast, they head back towards the Irish Sea along the North Wales coast. Two sites, the Point of Ayr and Great Orme, will produce moderate movements if conditions persist but views will be more distant.

➊Strumble Head - Dyfed
In strong W/NW winds, large numbers of Leach's Petrels will pass Strumble Head during September.

✱St Ives Island/Pendeen Watchpoint - Cornwall
Moderate numbers of Leach's Petrels move west past north Cornwall during fresh to gale force W/NW winds. The peak movement is in mid to late September, with records generally spanning the period 27th August to 4th November. Up to 100 can be recorded in good movements.

➊Sheringham - Norfolk
This is the premier site in eastern Britain for this species with up to 16 being recorded on days with strong NW/N winds. The peak period for sightings is between 11th September and 9th October, but in some years, November can be the peak month. Watch from the seated shelters along the west promenade, especially those on the lower seawall.

Leach's Petrels are particularly vulnerable to winds in excess of their normal flight speed and in autumn, when many are in an advanced state of moult, they can be blown about like autumn leaves. This can often lead to dramatic 'wrecks', with many birds, sometimes hundreds, being blown inland. Such events occurred in the autumns of 1881, 1899, 1908, 1917, 1952 and 1987. Under extreme conditions, birds can be recorded at virtually any inland water.

Common Bittern *Botaurus stellaris*

The Common Bittern is a scarce breeding species with less than twenty pairs nesting annually in Britain. It is more frequent in winter, when numbers are swollen by immigrants from Continental Europe, with between 50 and 80 birds noted in recent years.

It is generally one of the most difficult birds to see in Britain, primarily because of its preference for secluded reedbed habitat and also because of its superbly camouflaged plumage and secret behaviour. Male Bitterns announce their presence by 'booming', a far-carrying sound, somewhat reminiscent of the sound made by blowing into an empty bottle or a muffled distant foghorn. This is most often heard in spring or early summer and although it gives an idea of where the bird is, it is very unlikely that the sound will result in a view. The best chance of seeing Bittern in the breeding season is in early morning, when they regularly fly to and from fishing sites, or in late summer when they are preoccupied with feeding young. The best strategy is to find a good vantage point overlooking a favoured reedbed and just patiently wait. In late June and throughout July, it should not be too long before one is rewarded.

Bitterns fly with quick wingbeats with neck retracted, generally low over the reeds. They fly directly and purposefully and rarely afford observers prolonged views. Each male has a large territory and may have several different females with which to mate.

The best time of the year to observe Bitterns is in winter when the meagre British population is swollen by a variable number of immigrants from the continent. Although virtually any wetland habitat in Britain can harbour birds, it is those with extensive reedbeds that attract Bitterns on a regular basis. They are considerably susceptible to severe winter weather, as it becomes difficult for them to extract fish and other prey from beneath the ice. In such circumstances, they are forced out of their wetland retreats into more open areas, and may be found, weak and starving, in very unlikely locations, such as roadside verges, canals or on garden ponds. Many birds perish in such conditions.

The best site in which to observe Bitterns in Britain is Lea Valley Park, Hertfordshire, where, since at least 1987, 1-4 birds have wintered at the site. Elsewhere, Leighton Moss (Lancs), Cley NWT Reserve (Norfolk) and Minsmere RSPB Reserve (Suffolk) are most reliable.

✱Lea Valley Park - Hertfordshire
Up to three Bitterns frequent the small fragmented reedbed outside Waverley Hide, adjacent to Fisher's Green car park, between October and early March. They perform unusually throughout the day and offer observers breathtaking views, individuals often fishing to within fifteen feet of the hide. In freezing conditions, they perform even better and it is possible to see all three birds feeding within a few yards of each other. Fisher's Green car park can be easily reached by leaving the M25 at Junction 26 and following the A121 through the eastern suburbs of Waltham Abbey. Turn north onto the A112 and then leave the major roundabout northwards on the B194 towards Lower Nazeing. After about one mile, turn left by the garden centre at the first sharp right hand bend and park in the Lea Valley car park at TQ 377033. From the car park, proceed over the footbridge to Waverley Hide, 50 yards along the footpath to the right. Please note however, that this hide may well be locked outside of weekends.

✱Leighton Moss RSPB Reserve - Lancashire

This reserve lies on the east side of Morecambe Bay and holds the largest population of breeding Bitterns in Britain. They are best seen from late March to early May and primarily at dawn and dusk from either Lilian's, Jackson or Grizedale Hides on the reserve. Leave the A6 north of Carnforth (Junction 35 of the M6) on minor road to Yealand Redmayne and then continue further west towards Silverdale. The RSPB reserve car park and visitor centre are on the left just east of Silverdale Railway Station.

❶Cley NWT Reserve - Norfolk

At least one pair of Bitterns has bred successfully at Cley since 1992 and birds can be observed during feeding sorties from mid July to mid August. Daukes Hide tends to be the favourite site but to be sure of views, it is best to watch from the high vantage point at the visitor centre, which overlooks the whole reserve. Alternatively, watch from the raised East Bank or Bittern Hide. Access to the reserve hides is by permit only, these being obtainable from the visitor centre on the A149.

❶Minsmere RSPB Reserve - Suffolk

The extensive reedbeds of Island Mere and North Marsh hold viable populations of nesting Bitterns which can be encountered throughout the year. The optimum chance of viewing is from the Tree or Island Mere Hides during early morning or late evening, as birds fly from one part of the reedbed to the other.

Other recommended sites include Walberswick reedbeds (Suffolk) (in both summer and winter); Duddingston Loch, in central Edinburgh (Lothian); Rostherne Mere (Cheshire); Oxwich Marsh (West Glamorgan); Pugney's and Angler's Country Parks (West Yorks); Stodmarsh (Kent); Frensham Little Pond (Surrey); Rye Harbour NR (East Sussex); Titchfield Haven (Hants); Chew Valley Lake (Avon) and Kenfig Pool (Mid Glamorgan).

Little Egret *Egretta garzetta*

A southern vagrant on the verge of colonising Britain, with between 200 and 600 individuals being recorded annually between 1992 and 1995. Nesting has been attempted at several south coast localities and the species is now resident along the south coast between West Sussex and Cornwall. Numbers are further swollen in summer, when post-breeding dispersal from Brittany and western France sees large numbers arrive between mid July and September. Roosting flocks of up to 143 birds have now been recorded in late summer.

Little Egrets can be encountered at virtually any coastal wetland site between Dungeness (Kent) and Marazion Marsh (Cornwall), but generally congregate in the vast inter-tidal harbours in Sussex and Dorset and the estuaries in Cornwall and Devon. The following are a list of the most regular sites, flocks of ten or more birds being commonplace.

✱Pagham Harbour - West Sussex

A resident group of at least seven Little Egrets survives in Pagham Harbour, these being best observed from the coastal footpath running between Sidlesham Ferry and Church Norton along the south shore of the harbour. Birds can also be observed from the sea wall immediately west of Pagham village, in the NE corner of the harbour.

✱Chichester Harbour - West Sussex

A large number of Little Egrets wander the various creeks and inter-tidal channels in Chichester Harbour throughout the year. Favoured haunts are the mudbanks opposite Dell Quay (two miles SW of Chichester), the saltings and mud surrounding Pilsey Island and off West Itchenor. At dusk, birds depart westwards to the roost on Thorney Island.

✱Thorney Deeps - West Sussex

The largest consistent roost of Little Egrets in Britain is on Thorney Island, where between 60 and 140 birds roost almost daily. The chosen roost-site is an area of dense vegetation and bushes immediately south of the Great Deep. Although up to a dozen birds can be present throughout on the nearby channel, birds begin to arrive during the last two hours of light and fly in from the nearby harbours at continuous intervals until dusk, often in groups of up to twenty birds. The site can be accessed from the yacht club at Prinsted, following the seawall southwards to the Great Deep. Wait quietly on the seawall and watch the egrets flight in at dusk from both directions. This area is also excellent for wintering Pintail, Spotted Redshank, Greenshank and Common Sandpiper, Barn and Short-eared Owl, Merlin and Peregrine and Cetti's Warbler.

✱Langstone Harbour - Hampshire

Langstone Harbour lies immediately west of Chichester Harbour, with those birds roosting at Thorney Deeps being largely made up of birds dispersing widely in both harbours throughout the day. Main vantage points are from the oyster beds on the NW side of Hayling Island and from anywhere along the seawall bordering Farlington Marshes NR. Birds are best observed at low tide, feeding in the vast network of muddy creeks, but can still be seen at high tide, when they roost on the conglomeration of islands in the Longstone Channel. Birds are present all year but more numerous in winter and late summer.

✱Poole Harbour - Dorset
Poole Harbour is home to a large number of Little Egrets, particularly in late summer. Although birds disperse widely at low tide, many roost in the tall trees on Brownsea Island at high, and at dusk, fly into the trees at Littlesea, Studland, to roost. These can be clearly seen from the hide overlooking the south end of the lake.

➊Dawlish Warren/River Exe - Devon
The River Exe has long been known as a regular haunt for wintering and passage Little Egrets. Favoured haunts are at Dawlish Warren, Powderham Park at high tide or the flooded meadows at Bowling Green Marsh, Topsham.

➊Teign Estuary - Devon
The River Teign has, over the last five years, been a major attraction to Little Egrets from late July until early March. Birds congregate at the western end and at high tide can be seen roosting in the tall trees on the south side of the estuary, directly opposite the Passage House Inn.

➊Taw/Torridge Estuary - Devon
Up to fifteen Little Egrets are present in this area of north Devon throughout the year. They are best observed from the minor road between Bideford and Fremington and range between Penhill Point and the mudflats at Instow. Islay Marsh is a favoured roosting area.

➊Kingsbridge Estuary - Devon
Large numbers of Little Egrets congregate in the Kingsbridge Estuary from late July until February and can be seen from various vantage points around the site. Frogmore Creek and Southpool Creek are the favoured haunts and the raised bank just south of West Charleton Marsh is the best site from which to view them from.

✱River Lynher - Cornwall
Up to twenty Little Egrets winter on the River Lynher between July and April and can be seen from various sites along the southern shore. The best vantage point is at Wacker Quay, about a mile west of Anthony on the A374. Large numbers often roost in the tall trees at Jupiter Point, two miles north of Torpoint. This site is also good for wintering Spoonbill.

➊Camel Estuary - Cornwall
Large numbers of Little Egrets frequent the Camel Estuary throughout the year and are best seen from the Tregunna Hide on the south side of the river. This can be reached by walking west for two miles from Wadebridge. Alternatively, walk south from Trewornan Bridge, 1 mile NW of Wadebridge on the B3314, and view the river to the west from the north bank.

✱Hayle Estuary - Cornwall
The Hayle Estuary has long been known as one of the best sites to see Little Egret. Birds are virtually ever-present and can be viewed on both tides from the RSPB hide at the back of the Quay House Inn, in the far southern corner or from the B3301 which runs parallel to the estuary. In winter, birds often visit the nearby Copperhouse Creek, where they can be watched from the Library car park.

✱Tresco - Scilly
The population of Little Egrets reached fifteen on Tresco in autumn 1995 and looks set to increase even further as the species continues its northward spread from France. They favour the tidal channel between Tresco and Bryher and can be seen at high tide roosting on the rocks in the channel or in the pines on the south side of the western end of the Great Pool.

European Spoonbill *Platalea leucorodia*

The European Spoonbill is a regular passage migrant and summer visitor to Britain with between 30 and 70 birds recorded annually. The majority of records are coastal and restricted mainly to the south coast and East Anglia, with very few appearing inland or in Scotland. Small parties of birds may be recorded at almost any wetland site in southern England and East Anglia from mid April until mid August, with some years producing small summer influxes, when sites such as Minsmere RSPB Reserve scrape or Island Mere (Suffolk), Cley NWT Reserve Simmonds and North scrapes (Norfolk), Titchwell RSPB Reserve lagoons (Norfolk), Hickling Swim Coots lagoon (Norfolk) and Titchfield Haven NR pools (Hants) may have long-staying individuals present.

Since 1990, at least four Spoonbills have chosen to overwinter in Britain, showing great site fidelity. The chosen haunts are large inter-tidal channels in estuaries and harbours in SW England. All four birds were still surviving in January 1996.

✱River Lynher - Cornwall
1-2 Spoonbills have overwintered at this site for at least six years and are best observed from Wacker Quay, on the south side of the estuary. Wacker Quay is a public viewpoint signposted off the A374, about one mile west of Antony village. The Spoonbills are most reliable at low tide and favour the deep channels on the north side near Sconner Farm. Scan the northern shoreline to the north-east of the car park and the deep channel of the river. This site is also excellent for Little Egret.

✱Poole Harbour - Dorset
Another long-staying and regularly wintering Spoonbill survives in Poole Harbour between late October and April. It spends most of its time commuting between the marshes and river channels at Arne NR (in the NW of the harbour at SY 973 877 - Sheet 195) and the freshwater lagoons at the eastern end of Brownsea Island. These can be viewed at vast range from Lilliput Hill, south of Parkstone (Sheet 195: SZ 041 894). This latter site is also excellent for wintering Avocet, up to 200 normally being present. In January 1996, two Spoonbills were present, the regular adult and a new immature.

✱River Taw - North Devon
The extensive mudflats and shallow tidal channels of the Taw Estuary have been attractive to a long-staying Spoonbill since November 1987. It is generally present from late October until early May and can be found, especially at low tide, sifting through shallow water off Instow Power Station, on Islay Marsh or in the wide middle reaches of the estuary. Islay Marsh can be reached from the west end of Lower Yelland village. A narrow lane runs 300 yards north towards the estuary before being crossed by the disused railway track. Turn right along here for about 200 yards and Islay Marsh bay is immediately on the left. Either view the area from the path or, as you reach the bay, cross a stile to the left and take the coastal footpath around the western perimeter of the bay on a raised bank (the former power station waste tip site). Do not walk onto the marsh, which is an RSPB reserve.

Greater Flamingo *Phoenicopterus ruber*

The Greater Flamingo is an exceptionally rare vagrant to Britain which has appeared in a wild state in Britain on numerous occasions (see Evans 1994, *Rare Birds in Britain 1800-1990*, pages 455-46). However, records are tainted by escapes, with large numbers being present in captivity throughout Britain and western Europe. Furthermore, it is essential to correctly identify flamingos seen at large in Britain, most sightings involving the smaller and pink-kneed Chilean Flamingo much more abundant in captivity. Greater Flamingos of the European race *roseus* are large, rather white in appearance and longer-necked. The long legs are pale pink in adult and bluish-grey in immature, whilst the bill has a much more restricted area of black, being mainly confined to the tip below the curvature.

In late October 1994, a third-winter Greater Flamingo arrived in Kent from the continent and then took up residence in Britain until at least 31st December 1995. It quickly sought out a long-staying Chilean Flamingo that had been present in the Thames Estuary for at least six years and resided with this bird at Cliffe Pools (Kent) until mid May. It spent the long hot summer of 1995 at Breydon Water (Norfolk) and Minsmere RSPB Reserve (Suffolk) and then returned via Essex to Cliffe for its second winter. For much of this period, it accompanied the escaped Chilean Flamingo.

At Cliffe, the Greater Flamingo spends much of its time roosting on the Alpha Pool, north of the cement works, but at low tide, visits the adjacent Higham Bight, on the south bank of the River Thames. Both sites can be reached from Cliffe village: leave the B2000 westwards on the minor road just north of the village sign. Turn right at the T-junction and then left between the two quarries after half a mile. Follow this road to the end and park sensibly by the cement works. A long straight road then runs northwest to Cliffe Fort signposted to the watersports and sailing centre. Alpha Pool is the large pit to the west of the road and Higham Bight is the expanse of tidal mud immediately west of the pit. This area can be viewed from the public footpath that runs along the sea wall and is excellent for wintering Avocets and Dark-bellied Brent Geese.

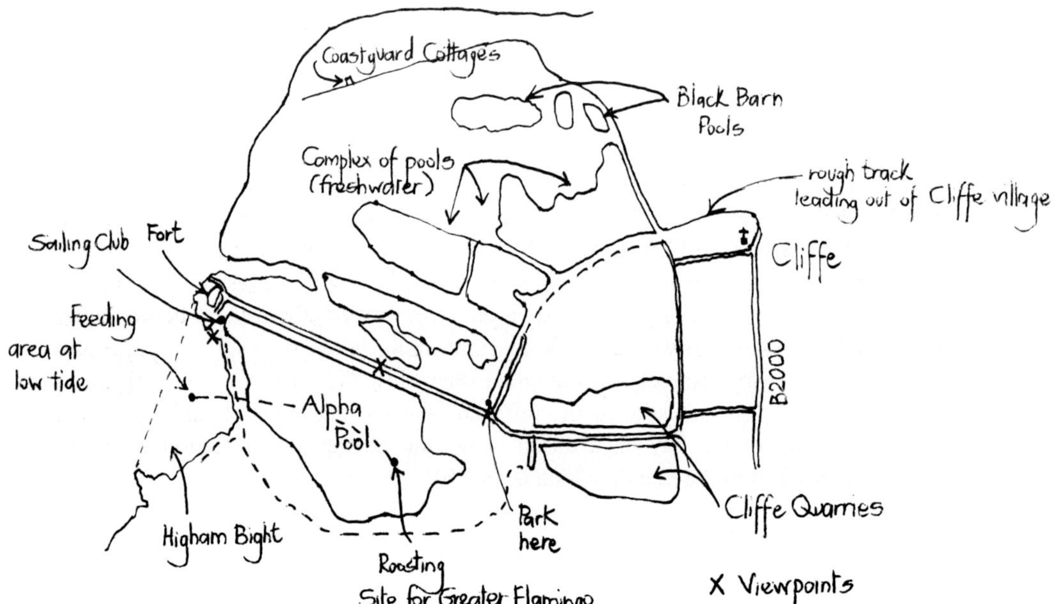

Bewick's Swan *Cygnus columbianus*

The Bewick's Swan is a species of the high Arctic, breeding on sparsely vegetated areas of grassy tundra in Canada, Alaska and Siberia. Apart from the odd injured straggler, they are entirely winter visitors to western Europe, with some 7,000 birds visiting Britain between late October and late March. The preference is for natural wetland habitats, mainly shallow freshwater lakes and ponds and slow-moving rivers adjacent to extensive grazing fields susceptible to flooding. In East Anglia, foraging on waste root crops, grain stubble and winter cereal is increasingly popular, and the rise in numbers wintering on the Ouse Washes has been nothing short of spectacular (from just a handful in the 1940's to over 1,000 in 1971 and an all-time peak of 5,227 in February 1985 - 3,668 were counted in December 1994). The reasons for this increase are believed to be the increased protection afforded to them at the RSPB and Wildfowl Trust refuges, and their ability to adapt to new food sources.

There are six main sites for observing Bewick's Swans, these being detailed below:-

✱Welney Wildfowl Refuge/Ouse Washes RSPB Reserve - Norfolk/Cambs
Between three to five thousand Bewick's Swans winter on the Washes and can be easily seen at Welney WWT Refuge or from Ouse Washes RSPB Reserve. Welney Refuge is easily reached from the A1101; at the south-east corner of the bridge take the minor road east signposted to the refuge. The car park and reception area are reached after about a mile and a half. Permits must be obtained from the office and a number of hides overlook the feeding areas, floodlit in winter.

The RSPB reserve is reached via Manea village. Follow signposts towards reserve along a series of minor roads. Pass Purls Bridge and continue south-westwards adjacent to river towards Welches Dam car park. A selection of hides overlook the washes on the opposite side of the river. Large numbers of Whooper Swans and dabbling duck also winter at these sites.

✱Slimbridge WWT - Gloucestershire
Up to 350 Bewick's Swans winter on the grazing meadows and feeding stations at Slimbridge and can be easily seen from the Holden Tower and numerous hides on the reserve between 20th October and 5th March. Admission is free to members and £4.70 otherwise. Access is from the M5 Junction 13 and then south along the A38. It is well signposted from thereon.

Birds also visit nearby Walmore Common.

✱Martin Mere WWT - Lancashire
Some 800 Bewick's Swans visit Martin Mere WWT between late October and March and can be seen from the reserve's hides. The refuge is situated on the minor road between the B5246 at Holmeswood (south of Mere Brow on the A565) and the A59 at Burscough Bridge, and is clearly signposted. It is open daily from 9.30am to 5.30pm.

✱Arundel Refuge - West Sussex

The Arun valley is a traditional site for wintering Bewick's Swans and, despite drainage and extensive agricultural intensification, still attracts up to seventy birds. During the day, the swans feed on flooded meadows along the Arun valley west of North Stoke village but in the evening, they roost at Waltham Brooks, situated along the minor road between the A29 at Coldwaltham and Greatham. There is limited parking space east of the railway.

✱Avon Valley, Ibsley Bridge - Hampshire

Between 48 and 300 Bewick's Swans winter in the Avon valley between mid November and February, commuting between Ibsley and Harbridge Green and Avon village to Christchurch. They wander the length of the valley but most consistently frequent the meadows to the north of Ibsley bridge immediately west of the A338.

✱Catcott Lows NR, West Sedgemoor, Curry Moor and Wet Moor - Somerset Levels

Though numbers depend on the extent of winter flooding, at least 65 Bewick's Swans winter on the levels between Bridgwater and Yeovil between December and February.

West Sedgemoor is situated on the A378, about two miles west of Curry Rivel. This is an RSPB reserve and has several hides overlooking the levels. Curry Moor can be viewed from the minor road between North Curry and West Lyng villages. For Wet Moor, park at the Yeo Bridge on the B3165 in Long Road and follow the footpath west along the south bank of the river.

Elsewhere, Bewick's Swans are scarce visitors, with less than 100 reaching Scotland in an average year (primarily at Vane Farm RSPB Reserve, Tayside, and Caerlaverock WWT, Dumfries & Galloway) and small numbers in NW England and Wales and passage visitors to sites in the Midlands and eastern England.

Whooper Swan *Cygnus cygnus*

The Whooper Swan is a common winter visitor to Britain, predominantly to Scotland, NW England and East Anglia, with up to 6,000 birds wintering annually. They favour small lochs and ponds and grazing marshes and grasslands, and feed on a variety of emergent and submergent water plants, grass and winter cereal. They also visit intertidal mudflats, where eel grass is a favoured food source. Although occurring in herds of over 600 in northern Scotland, most are seen in family parties or in groups of 50-100.

Whooper Swans bred in Orkney in the eighteenth century but since then, only sporadic nesting has taken place in the Northern Isles. However, a small number of individuals may linger throughout the summer.

Most Whooper Swans wintering in northern Britain (and Ireland) are probably from the Icelandic population, whilst those in Norfolk and elsewhere in southern Britain are possibly from Sweden and western Siberia. It is generally widespread in Scotland, with large wintering flocks in Shetland, Orkney, Caithness and Moray & Nairn, and is relatively easy to encounter. The following is a summary of the better sites in which fifty or more birds are normally present.

✱Insh Marshes - Inverness-shire, Highland Region
The extensive areas of wetland, marshes and flooded meadows at Insh Marshes attracts herds of up to 230 Whooper Swans between mid October and late April. There are numerous viewing areas, especially from the B970 and B9152, and birds can also be seen from the lay-by's on the A9, between four and five miles south of Aviemore.

✱Caerlaverock WWT - Dumfries and Galloway
Although more famous for its wintering flocks of up to 14,000 Spitzbergen Barnacle Geese, Caerlaverock WWT attracts a wintering population of up to 300 Whooper Swans. The Eastpark reserve is signposted from the B725, 1 mile south of Bankend (south of Dumfries). Visitors should report to the reception centre on arrival.

Small herds of birds can also be seen on Kirkconnell Merse, viewable from the B725 between Dumfries and Caerlaverock Castle. Park at NY 018 653 (sheet 84).

✱Martin Mere WWT - Lancashire
Up to 636 Whooper Swans winter at Martin Mere WWT between mid October and late March, favouring the extensive area of flooded meadows and pools. The refuge is open daily between 9.30am and 5.30pm and is situated on the minor road between the B5246 at Holmeswood and the A59 at Burscough Bridge.

✱Llanerchymedd, Llyn Alaw and Cefni Reservoir - Anglesey
Up to 40 individuals wander between these three sites on Anglesey from late October until mid March. The birds favour the flooded meadows and river valley SE of Llanerchymedd (view from B5111), the shallow fringes of Llyn Alaw (view from minor road between Rhosgoch and Llanddeusant) and the fields surrounding Cefni Reservoir (situated two miles NW of Llangefni).

✱Glaslyn Marshes - Caernarfon, Gwynedd
A large herd of up to 79 birds frequents the upper reaches of the Afon Glaslyn estuary at Minffordd from November to March and can be viewed from the A487. They tend to associate with the large herd of resident Mute Swans in this area.

✱River Tweed - Northumberland
A large wintering herd of up to 120 Whooper Swans can be found in the Tweed river valley to the west of Cornhill-on-Tweed. They are best viewed from the A698 east of Coldstream or north of Cornhill.

✱Lindisfarne NNR - Northumberland
A regular flock of some 50 or more Whooper Swans winters in the fields and flats to the west of the reserve. The best vantage points are Granary Point, Fenham, or Lowmoor Point nearby. Turn east off the A1 on a minor road to Fenham and park in designated car park at NU 088 408 (sheet 75). Lowmoor Point is reached by turning east off the A1 towards Fenham-le-Moor. Continue past Lowmoor railway crossing to car park at NU 097 399. Public footpaths lead north and south and high tide is best.

✱Druridge Pools - Northumberland
A large herd of Whooper Swans can be found in the Druridge Bay area from November to February. However, they are wide-ranging and wander to Warkworthlane and Linton Pond to the south. Drive north from Ellington and Cresswell along coast road to Druridge.

✱Bolton-on-Swale/Great Langton and Lower Derwent Valley - North Yorkshire
These two Yorkshire sites hold regularly wintering herds of Whooper Swans, with up to 30 and 48 birds respectively. The former flock ranges from Bolton-on-Swale in the north to Great Langton in the south and can be located from the B6271 south of Scorton. The latter are best observed from the hides at Wheldrake Ings NNR or from Bubwith Ings. Wheldrake Reserve lies on the east side of the minor road between Wheldrake and Thorganby.

✱Fairburn Ings and the Dearne Valley - West/South Yorkshire
Sizeable roosting herds of Whoopers may be encountered at Fairburn Ings RSPB Reserve during December to February and 20-30 are normally present in the Dearne Valley, near Barnsley, commuting between Broomhill Flash, Wath Ings and Edderthorpe Flash. Fairburn Ings is situated immediately west of the A1 along the Fairburn to Allerton Bywater minor road whilst Broomhill Flash lies east of Wombwell north of the A633.

✱Welney WWT Refuge - Norfolk
This refuge has possibly the largest single-site total of Whooper Swans wintering in Britain, with up to 900 birds visiting the site between December and February. The majority can be seen from the large, heated hide directly opposite the information centre. Welney Refuge is well signposted from the A1101 between Littleport and Upwell.

✱Horsey/Waxham - Norfolk
A regular wintering herd of up to 60 birds can be seen in the roadside fields between Horsey and Sea Palling. The favoured area is the ploughed field just east of Brograve Farm, although the fields south of Horsey Mill are often used. Drive north from West Somerton on the B1159 and scan.

Western Bean Goose *Anser fabalis*

The Western Bean Goose is a rare and localised winter visitor to Britain with up to 700 birds wintering annually, primarily in the Yare Valley in Norfolk and in the Avon Valley in central Scotland. Small numbers of the Russian Bean *rossicus* are also recorded each winter.

✱Prime Site: Yare Valley - Norfolk

The main wintering site in Britain is in the Yare Valley, where up to 400 birds remain from late November to February (416 present in January 1994). The flock selects cattle-grazed swards for feeding where the grass food plant *Poa pratensis* is abundant, but they are extremely wary and unapproachable. They spend the majority of the winter in fields north of the River Yare about a mile west of Cantley Beet Factory (at about TM 372 042) and are best observed from the high point on the public footpath running west from the village (TM 377 042). Please park sensibly in village and do not obstruct residents' driveways.

Later in the winter, the flock tend to move further upriver to Buckenham Carrs and can be watched at long range from the hide on the north side of the river at TM 354 045. The birds feed in the grass to the south of the railway looking north-eastwards from the hide. Park at Buckenham railway station and follow footpath to the hide. This is a designated SSSI reserve.

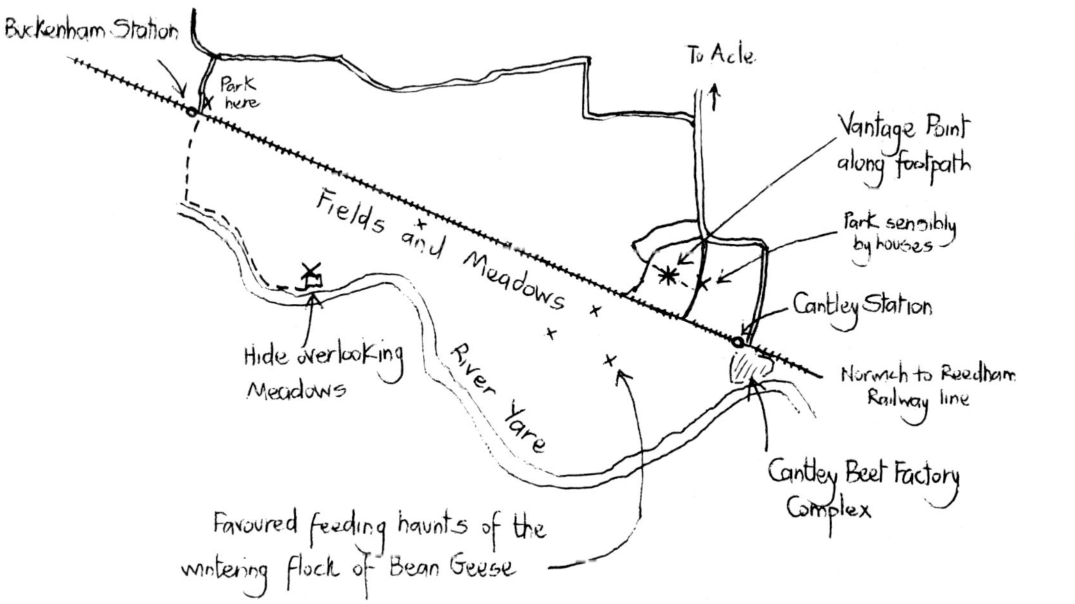

✱Prime Site: Avon Valley - Central Region

A wintering flock of up to 150 birds ranges widely between Cumbernauld and Falkirk. They commute between Garbethill Muir, Fannyside Loch and Loch Ellrig, and are best looked for by driving east on the minor road from West Fannyside to Threaprig and Jawcraig and then on to Loch Ellrig. They are generally present from late September until late February, but are at their easiest when they first arrive. During some winters, the flock choose to roost on the east Fannyside Loch. However, they often flight in after dark.

Pink-footed Goose *Anser brachyrhynchus*

Pink-footed Goose is a common winter visitor, especially to the agricultural farmland of North Norfolk, Lancashire and central Scotland, with an estimated wintering population of at least 120,000 birds. It is fairly abundant in Scotland and can be easily encountered in winter in Grampian, Tayside, Fife, Central Region and Dumfries and Galloway. Elsewhere, large numbers can be seen at Martin Mere and on the Ribble Estuary in Lancashire and along the north Norfolk coast.

Prime Sites

✱Strathallen and Strathfearn - Central Region
The A9 between Dunblane and Perth will produce guaranteed sightings of Pink-footed Geese during October to April. They feed in the large roadside fields and can be seen in flocks of 800 or more. The best section is that between Aberuthven and Forteviot.

✱Martin Mere WWT - Lancashire
Totals of up to 34,000 Pink-footed Geese have been recorded at Martin Mere WWT in early winter, with large numbers remaining at the reserve throughout the winter and into April. The refuge is situated on the minor road between the B5246 at Holmeswood and the A59 at Burscough Bridge. It is well signposted.

✱Snettisham RSPB Reserve, Scolt Head Island and Holkham Marsh - Norfolk
Co-ordinated roost counts in NW Norfolk in January 1994 revealed a total of over 68,000 birds wintering in the county, an unprecedented figure. By day, large flocks congregate in beet fields inland near Docking and in the fields due west of Holkham Gap, but in the evening towards dusk, huge skeins can be seen heading west over Scolt Head Island and west over Snettisham GP to roost in the Wash.

Lesser White-fronted Goose *Anser erythropus*

A rare vagrant which has declined considerably over the last twenty years, with a total of 105 recorded to the end of 1994. Two Scandinavian reintroduction schemes have complicated the status of this species in the last five years, with several individuals emanating from this source and others being of possible captive origin. A large percentage of the British records have come from just one site - the Wildfowl Trust reserve on the River Severn at Slimbridge. This is the best possible site in which to see one of these geese, grazing with the wintering flocks of European White-fronted Geese on the Dumbles. The peak period for vagrancy is from 3rd January to 12th March, with the Holden Tower being the best vantage point.

European White-fronted Goose *Anser albifrons albifrons*

Two quite distinct subspecies of White-fronted Goose winter in Britain, the dark plumaged and orange-billed Greenland White-front *A. a. flavirostris* and the paler-bellied, pink-billed European White-front. The European White-front breeds on the tundra of northern Russia, with many wintering in southern England and South Wales. The numbers wintering in Britain fluctuate depending on the severity of weather in Continental Europe, but range from 5,000 to 13,000 birds between October and April. They favour flat grazing land and coastal marshes in which to winter.

The following are the most regularly used wintering sites in central and southern England and South Wales:-

✱Slimbridge WWT - Gloucestershire
The largest single concentration of European White-fronted Geese in Britain with at least 1,200 birds wintering on the Dumbles between November and March. They are best observed from the Holden Tower or the series of South Finger Hides. Slimbridge WWT is well signposted from the A38 south of junction 13 of the M5. A Lesser White-fronted Goose is often discovered amongst the flock in January.

✱Holkham Freshmarsh - Norfolk
Up to 400 birds winter in the grazing fields immediately north of the A149 three miles west of Wells-next-the-Sea. Most congregate in the damp meadows to the west of Holkham Gap and can be viewed from the entrance road.

✱Buckenham/Cantley - Norfolk
Up to 300 birds winter in the Yare Valley and can be viewed from School Lane, Cantley, or from Buckenham Carrs SSSI reserve or railway station. They are generally distant.

✱Kessingland Levels, Reydon Marshes, Southwold Town Marshes, Eastbridge Marshes and Aldeburgh Marshes - Suffolk
A wintering flock of up to 300 birds winters on the Suffolk coastal strip between late November and late February. They are highly mobile and wander freely between sites. They can be encountered at Kessingland Levels (watch from A12 lay-by, 1 mile south of Kessingland, looking east), Reydon Marshes (south of A1095), Southwold Town Marshes (due south of town), Eastbridge Marshes and Minsmere Levels (check from sluice gates at Minsmere beach or along public footpath between sluice and Eastbridge) and Aldeburgh Marshes (viewed from beach car park at south end of town).

✱Capel Fleet and Leysdown Marshes, Isle of Sheppey - Kent
Flocks of up to 1100 European White-fronted Geese remain on Sheppey from October to March and are best observed from the minor road running south from the B2231 to the Harty Ferry Inn. The favoured area is the fields either side of the road about half a mile east of the Capel Fleet crossing. Large groups congregate on Leysdown Marshes to the north of the road.

❶Avon Valley - Hampshire
A small wintering population of up to 84 birds winters in the Avon Valley from mid November until March. Although highly mobile and dispersive, they invariably spend a large majority of the winter on the floodmeadows at Bisterne, between Ringwood and Christchurch. Watch from the B3347.

❶Dryslwyn, Carmarthen - Dyfed
This dwindling wintering population of up to 64 birds can be found in the water meadows about Dryslwyn and are best viewed from the B4300 or Dryslwyn Castle nearby. Dryslwyn is about six miles west of Llandeilo.

Greenland White-fronted Goose *Anser albifrons flavirostris*

The Greenland White-fronted Goose is a localised winter visitor which is found in internationally important numbers in Britain and Ireland. It can be separated from the European White-front in its much darker plumage and orange rather than pink bill. It breeds only in west Greenland and winters almost exclusively in Ireland, north and west Scotland and in mid-Wales.

A total of about 17,400 winter in Britain and up to 9,000 in Ireland, with the vast majority wintering at Wexford North Slobs (Co. Wexford) (7-10,000) and on Islay (Strathclyde) (9-10,000). Elsewhere, up to 1,200 winter at Rhunahaorine, Kintyre (Strathclyde), Machrihanish, Kintyre (Strathclyde) and on Tiree (Strathclyde), whilst 7-800 can regularly be found on Coll (Strathclyde); The Loons RSPB Reserve (Orkney) and in Caithness. Up to 350 still regularly occur on Loch Ken (Dumfries & Galloway) and up to 134 at the Dyfi Estuary (Dyfed).

Greenland White-fronted Geese reach their wintering grounds in Britain in October and remain until mid-late April. They are generally static throughout the period, although some wandering between sites does occur.

Prime Sites

✱Loch Gruinart RSPB Reserve, Islay - Strathclyde
Loch Gruinart holds the largest wintering population of Greenland White-fronted Geese in the world, with counts of in excess of 10,000 birds being recorded in October-November and March-April. It is situated on the north shore of Islay and can be reached from either Port Askaig or Port Ellen. The geese are best viewed from the B8017 or the unclassified road that branches north from Ardnave, whilst at Aoradh Farm, an RSPB visitor centre is open daily from 10.00am to 5.00pm. A small hide on the reserve gives ample views over the best feeding and roosting grounds.

Islay itself lies twelve miles off the Argyll mainland and is reached by ferry - Caledonian MacBrayne operate a car and passenger service from Kennacraig, seven miles west of Tarbert, to Port Ellen or Port Askaig. There are two sailings a day, Monday to Saturday, with a single sailing on Sunday, and takes just under two hours, weather permitting. For bookings, contact Caledonian MacBrayne, The Ferry Terminal, Gourock, PA19 1QP (01475 34531). It is also possible to fly from Glasgow Airport, *Loganair* operating two flights a day Monday to Friday, and one on Saturday. Enquiries on 0141 889 3181.

✱Loch of Mey, Caithness - Highland Region
A flock of at least 100 Greenland White-fronts winters in the Loch of Mey area to the east of Dunnet Bay. This area can easily be checked from the B855, north from Dunnet village on the A836 Thurso/John O'Groats road.

✱Loch Ken - Dumfries and Galloway
A traditional wintering flock of up to 300 Greenland White-fronts can be found in the Loch Ken area from December to February. The favoured areas are the fields at Mains of Duchrae or the pastures around Cogarth. These sites are reached from the A762 Laurieston-New Galloway road on the west bank. Turn east in Laurieston on the B795 to Townhead of Greenlaw and then head north, immediately west of Glenlochar Bridge, along the minor road (C50). This road affords views of the southwest part of

the loch. Please park considerately and do not obstruct farm traffic. The RSPB have built a car park at the entrance to the Mains of Duchrae Farm (NX 699 684) and from here visitors may walk to a hide on the foreshore offering views over the loch and adjoining farmland. The eastern shore between Crossmichael and Parton, particularly around Cogarth, is an excellent grazing area for the geese.

✱Dyfi Estuary - Dyfed

The Dyfi Estuary is one of the most enchanting estuaries in the whole of Britain, being bounded to the north by Cader Idris and its foothills, and to the south by the lesser slopes of north Cardiganshire. A dwindling flock of up to 134 Greenland White-fronted Geese still winters here but are under constant pressure from habitat destruction and disturbance from low-flying jet fighters from the Ministry of Defence. They can be encountered at Cors Fochno, an unmodified raised peat mire to the SE of the estuary or on the estuary itself. Cors Fochno is private and specially protected but can be viewed from a distance from the adjoining roads west of Llancynfelyn or from the Leri embankment, where a footpath follows the west bank from the railway bridge south. Visitors for Ynys-hir RSPB reserve should turn off the A487 at SN 685 952 (sheet 135) and follow signs for the reception centre. The reserve has a selection of hides offering ample views over the estuary.

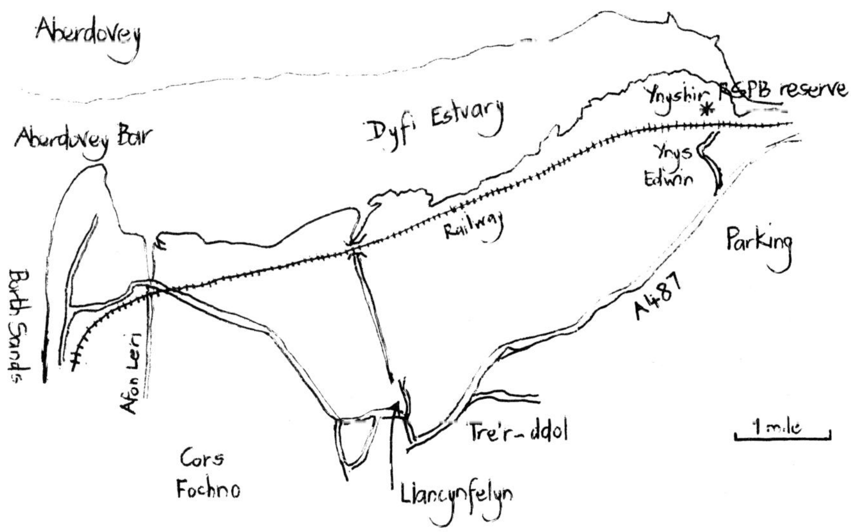

Greylag Goose *Anser anser*

The Greylag Goose is a regular winter visitor from Iceland (primarily to Scotland) with up to 100,000 birds wintering in Britain between mid October and May. A large feral population also exists in southern England and Wales. The main wintering concentrations are found in the Moray Firth area, Aberdeenshire, the central Southern Uplands and SW Scotland. A total of about 800 pairs of truly native stock still survives in NW Scotland and Orkney and these are largely resident.

Wild Greylags feed almost exclusively on farmland by day and roost on estuaries, reservoirs, lochs or river islands at night. The largest numbers to be found are in Inverness-shire, Moray & Nairn and Aberdeenshire. The following three sites should yield numbers well in excess of 7,000 birds:-

✱Loch Eye - Highland Region
Loch Eye is the only large eutrophic loch in Easter Ross and is winter home to over 15,000 Greylag Geese. The land surrounding the loch is strictly private but reasonable views can be obtained from the road to the north and west of the loch. The geese flight into roost between 3.30pm and dusk and during the day favour the area between Arabella and Balmuchy. The loch itself is situated east of Tain and the A9 and north of the B9165 at NH 830 795 (sheet 21).

✱Loch of Skene - Grampian
A wintering flock of up to 3,600 birds regularly roosts on the Loch of Skene between October and March. The loch is situated about ten miles west of Aberdeen and can be viewed from the A944. Similarly sized flocks can be encountered in the Haddo area, on farmland either side of the B9005 or on Haddo House Lake in the evenings, whilst autumn counts of up to 22,000 birds visit Dinnet Lochs, five miles east of Ballater north of the A93, in October-December.

✱Loch Spynie - Grampian
Several thousand Icelandic Greylag Geese winter in the Loch Spynie area and can be seen daily flighting over the B9016 at Clochan from late October to mid November. Roosting on Loch Spynie, they spend the day feeding in fields north of Aultmore or in the Roseisle/Coltfield area.

The vast majority of Greylag Geese seen in England are of feral origin but in winter, numbers in northern England are swollen by genuine immigrants.

✱Eden Valley - Cumbria
A wintering flock of up to 1100 Icelandic Greylag Geese frequent the Eden river valley between Culgaith and Langwathby from October to February. Leave the A66 four miles east of Penrith north on the B6412. The geese can be seen anywhere along the valley but favour the fields around Staingills and Watersmeet.

✱Morecambe Bay - Lancashire
A wild flock of up to 400 Greylags can be seen on Morecambe Bay marshes from December to March and are best seen in the Carnforth inner marsh area and roadside fields between Warton and Leighton Moss RSPB Reserve. This flock was formerly of wild origin but is now a mixture of wild and resident feral birds.

In severe winters, wild Greylag Geese from the Scandinavian and central European populations may visit southern Britain, although most of these birds winter in Spain.

Snow Goose *Anser caerulescens*

The Snow Goose is a rare, but annual vagrant to Scotland, with up to ten occurring in flocks of Icelandic Greylag or Pink-footed Geese or Spitsbergen Barnacle Geese. In England, at least 242 free-flying feral Snow Geese were recorded in 1995, with breeding taking place in at least four locations.

A flock of full-winged Snow Geese exists on the island of Mull (Strathclyde), this flock numbering 41 in December 1993. These birds winter on nearby Coll and successfully breed in most years.

In the winter of 1995/96, only one regularly wintering Snow Goose of presumed genuine origin could be found, this being an adult of the white morph on Islay (Strathclyde). This bird consorts with the wintering Greenland White-fronted Goose flock and can be best seen in the Port Charlotte area on the west side of the island, viewable from the A847.

Recent winters have also seen single white morph birds in the Loch Leven area (Tayside) (with wintering Pink-feet) and in fields south of the A9 near Forteviot. *Birdline Scotland* (0891 700234) carries full directions to Snow Geese as and when they occur.

Barnacle Goose *Branta leucopsis*

The Barnacle Goose is a common winter visitor to SW Scotland and Ireland with up to 33,000 birds appearing annually. Large numbers are also breeding ferally in southern England with flocks of up to 200 birds being commonplace.

There are three separate breeding populations of Barnacle Goose in the world and two of these winter exclusively in Britain and Ireland. Birds breeding on the east side of Greenland winter on islands off the north and west coasts of Scotland, in Ireland, and on Skomer Island and the Dyfi Estuary (Dyfed), whilst those from Spitzbergen, in the Svalbard, winter almost exclusively on the inner Solway Firth. The third population nests on the islands of Novaya Zemlya and Vaygach, off the Siberian coast, and winters in Holland and West Germany. Fluctuating numbers of these birds cross the North Sea to East Anglia and southeast England, particularly when the continent experiences freezing conditions.

Barnacle Geese are very gregarious in their winter quarters, both at roost and when feeding, and form vast flocks. They favour coastal sites and frequent salt marshes or agricultural pastures within ten miles of the sea, and are faithful to their chosen winter haunts.

The largest numbers are to be found on the outlying islands off the north and west coasts of Scotland, with some 30,000 birds wintering in Argyll. The biggest proportion of these winter on Islay, where 26,000 may be counted roosting at the head of Loch Indaal (watch from NR 335 620 or NR 324 627) or in the fields around Ardnave. However, for most birders, a visit to Caerlaverock WWT will reap rewards.

✱Prime Site: Caerlaverock WWT - Dumfries and Galloway
A total of up to 12,178 Barnacle Geese winter at Caerlaverock and can be seen from the reserve hides or at nearby Mersehead or in roadside fields in the area. The refuge is well signposted from the B725, one mile south of Bankend. This road also offers ideal viewing opportunities of the Merse and the eastern side of the Nith Estuary. There is a parking area at NY 018 653.

Later in the spring, many of these birds move east to Rockliffe Marsh (Cumbria), where they can be viewed from minor roads west of Rockcliffe Cross.

Small numbers of Barnacle Geese can be encountered at many other sites in winter south to Lancashire, and also, as stated earlier, large flocks may arrive in southern England from the continent.

Pale-bellied Brent Goose *Branta bernicla hrota*

The Dark-bellied Brent Goose breeds in northern Russia and is a common winter visitor to the Wash, the Thames Estuary and along the south coast, whilst the Pale-bellied Brent Goose is restricted to Northumberland, Scotland and Ireland.

Pale-bellied Brent breeds in Arctic Canada, Greenland, Spitsbergen and Franz Josef Land, with estimated populations of 100,000, 26,000, 2,500 and 4,500 respectively. The Greenland population winters on the eastern seaboard of North America and in Ireland, and the Svalbard and Franz Josef Land in Denmark and NE England.

It can be separated from Dark-bellied in its pale greyish-brown lower breast, belly and flanks, often shading to whitish in the centre and towards the rear.

The whole Siberian population of *hrota* arrives at its Danish wintering grounds in October, and from there fluctuating numbers move later in the autumn to their only regular British site at Lindisfarne, Northumberland. The variation in numbers at the site largely results from the severity of the weather in Denmark and ranges from as low as 300 to almost all of the population at 3,100.

✱Prime Site: Lindisfarne - Northumberland
A wintering flock, normally in the region of 1,300 to 2,150 birds, resides at Lindisfarne from mid October until late March, with occasional injured birds remaining throughout the summer. The best site in which to see them is from the main causeway crossing over to Holy Island. Leave the northbound A1 at West Mains and take the minor road eastwards to Beal. After three miles this road reaches the shore and causeway. At high tide, one can view from the public footpath which runs south along the shore, but the geese are best seen at low tide, feeding on the mudflats. They can be observed from anywhere along the causeway or from the road as it runs eastwards along the shore towards the village.

An alternative site in which to view the geese is from the hide overlooking Fenham Flats. Leave the A1 towards Fenham-le-Moor and continue over the railway crossing to the limited parking space on the shoreline at NU 097 398.

Egyptian Goose *Alopochen aegyptiacus*

The Egyptian Goose is an introduced species that has maintained a successful feral breeding population which is now in excess of 750 individuals (mainly Norfolk and Suffolk). Its native range is in Tropical Africa, where it is widespread and abundant. In the Western Palearctic, it is now restricted to the Nile Valley, Egypt, where it is still reasonably common in the Aswan Dam area.

In Britain, birds are restricted to East Anglia, with very small numbers being recorded elsewhere. They are found almost exclusively near fresh water and favour wooded park land containing areas of water, breckland meres, open pastures and broadland.

The following are the best sites in which to see this species:-

✱Holkham Park and Freshmarsh - Norfolk
A self-supporting flock of up to 100 birds survives in and around Holkham Park and can be best encountered in the fields north of the A149, between 1 and 2.5 miles west of Wells-next-the-Sea and from the access road to Holkham Gap or on grassland surrounding the southern end of the Holkham Park Lake. Park in the designated car park directly south of the Holkham Gap turning and follow public footpath along road to lake.

✱Cley NWT Reserve to Salthouse - Norfolk
Up to twenty pairs can be found between Cley West Bank and Salthouse Quags throughout the year and can be viewed from the A149 or from Cley reserve, Salthouse Duckpond car park or the Iron road, Salthouse.

✱Bure River Valley - Norfolk
A large percentage of the population survives in the Bure river valley, between Coltishall and South Walsham. Most are concentrated in the Wroxham, Cockshoot and Ranworth Broad areas. These sites can all be reached via the minor road running between Wroxham and Woodbastwick.

✱Blickling Hall - Norfolk
A breeding population of some 10-16 pairs survives in Blickling Park, situated two miles NW of Aylsham on the B1354. The Hall and park lands are managed by the National Trust.

✱Thetford Nunnery - Norfolk
Thetford Nunnery Lake and adjoining river valley holds a resident population of up to 24 Egyptian Geese. The site lies on the south side of Thetford, east of the A134.

✱Waveney Valley - Suffolk
A large population of some 97+ birds frequents the river valley between Burgh Castle and Beccles, with the largest concentrations roaming Haddiscoe and Burgh Marshes. Observations are best made from Somerleyton village, about five miles NW of Lowestoft. In summer, many of the birds move to Peto's Marsh, at the west end of Oulton Broad, and to Fritton Decoy Pond and Lound Sewage Works nearby.

✱Livermere Park - Suffolk
A resident flock of up to 17 birds resides in Livermere Park and can be seen in the fields and farmland surrounding the lake and on the lake itself. Leave the A143 five miles NE of Bury St Edmunds at Great Barton. Follow minor road north-westwards towards Great Livermere and park sensibly by entrance gates. A public footpath leads to the northern part of the lake. Nearby, small numbers can be seen in roadside fields at Ixworth Thorpe and at the Lackford WR, West Stow.

Away from East Anglia, Egyptian Geese are recorded as vagrants from this population or as feral escapes from captivity.

Mandarin Duck *Aix galericulata*

The Mandarin Duck is an introduced species from the Far East (western USSR, China and Japan) with a viable breeding population in Britain of at least 2,000 pairs. The strongholds remain in Surrey, Berkshire, South Buckinghamshire and Hampshire, with smaller numbers in Tayside, Borders, East Anglia, the Midlands and Wales.

It is a remarkably secretive species, particularly in the breeding season, and haunts secluded streams or ponds in mature open deciduous woodland, wooded river valleys and ornamental park land. Sites with extensive overhanging vegetation are particularly favoured. In winter, Mandarins flock together and at times of severe weather, large flocks may be encountered roosting on ice-frozen lakes.

The following is an exhaustive and comprehensive list of Mandarin sites in Britain:-

In **Scotland**, the breeding population in recent years has plummeted with less than two pairs remaining on the River Tay at Perth (from a peak of about 60 birds in 1984), 1-2 pairs in Inverness-shire, a pair at the northern end of Loch Eck on the Cowal Peninsula, Argyll (Strathclyde) (NS 134 963 - Sheet 56) and up to ten birds in Borders. In this last-mentioned county, the Eye Water between Houndwood and Reston are prime sites. Leave the A1 at NT 866 627 (Sheet 67) and check from river bridge after 80 yards or from nearby Shawbraes (NT 864 624) or from Houndwood (NT 842 638).

Lancashire: Mere Sands Wood - 1-2 birds may be found in the wooded river valley at Mere Sands Wood LNR. Access from B5246 one mile west of Rufford.

Martin Mere WWT - 1-3 birds present on the reserve and occasionally visible from hides. Refuge is clearly signposted from Burscough Bridge and Mere Brow.

Cheshire: River Dee, Eccleston - up to ten birds survive on the tree-lined River Dee between Eccleston Ferry and Iron Bridge. Park in Eccleston village and walk down to river and follow upstream for up to 200 yards. Mandarins frequent the river between here and up to two miles to the north. Another site nearby has regularly attracted small numbers in recent years on the River Gowy at Beeston, three miles south of Taporley, west of the A49. View from bridge over the river by the Shady Oak public house. Two pairs are also resident on Tatton Park Mere, although their origin is somewhat dubious.

Norfolk: a population in excess of 100 birds survives in Sandringham Park, although these were introduced from Windsor Park in 1973. The birds frequent the main lakes and adjacent water courses but can only be seen reliably from within the confines of the Royal Estate, access via entrance fee. Elsewhere, 1-2 birds can be encountered at Bayfield Lake or on the river at Letheringsett, south of Glandford/Cley; Narford and Sparham, and up to twenty birds of unknown origin at Felbrigg Hall Lake.

Cambridgeshire: up to 20 birds frequent the River Cam, on the south side of Cambridge, and are best observed at Bolton's Pit or the Botanical Gardens. Up to 30 frequent Milton Park, north Huntingdon.

Essex: although fairly widespread in the county, the most reliable sites with regularly up to 20 birds are Connaught Water and Ingatestone. Connaught Water is situated in Epping Forest at TQ 403 951 (Sheet 177), whilst Harding's Farm Reservoir and Howlett's Hall lake attract birds in the Ingatestone area.

Northamptonshire: noted regularly at various sites but most consistently at Castle Ashby engine pond and at Overstone Solarium Lake.

Bedfordshire: up to 34 birds survive in and around Woburn Park and can be seen roosting on the wooded island on the main lake in front of Woburn Abbey from September to April, whilst smaller numbers frequent Linden Lake, Eversholt, nearby. At least six pairs are present in Stockgrove Park and Luton Hoo, and 1-2 pairs on Harlington Mill Pond.

Buckinghamshire: A large population of at least 85 birds is present on the densely wooded section of the River Thames between Taplow Court and Cliveden, but much of this area is strictly inaccessible. Elsewhere, up to ten birds can be found at Littleworth Common and at Hitcham.

Hertfordshire: The most consistent site for sightings is Panshanger Park, near Welwyn Garden City. Elsewhere, 1-2 are recorded annually at up to twelve sites.

Greater London: the best sites are at Grovelands Park, Southgate (up to 35 birds) and Trent Park.

Kent: up to 50 birds are regularly recorded in the Broomfield area, particularly around Fulling Mill. The NE corner of Bough Beech Reservoir can also be very good for this species. View from the causeway at Winkhurst Green.

Sussex: a fairly widespread and common species with favoured sites being Swanbourne Lake, Arundel; Pulborough Brooks RSPB Reserve, Weir Wood Reservoir and Maynards Green.

Berkshire: Possibly the largest concentration in Britain inhabits Virginia Water in Windsor Great Park. Counts of up to 240 birds have been made at dusk as birds flight into roost. Visits between October and April are recommended as birds disperse in summer to breed on outlying ponds. The birds frequent the western half of the lake and are best approached from the car park at Blacknest Gate, on the A329.

Elsewhere, an increasing population can be found in the Wraysbury GP complex and on the River Thames, between Runnymede and Henley.

Surrey: Relatively abundant with well over 200 pairs breeding in the county. Prime sites are Cutmill Pond, Busbridge Lakes, Thursley Lake and Frensham Ponds.

Gloucestershire: a widespread breeding species, particularly in small wooded ponds in the Forest of Dean. It is best observed in winter when large gatherings can be found at Slimbridge WWT, Cannop Ponds and Soudley Ponds. Cannop Ponds are located immediately east of the B4234 south of Cannop and Upper Lydbrook, whilst Soudley Ponds are east of the B4227 south of Cinderford.

Hampshire: widely distributed in small numbers with Eyeworth Pond being the most reliable site in winter.

Isle of Wight: Favoured localities are Newchurch Pond and Marsh and Godshill Pond.

Wiltshire: Up to 16 can be found on Fonthill Lake, north of Tisbury, in winter, and up to 30 frequent the River Nadder, around Compton Chamberlayne.

Avon: The only consistent site is Tortworth Lake, where up to nine birds may be recorded. Tortworth is reached from the M5 Junction 14. Turn north off the B4509, 1 mile east of the junction.

Dorset: 1-2 birds may be encountered at Sherbourne Lake and Bryanston.

American Black Duck *Anas rubripes*

The Black Duck is an extremely rare vagrant from North America with just fourteen recorded in Britain and six in Ireland to the end of 1995. The Isles of Scilly has recorded four of these and is probably the best place in Britain in which to ever encounter this rather drab vagrant. At the time of writing (June 1996), a long-staying drake was still present on Scilly, having first arrived on 1st April 1994. This bird, like all the previous records on the archipelago, consorted with its Old World counterpart, the Mallard, and frequented Tresco Great Pool. It is best looked for at the western end of the pool and spends much of its time in the reed fringes. It can be seen from either the David Hunt hide, accessed from the footpath along the northern edge of the pool, or from the obscured hide at the SW corner of the pool.

Garganey *Anas querquedula*

A scarce summer visitor with between 40 and 60 pairs nesting annually (mostly in East Anglia and the Fens) and passage visitors between March and May and August and October. 1-2 birds occasionally overwinter. Peak periods in which to see Garganey are in April-May or from August-September.

Prime sites for observation are on the Ouse Washes (Cambs), Cley NWT Reserve (Norfolk) and Minsmere RSPB Reserve (Suffolk), but birds can be seen at virtually any wetland site in England north to Lancashire and Yorkshire during passage periods. It is very rare in Scotland.

In spring, the drake Garganey is very easy to identify with its broad white head stripe, brown face, pale flanks and heavily mottled breast, but in autumn, eclipsed-plumaged adults and newly-fledged juveniles are quite difficult to locate amongst hordes of roosting waterfowl. They can be distinguished from Teal by the longer and more stout bill, pale creamy spot at the base of the bill and dark cheek bar bordered on both sides by creamy-buff stripes, and in all plumages by the pale greyish-blue upperwing coverts and broad white trailing edge.

✱Ouse Washes RSPB Reserve - Cambridgeshire

The RSPB reserve at Welches Dam attracts a breeding population of about 3-5 pairs of Garganey and these can be seen from the reserve hides. In April and May, pairs can be located on the flooded meadows from Stockdale Hide, and are generally present in the area of washes to the south of there. The reserve is well signposted but can be reached by taking the B1098 from Chatteris towards Upwell. The road turns sharp left after two miles and after 200 yards crosses Vermuyden's Drain. Turn right immediately after the drain towards Manea and at the beginning of the village, turn right towards Purls Bridge and Welches Dam. At the bridge, the road follows the Old Bedford River to the RSPB car park and visitor centre. The Garganey are best seen from any of the first four hides as you walk north along the barrier bank footpath on the opposite side of the river.

❶Cley Marshes NWT Reserve - Norfolk
Small numbers of Garganey can be seen at Cley between early April and mid September and are best looked for on Pat's Pool, Simmonds Scrape or North Scrape. Watch from Avocet, Daukes, Teal, Irene or North hides. The reserve lies just east of Cley next the Sea. Park in the designated car park on the A149. The reserve is open every day except Tuesday.

Garganey can also be seen at nearby Salthouse Marshes and Weybourne Quags Pool.

❶Minsmere RSPB Reserve - Suffolk
Garganey are frequently recorded at Minsmere during April-August and can be seen from Island Mere Hide or from any of the hides overlooking the main scrape. The reserve is best reached from Westleton. At the north end of the village, leave the B1125 east on a minor road towards Dunwich and after half a mile turn right and follow signs to reserve. The reserve is open every day except Tuesday.

❶Pulborough Brooks RSPB Reserve/Amberley Wild Brooks - West Sussex
Between late March and mid May, Garganey can be reliably seen on the floodmeadows at these two sites. Amberley Wild Brooks are situated immediately north of Amberley, about four miles north of Arundel. Leave Arundel northwards on the A284 and after 2.5 miles turn right onto the B2139. Continue east for a further 2.5 miles to Amberley village, where a track leads northwards across the marshes from the north end of the village. The RSPB reserve is reached from a public footpath leading south from the A283 in Pulborough village.

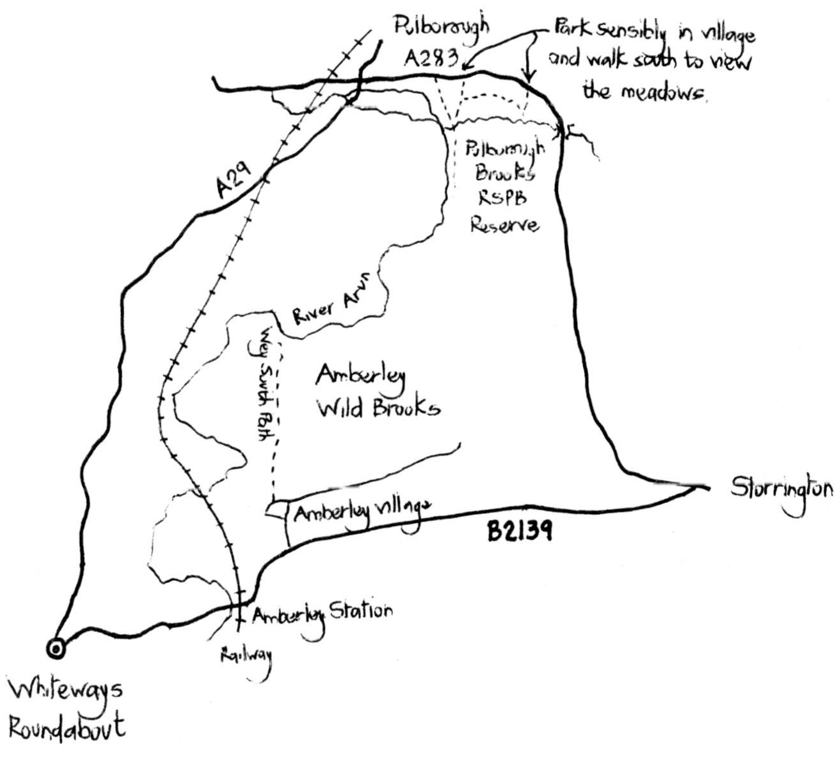

Red-crested Pochard *Netta rufina*

The Red-crested Pochard is an annual vagrant to Britain in small numbers, with most occurring in August-September or in winter. It is also widely kept in captivity and many individuals are suspected of being escapes. A feral breeding population of up to 60 birds exists in the Cotswold WP (Wilts) and up to 160 birds are recorded on an annual basis.

Red-crested Pochards favour reservoirs, lakes and flooded gravel pits and feed almost exclusively on aquatic vegetation, obtained by diving or more often, upending. The native range extends eastwards from the Black Sea into Asia, but there are also small and somewhat declining populations in the Netherlands, West Germany and Denmark, and more substantial numbers in France and Spain. Most west European birds winter in the north Mediterranean basin. For a period in the late 1950's/early 1960's, there was evidence of immigration into Essex of up to twenty birds from the late summer dispersal of moulting birds from NW Germany and the Netherlands. This pattern has somewhat ceased, mainly due to the reduction in numbers in the Danish population, but can still occur on a small scale if conditions arise (mainly in late August/September).

Apart from the well established feral population in the Cotswolds, the best place to see Red-crested Pochards is on the east coast, between Yorkshire and Kent. The nucleus of sightings in the Midlands region and within a 50-mile radius of London are almost certainly all of feral origin. However, under the terms of Category C species, these are all perfectly acceptable, and it is only those birds which show obvious signs of escape (i.e., residing in town parks, ornamental ponds, etc., and consorting with Mallard and Coot rather than with Pochard or other diving duck) that are unacceptable. A population of up to 42 free-flying birds are also known to be present in Paultons Park (Hampshire).

The following sites are divided into two sections: those for Red-crested Pochards considered to be of wild origin and for those of birds likely to be of feral origin.

❶Abberton Reservoir - Essex
A flock of up to four birds, possibly of wild origin, appears at Abberton Reservoir almost annually between late July and November, and consort with Pochard and large numbers of dabbling duck at the western end of the reservoir. They are best observed from the Layer Breton causeway, reached by leaving Colchester southwards on the B1026. The road crosses the Layer De La Haye causeway first and after 1.25 miles, a minor road leads north from a sharp left hand bend towards Layer Breton and Birch. The causeway is after half a mile.

Elsewhere in Essex, birds are regularly seen at Hanningfield Reservoir and at Coleman's Farm Reservoir.

❷Dungeness RSPB Reserve - Kent
Almost every autumn, Red-crested Pochards (mostly immatures) appear on the complex of flooded gravel workings at Dungeness between mid August and early October. They favour the largest (Burrowes Pit)and this is viewable from the information centre and hides 1-3. The reserve is open daily except Tuesday and is signposted off the Lydd to Greatstone-on-Sea road. About 1.5 miles south of the Lydd

roundabout, turn right at Boulderwall Farm and follow shingle road to reserve car park.

✱Cotswold Water Park - Gloucs/Wilts Border
Cotswold Water Park holds the largest concentration of this duck in Britain, with up to 65 birds being present in 1995. Several pairs breed in the park and the favoured site is Pit 64, although in autumn, Pit 29 is used as a moult gathering. The Park lies about four miles south of Cirencester. Turn right off the A419 on the B4696 towards Ashton Keynes and after two miles turn right towards Somerford Keynes. The series of pits on either side of this road are the favoured haunts of the pochards - visits between September and April are best.

✱Rutland Water - Leicestershire
Up to four Red-crested Pochards visit Rutland Water between mid July and February and are best observed from the Egleton Reserve at the western end of the reservoir. Leave Oakham south on the A6003 and after a mile turn left to the reserve information centre and car park. They can be seen from the main hide at reception or from one of five more along the nature trail. It is not unusual for birds to commute between Rutland and nearby Eye Brook Reservoir, where birds are best looked for from the Stoke Dry end along the western shore.

➊Pitsford Reservoir - Northamptonshire
A resident group of up to three birds tours various reservoirs, gravel pits and lakes in Northamptonshire throughout the year and generally visit Pitsford between July and February. The reservoir is situated five miles north of Northampton and can be easily reached from the A508. Leave Brixworth on a minor road to Holcot village and park in the narrow lane just NW of the causeway. The Red-crested Pochards can be identified amongst vast numbers of wintering duck in the Scaldwell or Walgrave arms of the reservoir. A bird club hide on the western shore gives ample views over the Scaldwell Arm.

Elsewhere, birds are regularly encountered on Sywell Reservoir, Hollowell Reservoir, Ravensthorpe Reservoir, Ditchford GP and Earls Barton GP.

➊Draycote Water - Warwickshire
During some winters, 1-3 Red-crested Pochards visit Draycote Water, particularly in Toft Bay or off the Dam. The site is permit access only however and is reached from the A426, 1.5 miles south of Dunchurch.

Elsewhere in the West Midlands, Chasewater and Blithfield Reservoir in Staffordshire regularly attract the species.

➊Stanton Harcourt GP - Oxfordshire
The large complex of gravel pits southwest of Stanton Harcourt regularly hold wintering Red-crested Pochards. The favoured area is Dix Pit, a large deep water site with two islands and good bankside vegetation. Entrance to the area is about a mile south of the village on the B4449. This is a working ARC site and visits outside working hours will see the gate locked. A limited area of parking can be found adjacent to the pit south of the weighing bridge.

Birds can also be seen elsewhere in the Lower Windrush Valley and on Farmoor Reservoir to the east.

Ring-necked Duck *Aythya collaris*

The Ring-necked Duck is an annual vagrant to Britain from North America, with several birds returning to favoured sites for up to fifteen years. A total of at least 220 had been recorded in Britain by the end of 1995 and a further 60 in Ireland. All sightings involve birds consorting with Tufted Duck or Pochard.

At the time of writing (January 1996), only one out of seven long-staying birds was being regularly reported, with drakes in Cornwall, Norfolk and Hampshire failing to appear. Details follow of sites

❷Inverdruie Fish Farm/Loch Morlich - Highland Region
A single drake appeared with Tufted Ducks at these two sites from 1980 until at least 1994. It spent much of the winter period commuting between Loch Insh and Loch Morlich and April-May on the freshwater pools at Inverdruie Fish Farm. These sites are all within close proximity of the A9 and Aviemore - the fish farm about a mile east on the B970 and Loch Morlich four miles east towards the Cairngorms Ski-Lift complex and Loch Insh south of Feshiebridge.

❷River Kent, Kendal - Cumbria
A single drake has consorted with Tufted Ducks in Cumbria since 25th November 1992 and was still present in January 1996. It seems to favour the River Kent between late November and early April and Killington Reservoir throughout the summer. At Kendal, it is best looked for at Scrogg's Weir or at the River Kent's confluence with the River Mint. Occasionally it visits Fisher Tarn Reservoir to the east. At Killington Reservoir, it favours the NW corner and can be best observed from the M6 service station at Killington or the access road leading north out of the back of the services.

❷Wayoh Reservoir - Lancashire
Until at least December 1994, a long-staying drake returned on a regular basis to Wayoh Reservoir, Edgeworth each winter. However, this bird has always been highly mobile in its movements and can also be seen at Holden Wood Reservoir, near Haslingden, Jumbles CP, north of Bolton, and at various sites in Greater Manchester.

❶Fairburn Ings/New Swillington Ings/Allerton Bywater - West Yorkshire
A regularly returning drake visits the Calder river valley in West Yorkshire between March and July. It is a mobile individual and rarely spends longer than two weeks at any one site. It visits the western lagoons at Fairburn Ings RSPB reserve (north of Castleford and accessed from the A1 at Fairburn), the large roadside pools at Allerton Bywater (adjacent to the A656) and at Astley Lake and New Swillington Ings, west of Great Preston.

❶Timsbury GP, Test Valley - Hampshire/Poole Harbour - Dorset
A single drake has returned since December 1985 to Timsbury GP, in the Test Valley (Hants). With Tufted Duck and Pochard, it frequented the flooded pit immediately west of the A3057, one mile north of the B3084 junction. The pit itself is strictly private but can be viewed from the hedgerow at the side of the road. Please park sensibly at this site. In some years, the bird commutes between here and floodmeadows at Fishlake, immediately north of Romsey.

The regular bird present in Cornwall from October 1990 until March 1995 was not reported in winter 1995/96. This individual had commuted between Drift Reservoir, Marazion Marsh and Helston Loe Pool.

Tresco Great Pool on the Isles of Scilly has also produced Ring-necked Ducks on a consistent basis, normally in September-October.

Ferruginous Duck *Aythya nyroca*

The Ferruginous Duck is a rare vagrant to Britain with between six and ten birds appearing annually, with just one surviving regularly returning bird. The main breeding range is in Russia, with large numbers wintering in Turkey, the Middle East, Pakistan and India, and smaller numbers in the Mediterranean basin. Vagrants to western Europe are believed to come from the eastern population, although those in Britain are most likely escapes from captivity. However, most records in Britain involve arrivals with large numbers of Pochards in late autumn, coincident with natural migration from central Asia to the Danube Delta and Black Sea.

The only regular individual in Britain is that which winters in the West Midlands between November and March. A further individual in SW Wales has returned for a short period of time for the past four autumns.

✱Lea Marston Balancing Lakes - Warwickshire
A drake has been present in the Lea Marston area for the past six winters and arrives with large numbers of continental Pochards between August and November. It normally remains until early March and appears to favour Balancing Lake 1B. Access to the lakes is from junction 9 of the M42. Drive northeast towards Kingsbury on the A4097 and at the first roundabout turn right towards Whitacre Heath. The Balancing Lakes 1A and 1B are on the right after 600 yards, whilst on the opposite side of the road is Coton Pool. The Pochard flock move regularly between all three sites and the area can be easily checked from a series of footpaths.

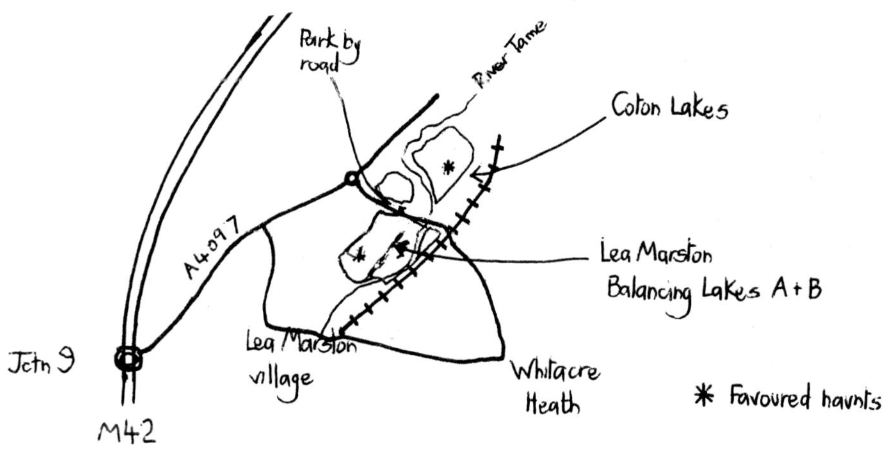

Scaup *Aythya marila*

The Scaup is a regular wintering species to the Firth of Forth, the Inner Solway and Islay with up to 5,000 birds appearing between September and April. It is less frequent elsewhere, but present off most coastlines during the winter months, with between 10 and 100 birds penetrating inland.

Scaups breed in Iceland and western Siberia and favour estuarine habitat and areas of relatively sheltered water. They dive to obtain food at depths of 20 ft and feed chiefly on molluscs, as well as crustaceans, invertebrates, worms and occasional small fish. In Britain, a preference for sewage and distillery outfalls has often been prevalent, but as many coastal sites are now environmentally cleansed, this type of habitat is difficult to find. These outfalls presumably contain nutrients which are distributed by tidal currents, benefiting the mollusc and marine life in the area. The following are the most productive sites in Britain for wintering Scaup.

✱Cromarty Firth, Inverness-shire - Highland Region
Up to 400 Scaup winter in the Cromarty Firth and are best observed from Invergordon, numerous stopping sites along the B817 towards Nigg Bay or from Udale Bay at Jemimaville, on the B9163 to Cromarty. Birds are present October to March.

Further north, the Dornoch Firth occasionally attracts the same flock, mainly in October-November. Edderton Bay is the favoured site.

✱Firth of Forth - Fife
The Firth of Forth regularly attracts a flock of up to 500 Scaup, with the birds being widely distributed between Kirkcaldy and St Andrews Bay. The largest concentrations are generally to be found in Largo Bay and can be 'scoped from the car parks at Levenmouth (opposite Methil Power Station) or from Lower Largo. Similar numbers can be counted in St Andrews Bay, along with vast numbers of scoter and other sea duck. View from the golf links to the north of the town, the birds concentrating at the mouth of the River Eden. At Kirkcaldy, the Tiel Burn area appears to be the most favoured feeding area.

✱Loch Indaal, Islay - Strathclyde
A wintering population of up to 1200 birds remains on Islay from late August to May and peaks in November to January. The birds favour Loch Indaal, on the west side of the island, and congregate particularly in Bridgend Bay, probably because of the sewage outfalls there and the distillery effluent. This gathering is often divided into 3-5 rafts and may be aligned along the River Sorn channel. Small numbers are sometimes recorded off other sections of the Upper Loch Indaal coast, between Bruichladdich and Port Charlotte on the A847. Loch Indaal can be easily reached from Port Ellen or Port Askaig, MacBraynes ferry services operating twice a day (Monday to Saturday) from Kennacraig, 7 miles west of Tarbert (☎ 01475 34531 for bookings).

✱Loch Ryan - Dumfries & Galloway

An increasing wintering flock of over 820 birds visits Loch Ryan between September and March, and is now the third most important site in Scotland. Loch Ryan is a very large and mostly enclosed sealoch situated in southwest Galloway. The Scaup favour the relatively shallow southern end and can be seen from the main A77 at the eastern end of Stranraer.

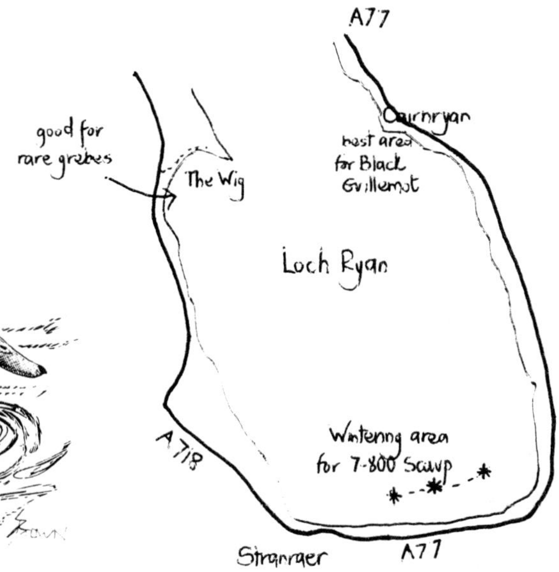

✱Blackshaw Bank - Dumfries & Galloway

The largest concentration of Scaup in the British Isles congregates in the Inner Solway, with in excess of 1,000 (occasionally up to 4,480) birds wintering in the shallow waters between the River Nith and Blackshaw Bank. These birds are best looked for from Carsethorn village or Southerness Point and in late winter, off Powfoot in the east.

A percentage of this same wintering flock can also be observed from the Cambrian coast under suitable conditions. The coastal road between Silloth and Skinburness and Campfield Marsh, west of Bowness-on-Solway, are the best observation points.

✱Pensarn - Clwyd

A wintering flock of up to 400 birds can be found in the vicinity of the sewage outfall off Pensarn beach. Park on the shingle ridge behind the railway station and 'scope offshore. The birds are invariably distant and are best viewed on calm days.

Elsewhere in Britain, Scaup are found in smaller concentrations, with up to 100 wintering in Wales (mainly in the Dee Estuary off Wirral CP, Cubbins Green, Caldy, Cheshire, and in the Glaslyn Estuary, Merionnydd), and up to 30 in the Wash (off coastline between Snettisham and Holme). A group of up to 220 also regularly winters on Scotney GP, four miles west of Lydd on the Camber Sands road (East Sussex).

King Eider *Somateria spectabilis*

The King Eider is an annual vagrant with at least 76 recorded in Britain and a further eight recorded in Ireland. Certain individuals have joined resident Common Eider flocks and have remained for periods of up to twenty years, but at the time of writing (January 1996), only four birds appear to be surviving. Scotland is obviously the best place in which to encounter this beautiful seaduck, and Shetland attracts the largest numbers.

✱Ythan Estuary - Grampian

Up to three drake King Eiders have been recorded on the Ythan Estuary in recent years and it is very unusual for not one to be present between April and July. The estuary is composed of inter-tidal mud and sandflats, with extensive mussel beds confined to the deeper channel. This in turn attracts vast numbers (up to 3,000) of moulting and breeding Eider. When present, the King Eiders appear to favour three distinct areas: at low tide, the shallow water between Inches Point and Waterside Bridge is best with the bird being visible from the Point (NK 005 256 - Sheet 38) or from the public footpath on the east side directly opposite, or occasionally the sandy dune beach at NK 008 250 is used for roosting. At high tide, King Eiders occasionally wander as far as the river mouth and can be looked for from the beach at the SE end of the village (NK 005 246). The estuary itself can be reached by taking the A92 north from Aberdeen for about ten miles, both Newburgh and Waterside Bridge being located on the A975.

✱Tayport - Tayside

A single drake King Eider, almost certainly the one that summers annually on the Ythan Estuary, winters at the mouth of the Firth of Tay between October and February. This bird is extremely difficult to locate amongst the 5,000 or so Eiders wintering here and is best seen at high tide, when the Eiders stream westwards into the mouth of the river as far west as Tayport Harbour. It is best looked for from the harbour car park or from the caravan site at Tayport (NO 465 285 - Sheet 54). At low tide, the bird can be identified roosting with Eiders on one of the sand banks in the middle of the river.

✱Tresta Voe - Shetland

For a number of years, a drake King Eider has spent most of the winter at Tresta Voe, between Veensgarth and Bixter. It can be looked for from the A971 at Tresta or from the minor road running south towards Sandsound.

The long-staying drakes in the Dornoch Firth and in Ayrshire/Galloway appear to have succumbed to old age and have not been recorded for several years.

Long-tailed Duck *Clangula hyemalis*

The Long-tailed Duck is a regular winter visitor to the British Isles with up to 14,000 birds present between November and April, primarily in Northeast Scotland and the Western Isles. It is regularly recorded inland with influxes up to 120 birds occurring during periods of severe weather. The main winter flocks occur on open coastal waters and feed mainly on a diet of molluscs, crustaceans and small fish, and are considered to be part of the Norwegian and western Siberian population.

They are found scattered in small numbers off much of the northern and eastern coastline of Britain and are most numerous in Shetland, Orkney, Highland, Grampian, Fife and Lothian. However, numbers wintering in the Moray Firth are internationally important.

✱Moray Firth - Highland Region
A staggering total of up to 12,000 birds winters in the Moray Firth and can be seen in large concentrations from various sites encompassing this huge area. In the north, large numbers can be seen from Golspie or Embo Pier (four miles north of Dornoch) and in the south, counts of 3,000 or more are regularly made from Burghead and large numbers can be seen offshore between Nairn and Culbin Bar, Findhorn beach, Lossiemouth and off Spey Bay from Tugnet car park.

✱Dunnet and Sinclair's Bays, Caithness - Highland
Both of these sites hold large numbers of wintering Long-tailed Ducks. Dunnet Bay lies to the east of Thurso and can be overlooked from the A836 and Sinclair's Bay is north of Wick and can be viewed from the beach car park at Ackergill.

✱Firth of Forth - Fife
A total of about 600 Long-tailed Ducks regularly winters in the Firth of Forth and can be best observed from Levenmouth, Lower Largo or from Ruddon's Point, west of Earlsferry.

Much smaller numbers winter on the Lothian side of the firth, but can be seen from Gosford Bay or Gullane Point.

✱Ross Back Sands and Bamburgh - Northumberland
A wintering population of between 130 and 160 birds remains off the north Northumberland coast during October to March. The favoured area is the stretch of coastline between Ross Back Sands and Bamburgh, Skate Road at the mouth of Budle Bay attracting the largest numbers. Walk east from Ross village across the golf links to view Skate Road or alternatively, watch from Bamburgh Lighthouse.

✱Brancaster Bay - Norfolk
A wintering flock of up to 120 Long-tailed Ducks can be found in Brancaster Bay between October and April. They are relatively mobile and move about depending on tides, but are best seen at high tide from Hunstanton Cliffs car park adjacent to the golf course or from Holme Dunes, or at low tide from Titchwell Beach or Thornham Point.

Surf Scoter *Melanitta perspicillata*

Surf Scoter is an annual vagrant from North America/NE Siberia with at least 200 recorded in Britain and 65 in Ireland. The majority of British records are from north and east Scotland, where, since 1979, a total of at least ten individuals have been resident. As to be expected, birds join up with wintering flocks of Common and Velvet Scoter and in most cases, it has become evident that they remain with these flocks throughout the year.

There are three main sites used in Scotland, the Moray Firth, St Andrews Bay and the Firth of Forth.

❷Moray Firth - Grampian/Highland
At least one drake Surf Scoter has been identified amongst the wintering Velvet and Common Scoters in the Moray Firth for the past twenty years. The birds are highly mobile in their movements and are only likely to be identified on calm days. The best sites in which to search are from Burghead Harbour, the dunes east of Lossiemouth or from Kingston or Spey Bay beach car parks.

❶St Andrews Bay - Fife
A large wintering flock of Common and Velvet Scoter congregate at the mouth of the Eden Estuary in St Andrews Bay and in late autumn/early winter are regularly joined by a flock of up to eight Surf Scoters. Calm sea conditions are essential for viewing at this site and the area can be overlooked from the metalled road running east of the golf links and parallel with the bay. Occasionally the flock gathers off the Sea Life Centre on West Beach.

✱Largo Bay - Fife
A traditional site which regularly attracts up to eight Surf Scoters between October and May. They loosely associate with the Common and Velvet Scoters in the bay and are best observed from Ruddon's Point, at the east end of the bay. They are very often feeding just west of the point and can sometimes afford excellent views. Ruddon's Point can be reached from the A917 between Lower Largo and Elie. About a mile north of Elie, turn right towards Shell Bay caravan park and follow to end. Between 9.00am and 4.00pm, the gate is open and one can drive to within 600 yards of the point, but at other times, a walk of one mile must be made from the gate. These birds commute between this site and St Andrew's Bay and in some years are seen from the south side of the Firth of Forth, off Musselburgh and Aberlady Bay.

Three further sites, used regularly in the past but not in 1995, were as follows.

❷Murcar Golf Course/Blackdog Rocks - Grampian
During most summers, 1-2 Surf Scoters are identified amongst moulting flocks of scoter offshore. They are regularly seen between June and August and presumably involve birds from wintering sites further south in Largo Bay and St Andrews Bay. The area can be checked from the dune car park at Blackdog, east of the A92 about two miles north of Aberdeen.

❷Abergele - Clwyd
For several winters, a drake Surf Scoter wintered with Common Scoters off the north coast of Wales. The favoured areas were offshore of Abergele beach, Pensarn beach and Llandullas beach, although in some years, it was seen as far west as Llanfairfechan beach.

❷Pendine - Dyfed
The Common Scoter flock which winters between Ragwen Point and Gilman Point, Pendine, occasionally attracts a Surf Scoter. Check from Pendine village.

Velvet Scoter *Melanitta fusca*

The Velvet Scoter is the least abundant of the regular wintering sea duck in Britain and is largely confined to the Firth of Forth and Moray Firth, with smaller numbers being recorded along the east coast to Kent and along the south coast to Cornwall. It is rare inland and is mainly found amongst flocks of the more abundant Common Scoter offshore.

The birds which winter in Britain are considered to come from the western Siberian and Norwegian populations and may number up to 6,000 birds in some winters. The principal wintering site is in the Moray Firth.

✱Moray Firth - Highland Region
A regular wintering flock of up to 1,000 birds frequents the Moray Firth between October and April. The principal observation points are from Burghead and Findhorn village in the south and Embo Pier, near Dornoch, in the north.

✱St Andrews Bay - Fife
Up to 1,200 Velvet Scoter join the raft of 3,000 Common Scoters in St Andrews Bay between November and March. They are best seen at the mouth of the Eden Estuary and form large flocks off the golf links road to the north of the town.

✱Firth of Forth - Fife/Lothian
A relatively small wintering population of up to 300 birds survives the season in the Firth of Forth. The largest numbers congregate in Largo Bay and are best observed from Lower Largo or Ruddon's Point. During some winters the birds drift south and can be seen offshore of Musselburgh and Gullane Point.

Elsewhere in Britain, Velvet Scoters are recorded in only very small numbers, with 10-30 being regular off Pendine (Dyfed) (see Surf Scoter), Llandullas beach (Gwynedd) and between Titchwell beach and Hunstanton Cliffs (Norfolk). However, large numbers are recorded on passage past several east coast locations, including Dungeness (Kent) and Flamborough Head (Humbs).

Smew *Mergus albellus*

The Smew is a scarce winter visitor to Britain with between 150 and 350 birds arriving during an average winter, primarily in East Anglia and Southeast England. However, severe freezing conditions on the continent, where up to 10,000 birds winter on the Ijsslemeer and Rhine delta in the Netherlands, can result in up to 600 birds arriving. It is very scarce in Scotland, mainly because of the lack of ice-free waters in the winter months, and also in SW England and exceptionally rare in Ireland.

Smews prefer much the same habitat in winter as Goldeneye and Goosander, namely moderate-sized bodies of fresh water where they can dive for small fish, and show great site fidelity. All regular sites attracting more than four birds are detailed below.

✷Dungeness RSPB Reserve - Kent
The complex of gravel pits at Dungeness RSPB reserve are the most reliable site in Britain for wintering Smew. An average year will produce up to 20 birds but in January 1996, an outstanding 50 or more birds were present. They commute widely between the pits but generally prefer the largest, known as Burrowes Pit. This is overlooked by a series of well-positioned hides and from the information centre. The reserve lies to the west of the Lydd-Greatstone road and is well signposted. A gravel track runs for about a mile to the reception centre and car park. The New Diggings and ARC Pit further south are also worth checking.

✷Wraysbury GP - Berkshire
The second most reliable site in Britain for seeing Smew is the complex of gravel pits surrounding Wraysbury GP. Up to 16 birds range widely over the area between December and March, and favour Heron Lake in the east, the main village pit and the Horton pits. Heron Lake lies immediately north of the disabled water sports pit and can be reached via the public footpath from the B376 or from Wraysbury Station. The north end of the village pit can also be checked from the footpath leading south from the station, and Horton Pits lie immediately behind the shop and sailing club south of the railway bridge on the B376.

✷Tilehouse North Pit - Buckinghamshire
A regular flock of up to six Smew can be found on Tilehouse North Pit from January to March. They are fairly elusive here as they often frequent the area at the back of the island or the edges covered by overhanging vegetation. The site is situated on the A412 between Denham and Maple Cross and may be viewed from a footpath leading to the pumping station at SU 038 898 (Sheet 176) or from some raised earth mounds at the side of the A412 at SU 035 900. These birds commute to Broadwater Lake immediately to the east (view from cement works at SU 046 895) and Stockers Lake, Rickmansworth, at SU 043 934.

✷Amwell GP - Hertfordshire
Up to ten Smew have been regularly recorded in the Lea Valley at Amwell over the last ten years. They favour the large pit at the south of the complex and can be viewed from the footpath adjacent to the pit and the river. Park at TQ 374 134, on the minor road between Stanstead Abbots and Ware, and follow the footpath to the pits.

A further ten localities also regularly attract Smew but in generally smaller numbers. They may be totally absent in mild winters.

❶Loch of Strathbeg RSPB Reserve - Grampian
This is the most reliable site in Scotland in which to record Smew, these being present on the loch between December and February. The reserve has a visitor centre, several hides and a lochside boardwalk and can be reached from the A952 Peterhead to Fraserburgh road, some seven miles NW of Peterhead, at Crimond.

➊Druridge Bay area - Northumberland
Up to five birds commute between the various freshwater sites in the Druridge Bay area between December and March. Most frequented are Ladyburn Lake, Hauxley Reserve, Druridge Pools and Cresswell Pond.

➊Langtoft Pits - Lincolnshire
A regular group of up to seven Smew visits Langtoft Pits between January and March. The pits lie three miles north of Market Deeping. Leave the A15 at Langtoft on a minor road towards Cradge Farm and at the first crossroads turn left. The pits are on the left and can be viewed from the road.

Toft Newton Reservoir, four miles west of Market Rasen, also regularly attracts 1+ Smew.

✱Eyebrook Reservoir/Rutland Water - Leicestershire
Up to six birds regularly winter at Eyebrook Reservoir or Rutland Water and often commute between the two sites. At Eyebrook, the birds are best observed from the western shore, whilst at Rutland Water, information is best sought from the information centre at Egleton Reserve, one mile south of Oakham.

✱Pitsford Reservoir - Northamptonshire
Small numbers of Smew regularly appear at Pitsford Reservoir and can be watched from the causeway between Brixworth and Holcot or from the dam at the south end.

Elsewhere in Northamptonshire, regular sites attracting Smew include Earls Barton GP, Stanwick GP, Stanford Reservoir and Daventry CP.

➊Dosthill GP - Warwickshire
The gravel pits at Dosthill, to the north of Kingsbury Water Park, regularly attract up to five Smew between December and March. From the M42 at junction 9, take the A4097 towards Kingsbury and turn left at the first roundabout. Follow the lane round to the left and the main entrance to the Water Park is on the right after half a mile. Continue past this and over the motorway and then turn immediately right into a tiny lane and take the entrance on the left to Broomey Croft. Follow this road, forking right to leave the caravan site on your left, pass through the lifting barriers and continue to the car park at the end. Walk north along the towpath and the gravel pits are on both sides after one mile.

✱Little Paxton GP - Cambridgeshire
The complex of gravel workings north of Little Paxton village attract small numbers of Smew in December to February. They favour the larger pools to the north and can be viewed from the nature trail running northwards beside the Great Ouse river valley. Leave the A1 at Little Paxton and turn immediately left towards the nature reserve. Follow the road to the designated car park at the end and follow the Heron Trail to the east of the pits.

➊Abberton Reservoir - Essex
In most winters, Smew are present in small numbers on the reservoir. They are best observed from the Layer Breton causeway at the west end of the reservoir and loosely associate with the more abundant wintering Goldeneye and Goosander.

➊Holyfield GP, Fishers Green, Lea Valley Park - Hertfordshire
Holyfield GP lie to the north of Fishers Green car park and regularly hold up to six Smew. They can be reached via the public footpath from the car park, the site being located west of the B194 at TL 376 033 - Sheet 166. Turn left off the B194 at the turning to Hayes Hill Farm.

➊Chew Valley and Blagdon Lakes - Avon
A small number of Smew (1-4) normally winters in Avon, commuting regularly between the two large reservoirs at Chew Valley and Blagdon. They favour the north end of Chew, where they can be observed from Villice Bay, Herons Green Bay and from the road running between Chew Stoke and Bishop Sutton. Blagdon Lake can be viewed from the dam at the west end and from Rugmoor Bay.

Honey Buzzard *Pernis apivorus*

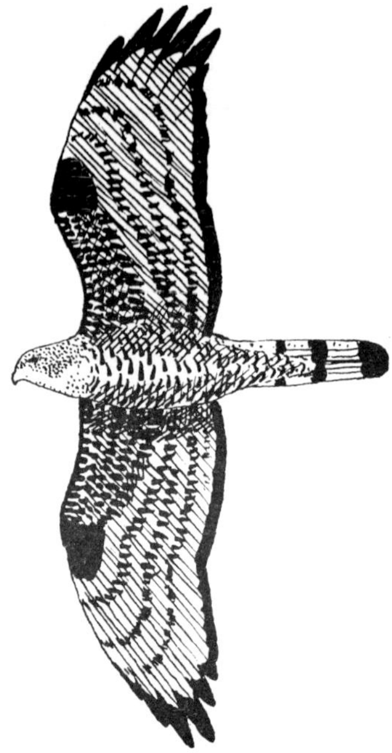

The Honey Buzzard is a rare summer visitor to Britain with a summer population of no more than 70 birds. About 26 pairs nest annually, mostly in southern England, with birds arriving in the third week of May and departing in the first week of September. They prefer mature deciduous woodland, especially beech, in which to nest, interspersed with wide open tracts over which they can search for food. They have vast territories, ranging up to five miles from the nest to obtain food, and regularly return to the same breeding sites year after year. The main food source consists of wasp larvae and pupae, which is dug out from ground nests, but amphibians, earthworms and other insects also form part of the diet throughout the summer.

Honey Buzzards are extremely secretive birds of prey and can be exceptionally difficult to locate, even at their traditional sites. There are four main sites in which to see them in Britain and with a little patience and time, should result in at least one view.

Honey Buzzards are best seen on clear, sunny days with little wind, and perform mainly between 10.30 and 11.45am and 1.30 to 3.00pm. Between 20th May and 20th June, males can be seen in acrobatic display. This takes the form of a tumbling display, with the male gathering speed whilst circling and curving steeply upwards on spread but motionless wings. He then stalls and performs a clapping action, raising the wings vertically over the body and trembling them. This can be repeated up to 12 times in quick succession before the bird resumes normal soaring mode. This display generally follows the presence of both birds, but can occur when just the male is in flight.

Honey Buzzards show great site fidelity and use 'rotational' nests in which to breed. With such regularity, it is possible for certain sites to be protected and permanently managed. However, Honey Buzzards are extremely susceptible to disturbance and are still on many an oologists 'wants list' and so site information has to be severely restricted for protection reasons. I have detailed six sites below which are permanently wardened and allow the birds to be viewed at a safe distance, and should provide, under suitable conditions, successful viewing of the species.

Prime sites

The following five sites are the most reliable in Britain for seeing soaring and displaying Honey Buzzards. It is imperative that birders do not approach or enter woodland breeding areas and view only from a distance. Information on other known sites cannot be published because of the sensitivity surrounding this species' breeding success and its threat from egg stealers.

✱Swanton Great Wood - Norfolk.
Apart from Devon, this is the best site in which to see Honey Buzzard in Britain. Birds arrive from 20th May onwards and if successful, remain in the area until 8th-15th September (adults leave in mid August). They are best seen from 10.30 to 11.30am and 1.15 to 3.00pm and afford some often extremely good views as they fly over the observation point.

Swanton Great Wood is an extensive woodland consisting of mature deciduous and coniferous trees. It is situated due south of Swanton Novers village and lies about six miles east of Fakenham. It can be reached by leaving the B1110 between East Dereham and Holt at the Fulmodeston turning, three miles north of the junction with the A1067. There is a designated *English Nature* car park at *TG 011 303* (OS Landranger Map 133) which allows a clear view of the entire woodland. Look north and scan the ridge just above the woodland, most birds spiralling up from the left hand side of the wood and often flying right overhead as they head for Fulmodeston Severals in which to feed. In 1994, up to five birds were present.

✱Great Haldon Forest - Devon
Prior to the Norfolk discovery, this was the best site in Britain in which to view this majestic raptor, with normally up to five birds present between 18th May and 10th September. Great Haldon Forest is a vast tract of woodland on the south Devon ridge and is easily accessed from the M5 SW of Exeter. Continue from the M5 on to the A38 and follow south for four miles ignoring turn-offs for Kennford and Torquay. Continue up the very steep incline until level with the race course on the left. Leave the A38 at the apex and turn right towards Dunchideock on the minor road at *SX 898 838* (OS Landranger Map 192). Follow this road through woodland north-westwards for two miles and then park at the *Haldon View watchpoint* clearly signposted on the left (at *SX 881 853*). Walk west from the car park to the viewpoint and look south into the steep valley. Birds appear daily between 9.30 and 11.00am and 2.00-4.00pm but prefer clear days with below force 4 wind. They can appear almost anywhere in the panoramic vista, but generally over Buller's Hill to the left or almost above the car park.

Common Buzzard, Hobby, Peregrine, Nightjar, Woodlark, Tree Pipit and Common Crossbill can also be seen at this site in summer.

✱Wykeham Forest - North Yorkshire
Up to five birds frequent the large mixed woodland at Wykeham from about 28th May to 27th August. Unlike other sites in Britain, these birds perform best between 9.00am and 1.00pm and rarely put in appearances during the afternoon. This is one of the better sights to actually study Honey Buzzards, as views here are often nothing short of spectacular, particularly when individuals head up Troutsdale or pass along the ridge above the viewpoint. However, birds here often gain tremendous height very rapidly and can be quickly lost from view.

The North York Moors Forest Enterprise kindly provided an access track and observation platform at this site, allowing easy observation over the entire forest. It can be reached by turning north on to a minor road off the A170 Pickering to Scarborough road at the Downe Arms Hotel in Wykeham. This should be followed for about four miles until one reaches the cross-roads at the top of the valley marked 'High Wood Brow'. Turn left here and the designated car park is on the right after half a mile at *SE 940 888* (OS Landranger Map 94). Walk north then past the forest pole barrier for 300 yards to the viewpoint. A further 'overspill' car park is located 200 yards further west along the road if required. Please note that extreme care is needed as the approach road to High Wood Brow cross-roads is extremely steep and narrow. For further information, read John Harwood's account of the site in *Yorkshire Birding* 3: 10-13.

✱Clumber Park - Nottinghamshire
For over twenty years, Honey Buzzards have returned to Welbeck woodlands in Clumber Park in which to breed or summer. They are notoriously difficult and are often at long range, but can be viewed from the minor road that runs between Norton and Carburton. They are generally present from 25th May until 20th August, but in recent years have been joined by 3-4 Common Buzzards in the area, making identification more difficult.

Clumber Park can be accessed from the A1 roundabout due east of Worksop. Turn south on the A614 and take first right after less than one mile. Follow minor road SW through Clumber Park and head towards Carburton village. At B6034 cross-roads, head straight across towards Norton and follow for 1.5 miles passing monument on right. Follow to highest point of road and park sensibly on verge at approximately *SK 583 721* (OS Landranger Sheet 120). View extensive woodlands to north of road, but be prepared for long periods of inactivity. Under no circumstances, leave the public road - the woods are strictly private and heavily keepered.

✱Acres Down, New Forest - Hampshire
The New Forest is an excellent area for Honey Buzzards with at least twelve birds returning on an annual basis. These birds are highly volatile and in many years, do not attempt to nest. The numbers are also swollen by sub-adult birds and these roam widely throughout the entire area during the summer months. Honey Buzzards are present in the forest from 14th May until late August and although several sites are known, the only 'publishable' site is that at Acres Down where birds can be viewed at a distance with no fear of disturbing them. Acres Down is not a regular nesting site but an excellent place where Honey Buzzards gather to feed and an area where sub-adult birds linger. Also, it is a superb vantage point from where many acres of New Forest can be viewed.

Leave the A31 west of Cadnam (at western end of M27) at Stoney Cross (signposted Emery Down). Follow SE for two miles and then turn right to Acres Down picnic site clearly signposted. Follow the track down crossing the ford and then turn right past the house and park in one of the two designated car park areas at *SU 267 099* (OS Landranger Sheet 195). Follow footpath to west along ridge for about half a mile to obvious viewpoint and look from heather- covered slope into surrounding valleys. Birds can be seen as they rise on thermals from the valley to the right of the ridge, anytime between 8.30am and 1.30pm. Afternoons are rarely productive at this site, unless a breeding pair is in residence.

Honey Buzzard Identification

Honey Buzzards are notoriously difficult to identify, being most easily confused with Common Buzzard. They are extremely variable in plumage, appearing in both pale and dark phases. In flight profile, they can be distinguished from Common Buzzard by their narrower and longer wings, more protruding and narrower head and neck and longer and more rounded tail. Honey Buzzard flight is more pronounced and less soaring, with deeper and more regular wing-flapping, whilst when gliding, Honey keeps its wings more right-angled to its body. Common Buzzard glides on flat or curved wings and soars with them raised. Plumagewise, the well-spaced dark barring on the undertail is diagnostic on Honey and against the light, Honey Buzzard always shows a large translucent patch confined to the inner primaries. Also, the barring on the underwing coverts is diagnostic and Honey is more heavily and evenly barred on the underparts. The black trailing edge to the secondaries and primaries is generally broader and more clear-cut than Common and the black on the wing-tip is confined to just the outer tips.

Immature Honey Buzzards are much more difficult to separate, being generally dark and unbarred below, but have more pronounced barring on the underwings and four dark tail-bands on the undertail. Flight mode and silhouette are also good characters.

During the breeding season, the male's distinctive 'tumbling' display flight is also diagnostic.

Summary of characters in order of importance

1.Long tail with three characteristic tail-bands.
2.Translucent primary patch contrasting with dark inner wings.
3.Strongly barred underwing coverts.
4.Flat-winged soaring profile, not raised as in Buzzard.
5.Protruding Cuckoo-like head.
6.More pronounced, undulating flight than Buzzard, with pinched in more accipiter-like wing profile.

For further information, it is essential to read Gensbol, *Birds of Prey of Britain & Europe, North Africa and the Middle East*, pages 294-297 and Porter *et al*, *Flight Identification of European Raptors*, pages 32-35 and plates 13-15.

Red Kite *Milvus milvus*

The Red Kite is a scarce breeding resident in Britain with at least 77 pairs nesting annually in Wales and a wintering population of at least 284 birds. Small numbers also appear on the south and east coasts between March and May and a large reintroduction programme is in operation in Scotland and Oxfordshire. The wardening of nests to protect them from disturbance and egg collecting has been an essential ingredient to their slow recovery, following their demise to just 3-4 pairs at the turn of the century. Despite recent successes and a general spread in range, the species is still limited to a restricted area of Wales.

Red Kites breed in mature Sessile Oak woodland, often on steep valley-sides, and feed on small mammals, birds, earthworms, beetles and carrion. They remain in the breeding valley from late February to October and then join other birds to scavenge refuse tips, abattoirs and feeding sites throughout the winter months. It is at these large communal feeding sites that Kites are best observed, especially as many nesting sites are extremely vulnerable to disturbance. The location of such breeding sites cannot be publicised here, but those offered will afford observers of unrivalled views of this majestic and endangered raptor.

Summer Sites

✱Dinas RSPB Reserve - Powys

A small population of breeding Red Kites frequents the reserve at Dinas Hill and are best observed on clear, sunny days in April-June. Leave Llandovery north on unclassified roads to Rhandirmwyn and continue towards the Lynn Brianne dam. The RSPB information centre is located at Nant-y-ffin, four miles to the north, and is open 10.00am until 5.00pm April-August inclusive (closed Fridays). Red Kites roam widely in the valley but are often seen around Dinas Hill.

✱Devil's Bridge and the Cwmystwyth Valley - Powys

A large proportion of the Welsh Red Kite population survives in the area between Devil's Bridge and Rhayader and can be easily seen from the maze of unclassified roads in the area. On warm, windless days in March-May, large numbers of Kites can be encountered thermalling low above the valleys, particularly from the A4120 1-5 miles west of Devil's Bridge. Furthermore, the B4574 between Devil's Bridge and Cwmystwyth will produce sightings and the hills and valleys surrounding Elan Valley many more. The twelve mile drive on the unclassified road between Cwmystwyth and Rhayader is particularly productive, as well as for other moorland species, Dipper, Common Redstart, Grey Wagtail, Raven and Peregrine.

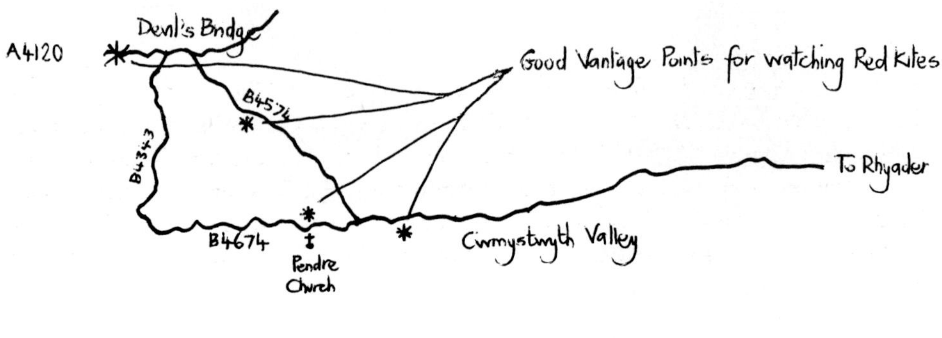

Winter Feeding Sites

✱Tregaron Bog - Dyfed

Tregaron Bog is a traditional winter site for Red Kites with up to 60 birds regularly wintering in the area. The raised bog is covered with moss and heather and patches of birch and willow carr and surrounded by a mixture of deciduous woodland and farmland rising up to more rugged hills, and lies immediately north of the town. Access is restricted, but a nature trail leads along the disused railway line to an observation tower, where Red Kites, Hen Harriers and Merlins may be watched between October and April. The trail starts from the B4343 some two miles north of Tregaron, immediately north of Maes-llyn Farm. Park in the designated lay-by. Red Kites, however, are often observed soaring over the road or the hills and woods to the east of the road.

A few miles south of Tregaron, on the A485, large numbers of Red Kites gather each evening between December and February to be fed by local inhabitants. A number of carcasses and bones are thrown out to the east of the road, with Red Kites arriving from 2.30pm onwards.

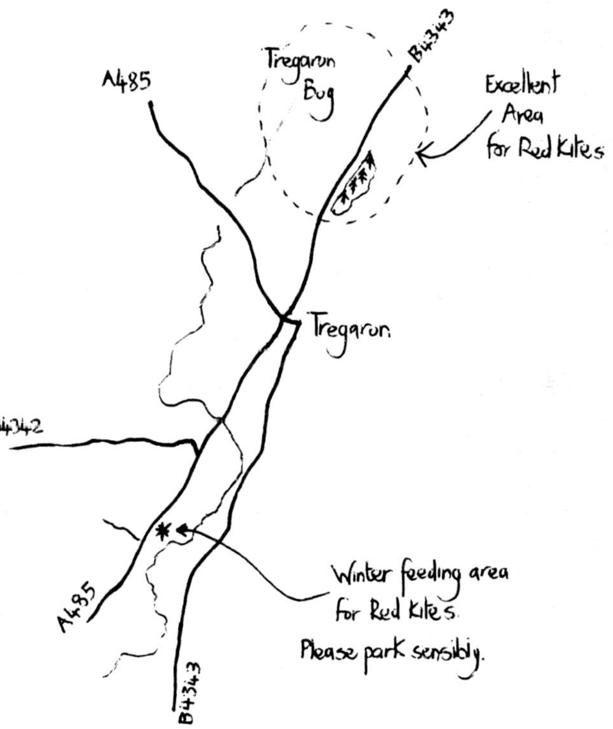

✱Gigrin Farm, Rhayader - Powys

A winter feeding site for Red Kites has been established at Gigrin Farm, adjacent to the RSPB reserve at Dyffryn Woods. Discarded meat and entrails are put out each afternoon to lure in the Red Kites and between November and March, up to 30 of these majestic birds can be seen at extremely close range from the observation hide. Feeding commences at 2.00pm and costs just £2.00 to witness. Leave Rhayader southwards on the A470 and after less than one mile turn left to Gigrin Farm (SN 980 678 - Sheet 147).

White-tailed Eagle *Haliaeetus albicilla*

The White-tailed Eagle is a rare vagrant to Britain with just fifteen birds recorded between 1950 and 1995. It is also a rare breeding resident, although this results from a successful reintroduction programme in northwest Scotland, initiated in 1975. By 1994, some eleven pairs had been established on territories producing an average yield of five young per year. The large majority of these reintroduced birds are ringed and wing-tagged, most recently with yellow upperwing tags with black numbering. The success of this latest reintroduction attempt (previous unsuccessful attempts were made in 1959 and 1968) has led the *UK400 Club* to recognise this population in terms of 'countable status' as of January 1996. However, birds are restricted to a handful of remote Scottish islands and it is generally advisable to refrain from causing undue disturbance to this tenuous population, particularly in summer.

It is desirable, if possible, to record White-tailed Eagle as a genuine vagrant, but as occurrences are currently running at one bird every six years this may not be conceivable for the average British birdwatcher (news on such an event being widely broadcast on *Birdline* 0891 700222 as and when it occurs). Alternatively, White-tailed Eagle may be encountered in remote NW Scotland, especially on Skye or on North Uist, and occasionally in the Beauly Firth or in Findhorn Valley, although the latter are mostly first-year birds, supplemented by fresh reintroductions.

❷Rhum - Highland
A total of at least 100 birds of Norwegian origin have been released on the island of Rhum since 1975, with some three quarters of these surviving therein or on nearby Hebridean islands. The island is a National Nature Reserve covering 26,400 acres and can be reached by passenger ferry from Mallaig. The boat sails on Monday, Wednesday, Thursday and Saturday, and takes 3-4 hours depending on sea and weather conditions. Both White-tailed and Golden Eagle can be seen from Kinloch, but sightings are highly dependent on weather conditions and are only about 10% reliable.

❷On nearby **Skye**, birds may be seen from Elgo at the end of the B8083 or on nearby islands of Soay and Raasay. However, sightings are highly unpredictable.

❶Gruinard Island - Highland Region
Gruinard Island is the site in which anthrax was purposefully introduced in the 1940's and for which reason is inaccessible and prohibited. This has led to its regular use by White-tailed Eagles as a wintering and roosting site and several birds, mainly immatures, can be seen at the south end of the island roosting on the beaches or on the slopes of the eastern side. Reasonable 'scope views of the birds can be obtained from the layby that overlooks the island on the A832 at Mungasdale, but remember this area is often shrouded in low cloud in winter and viewing may not always be possible. Gruinard Bay itself is an excellent area for wintering Great Northern and Black-throated Divers, as well as for good numbers of Black Guillemots.

Hen Harrier *Circus cyaneus*

The Hen Harrier is a scarce bird in Britain with a breeding population of about 630 pairs, confined mainly to north and west Britain and the Isle of Man. They are particularly widespread in Scotland, with breeding strongholds in Orkney, the east and north Highland, Strathclyde north and west of the Clyde and the Southern Uplands. Further nucleus's of populations occur in Lancashire and North Wales. The winter population is swollen by immigrants from the Netherlands and from Scandinavia and range between 170-800 birds, depending on the severity of the winter. The most recent of winter surveys revealed the presence of 77 different roost-sites during the 1983/84 winter. 50% of these were in reedbeds or on saltmarshes, the rest on rough grassland, lowland heath, heather moor and young conifer plantations. At almost all of the sites, birds roosted communally on the ground, and rarely exceeded more than six birds.

The following sites are the most regularly used roost-sites which should guarantee ample views of these beautiful raptors. The roosts are used between November and March and generally attract birds up to two hours before dusk:-

✱Ballaugh Curraghs - Isle of Man

The best and most heavily-populated roost-site in Britain is on the Isle of Man, where up to 161 birds have been counted coming into roost at Ballaugh Curragh, three miles northeast of Kirk Michael on the west coast. From the A14, take the minor road west to The Curraghs and then the first minor road on the left to SC 363 954 - Sheet 95.

❷West Freugh Airfield - Dumfries & Galloway

Up to four birds roost regularly at this site between October and April. Leave Stranraer south on the A77 and turn SE on to the A716 and then take the A757 towards Glenluce. Finally, after four miles, turn SW onto the A715. The harriers can be seen on the north side of the road about one mile south of the junction with the A757. Do not leave the road. This area is also good for roosting Merlin, Sparrowhawk, Peregrine, Short-eared Owl and large numbers of Greylag and Greenland White-fronted Geese.

❶Parkgate Marsh - Cheshire

A once-regular site used by up to eight birds but struggling to attract more than one during the winters of 1993-5. Park at the Old Baths car park at the northern end of the promenade at Parkgate and scan the saltmarsh. Leave the A540 westwards on Boathouse Lane (B5135) and turn right towards the Old Baths at the Boathouse Restaurant at the sharp left hand bend.

❶Scaling Dam Reservoir - Cleveland

A roost of up to six Hen Harriers is regularly encountered on the moorland surrounding Scaling Dam Reservoir. The birds arrive in December and remain until March and are best seen from the hide at the west end of the reservoir. The reservoir is situated about twelve miles west of Whitby on the A171.

✱Blacktoft Sands RSPB Reserve - North Humberside

The extensive reedbed at Blacktoft RSPB reserve attracts up to six roosting Hen Harriers between November and March. They can be seen from a selection of hides overlooking the reedbed from mid afternoon onwards. Leave Goole on the A161 to Swinefleet and Crowle. Fork left in Swinefleet onto the minor road signposted Reedness. Continue eastwards through Reedness, Swinefleet and Whitgift and about 0.5 miles east of Ousefleet, turn left into the reserve car park, immediately before the road bends sharply right. The reserve is open daily (except Tuesdays) from 9.00am until dusk.

✱Warham Greens - Norfolk
Up to five Hen Harriers roost on Warham saltmarshes and can be seen each evening between November and March flighting in from the east and west from 3.00pm onwards. Park in the coastal car park at the west end of Stiffkey Meals and walk west along the footpath to the metal gibbet. View the saltmarsh to the north and west of here. Turn north off the A149 at the west end of Stiffkey village along a minor road past the camping site.

❶Roydon Common - Norfolk
A limited number of Hen Harrier and Merlin still use the traditional site at Roydon, although disturbance from model aircraft and other leisure pursuits has seriously curtailed the attraction of this site. At the northern roundabout on the King's Lynn bypass, take the A148 east and after 50 yards, turn immediately right along the minor road to Roydon and then turn right across the common to view the bog from the raised hillock at TF 680 224.

✱Horsey Broad - Norfolk
The largest raptor roost in East Anglia is the extensive reedbed bordering the west side of Horsey Mere. This regularly holds up to ten Hen Harriers, eight Marsh Harriers, several Merlin and a resident flock of nine Common Cranes. The area can be viewed from two vantage points - from Horsey Mill on the east or Stubb's Mill on the west side. Leave Martham on the B1152 and head north towards Horsey on the B1159. Park at Horsey Mill and take the public footpath west along the north side of the mere to the disused mill on the raised bank after about a mile. Alternatively, leave the A149 eastwards to Hickling about one mile north of Potter Heigham. In Hickling Green, turn right along the minor road to the NWT reserve and after one mile, park in the car park at TG 420 220. The observation point at Stubb's Mill is at the eastern end of the road, about three quarters of a mile away. This is a sensitive area and it is imperative that one keeps to the public right of way.

✱Westwood Marshes, Walberswick - Suffolk
The vast reedbed at Westwood Marshes is a traditional site for roosting Marsh and Hen Harriers and regularly holds up to ten birds. The area can be viewed from Westwood Lodge, reached via the minor road leading south-westwards from Walberswick village. Park in the open space at the end of the road and 'scope the extensive reedbed to the south.

✱Wicken Fen - Cambridgeshire
A small wintering population of up to seven Hen Harriers traditionally roosts at Wicken Fen. For some reason, they are predominantly males and normally arrive about one hour prior to dusk. They can usually be seen from the Tower Hide on the reserve. Leave Cambridge north on the A10 and turn right onto the A1123 at Stretham roundabout. After just under four miles, turn right in Wicken village to the Fen. Park in the large car park on the left and advance on foot to the William Thorpe Building, where a permit can be obtained. An anticlockwise trail around the Sedge Fen leads on to the Tower Hide where ample views over the entire reserve and the adjacent Adventurer's Fen may be obtained.

✱Capel Fleet, Isle of Sheppey - Kent
The area around Capel Fleet and Leysdown and Harty Marshes are an excellent area for wintering geese, swans and wildfowl and for numerous birds of prey. From mid afternoon onwards between December and March, good numbers of Marsh and Hen Harriers and 1-3 Peregrine, Merlin and Short-eared Owl can be seen. From the B2231 west of Leysdown, turn right on a minor road signposted to 'Harty Ferry Inn'. This single track road crosses Capel Fleet after one mile and continues on through Harty Marshes. The extensive area of farmland and marshland either side of the road can be checked from numerous stopping places.

✱Stodmarsh - Kent

The extensive reedbeds at Stodmarsh have traditionally attracted Hen Harriers in winter, with up to eight being regularly recorded in the 1995/96 winter. The birds can be viewed from the Lampen Wall or from any of the stilt hides on the reserve, and flight in from 3.15pm onwards. Stodmarsh is a National Nature Reserve lying in the Stour valley five miles NE of Canterbury. Leave Canterbury on the A257 towards Sandwich and after one mile, turn left on the minor road opposite the Bekesbourne turning. This leads to Stodmarsh village after about four miles. In the village, turn left immediately past the Red Lion public house and follow the narrow and rather rough track to the reserve car park. From the car park continue on foot to the end of the lane, turning sharp right over a stile and onto the Lampen Wall. The path runs alongside woodland and scrub and then opens out into extensive reedbed and tracts of open water. From this open section, one can view the reedbeds on either side and watch raptors, Water Rails, Bearded Tits and the occasional Bittern.

✱St David's Peninsula - Dyfed

The St David's peninsula is one of the best sites in Wales for wintering Hen Harriers with up to 14 birds wandering the area between October and April. The favoured roosting site is Dowrog Common, where birds flight in from 3.00pm onwards to the cluster of willows in the middle of the heath. An unclassified road leaves the A487 St David's to Fishguard road at SM 782 268 (Sheet 157) and crosses the common to the west. There is a small car park at SM 772 277 and a number of bridle paths criss-cross the common from here.

❶Bursdon Moor - Devon

Up to five birds roost at this site from November to February as well as 1-2 Merlin. The site lies west of the A39, about thirteen miles west of Bideford. Take the minor road right towards Lutsford and Tosberry and view the moor from the road after 500 yards. It is best to watch these birds from inside the car as noise and disturbance may unsettle the birds.

❶Tregonetha Downs - Cornwall

A very small roost of 1-3 birds survives at Tregonetha Downs, north of the A30. Barn Owl, Peregrine and Merlin are also occasionally seen here, with December to February being best. Leave the A30 north on the B3274, four miles east of Indian Queens. Follow for just over two miles and over brow of hill, park carefully on the side of the road near the turning to Kernick. The harriers favour an area of heather-clad slope to the west of the road and usually fly in from a northerly direction up to an hour before dark. Scan area from raised verges and under no circumstances enter the down as this will cause serious disturbance to the roosting birds.

Montagu's Harrier *Circus pygargus*

The Montagu's Harrier is a rare summer visitor to Britain with a breeding population of no more than 28 pairs. It breeds largely in cereal crops, and occasionally heather-clad habitats, and is restricted to an area encircling the Wash and southern counties north to Oxfordshire. Arrival in Britain takes place between mid April and mid June, with departure in August and September. Numbers are swollen in both passage periods by migrants from southern and central Europe, where the population is in the region of 5,575 pairs (3,000 in Spain; 1,000 in Portugal; 372 in France; 80+ in West Germany; 55 in Sweden; 50 in Denmark; 15-25 in the Netherlands and 3-5 in Belgium).

Strict secrecy surrounds the whereabouts of most nesting Montagu's Harriers in Britain and rightly so, as past persecution by egg-collectors and game-keepers almost drove the species into extinction. Due to this rarity, nesting pairs arouse intense interest amongst birdwatchers and many hundreds of visitors may visit a pair throughout the breeding season. Apart from the risk of disturbance, this could cause problems to landowners who might then seek to discourage birds from nesting on their land. Where possible, the RSPB and other protection groups mount 24-hour wardening schemes, so that disturbance can be kept to a minimum. At one such site in Norfolk, a special observation area has been negotiated with local landowners so that pressures can be taken off other nesting pairs and birders can be rewarded with views of this rare species. It is recommended that all would-be observers visit this site and refrain from passing on details of other sites.

✱Estuary Farm, North Wootton - Norfolk
In the flat, wide expanse of low-lying cereal fields just east of the Wash, Montagu's and Marsh Harriers may be observed between late April and September. Up until late May, birds can be seen displaying or food-passing at regular intervals, but throughout June and into early July, sightings become much more difficult to obtain. The young birds are fed at irregular intervals throughout July and into mid August and once fledged, remain in the vicinity of the nest until the first week of September. At all times, the birds are best viewed prior to 10.00am or after 6.00pm, and are rarely seen during the heat of the day.

Leave King's Lynn north on the A149 turning left to Castle Rising a mile north of the A148 roundabout. Continue west to North Wootton and turn right just past the shop and phone box and right again by the church and follow a long winding road towards Estuary Farm. At the sharp left hand bend, park sensibly on the grass verge at the side of the road (making sure to leave plenty of room for tractors and farm vehicles) and view from corner. The Marsh and Montagu's Harriers favour the barley and rape fields to the southwest and can be seen quartering low over the fields hunting for food. This site is heavily wardened and visitors are requested to make a token donation for this cause in the security box provided. Under no circumstances leave the road as the farmland is strictly private. Any misbehaviour is likely to have serious repercussions on the nesting harriers and the public access to the site.

Migrant Montagu's Harriers can be encountered virtually anywhere in southern Britain between mid April and mid September, and are seen annually at Titchwell and Cley reserves (Norfolk) in May and at Elmley RSPB Reserve (Kent) throughout the summer.

Northern Goshawk *Accipiter gentilis*

The Northern Goshawk is a rare breeding resident in Britain with a population of between 312 and 360 pairs. It is seldom seen, however, because of its rather secretive behaviour. They breed at rather low densities in primarily larch and coniferous woodland and are predominantly sedentary. Food is obtained by hunting in both woodland and open country and generally consists of birds (mostly Stock Dove and Wood Pigeon), Grey Squirrels, Rabbits and other medium-sized mammals. The current British population of Goshawks is largely derived from imported birds which have escaped from hawk keepers or which have been deliberately released into the wild. The natural population, which was last recorded nesting in Yorkshire in 1893, was heavily persecuted by game-keepers and estate owners throughout the last century. It is probable that small numbers survived in Scotland and during 1938-1951, a small nucleus of birds was present in Sussex. Widespread releases and escapes took place between 1968 and 1976 and the population dramatically increased to up to sixty pairs in 1981 and this increasing trend still continues today.

The Goshawk is a highly desirable species for falconers and egg collectors alike and for this reason it is only possible to publish a few known sites. Nest-robbing and persecution are rife and even organised wardening and protection can often be ineffective.

I have chosen four sites at which it is possible to see Goshawks, but with a minimum risk of disturbance to the birds. Report any suspicious activity to the police and nearest RSPB office.

Goshawks are remarkably secretive species and are only reliably seen in March and April. At this time of year, pairs partake in display and can be seen circling high up over woodland for periods up to fifteen minutes. They show best on days of negligible wind and clear, blue skies, and normally appear between 9.30 and 11.00am and 2.15 and 3.00pm.

Identification is difficult as many female Sparrowhawks can appear large from a distance. It has a heavier and more powerful flight than Sparrowhawk and in some ways, recalls the appearance of a male Hen Harrier, especially when soaring and in flight mode. The wing-beats are slower and interspersed with long straight glides on flat, rather pointed wings giving a more pronounced curved trailing edge to the wings. Female Goshawks are generally Common Buzzard-sized but many males approach female Sparrowhawk in size. These individuals are about the same size as Carrion Crow, but can be separated by their shorter and more rounded tails (not square-cut or notched as in Sparrowhawk), broader and longer wings, deeper body structure producing a heavier breast profile, hooded appearance formed by distinct dark eye-streak and ear coverts and dark crown and more extensive and bulging gleaming white undertail coverts.

The following are the best publishable sites:-

❷Craigvinian Forest, Dunkeld - Perthshire
Craigvinian Forest is a huge area of coniferous woodland bordering the A9 west of Dunkeld. It is productive for both Goshawk and Sparrowhawk and birds may be seen between March and mid May displaying high over the plantations. The woodlands are strictly private and well keepered and observation can only be made from a selection of A9 lay-bys up to two miles north of Dunkeld or from the B898. Birds can be present on either side of the road.

✱Bakethin raptor watchpoint, Kielder Forest - Northumberland
An excellent site for displaying Goshawks with full protection being afforded to these heavily persecuted birds. Up to two pairs can regularly be seen from the viewpoint between 25th February and 30th April, but remember to visit on calm, clear days. The viewpoint is on the south side of Kielder Water at the NW end at NY 635 910 (Sheet 80) and can be reached from Saughtree in the west or from Bellingham in the east.

✱Derwentdale - Derbyshire
A relatively healthy population of Goshawks survives in the Upper Derwent Valley and can be seen under suitable conditions between mid February and mid April. The most reliable area for sightings is at the north end of the valley, adjacent to Howden Water. The best vantage point is from Windy Corner at SK 168 927 (Sheet 110). Park sensibly at the side of the road and 'scope the wooded valley to the north and the uplands to the east. Mornings are best at this site. Leave Sheffield west on the A57 to Ladybower Reservoir and then turn north up the minor road immediately as you cross the causeway. Follow this road north for about seven miles.

Nearby, birds may be seen at Holling Dale Plantation, Strines. Park at SK 221 908 and walk north along the footpath to a point at which one can view the woodland. Strines is north of the A57, along a minor road two miles east of Ladybower Reservoir.

❶Mayday Farm - Suffolk

The viewpoint at Mayday Farm is an excellent vantage point in which to see Goshawks between March and May. The large clearings at the site can often attract up to five different birds as well as a sizeable population of Sparrowhawks. It is essential to visit on calm, crisp and clear blue mornings and is also an extremely reliable site for Golden Pheasant, Common Crossbill, Siskin, Woodlark and Tree Pipit and often in spring, Great Grey Shrike.

The site is situated west of the B1106, two miles south of Brandon. Park sensibly by the entrance track and walk SW for about 1.25 miles to an obvious viewpoint, where four tracks converge. The Goshawks can be seen soaring high over the heath or flying low just above the fir woods on feeding sorties.

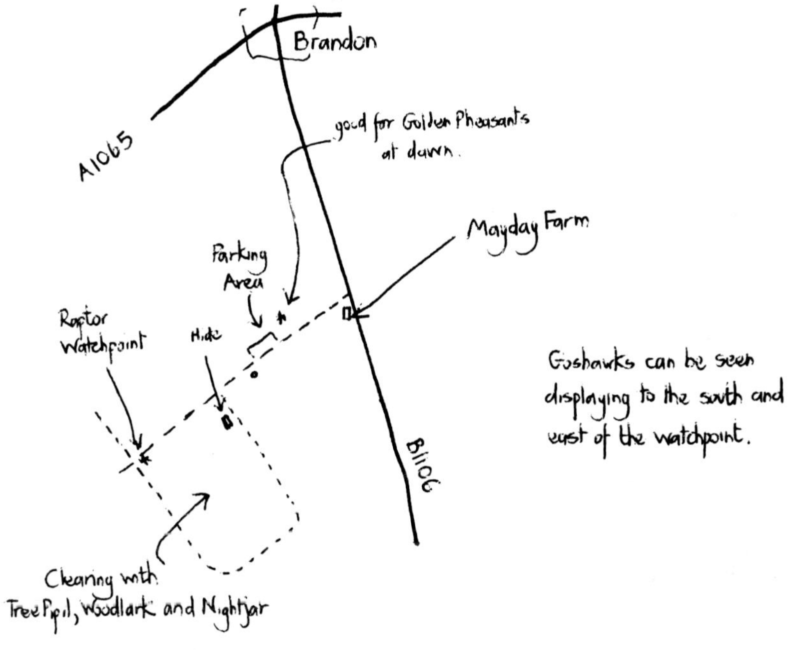

❷Crychan Forest - Dyfed

Goshawks are present in the massive upland block of conifer forest at this site and may be seen on bright, clear days in March and April. Turn off the A483 at Talgarth public house (SN 809 396) and continue along the minor road to the picnic site at SN 838 412. Walk along the main footpath until you reach the open clearing, where the woodland can be viewed. Further access points are at the Sugar Loaf picnic site at SN 837 428 or at Banc Cefngarreg at SN 820 390.

❷Brechfa Woods - Dyfed

This heavily wooded valley with extensive conifer plantations is another good site in which to see Goshawks between March and early May. Leave Carmarthen on the B4310 and park at the Abergorlech picnic site at SN 587 338. A forest track leads north into the forest and after about a mile opens up into an area that overlooks the woodland.

Elsewhere, Goshawks are particularly vulnerable to egg thieves and persecution and it would not be in the species' interest to publish site information. With well in excess of 100 pairs, parts of Mid and Central Wales can also produce results.

Rough-legged Buzzard *Buteo lagopus*

The Rough-legged Buzzard is a scarce winter visitor whose appearance depends largely on the availability of food and population levels in Scandinavia. In years of good Lemming availability, Fennoscandian populations of Rough-legged Buzzards thrive and produce many more offspring than average. This generally in turn sees large numbers dispersing south-westwards in autumn, with large numbers, often involving up to 400 birds, appearing in Britain and the Netherlands. This phenomenon however, tends to happen only once in every six years and numbers involved in any movement appear to fluctuate immensely. As expected with a Scandinavian migrant, the British occurrences are distinctly east-coast biased, with Yorkshire, Norfolk, Suffolk and Kent attracting the largest percentage of sightings. There are no regular sites in which to see this species in Britain, as in each winter the 1-5 wintering individuals frequent different sites from those the winter before. However, several sites are repeatedly visited, these being summarised below.

❷Fritton/Chedgrave Marshes - Norfolk
Rough-legged Buzzards favour this area of low-lying farmland, scattered trees and extensive marshland on a fairly regular basis and can be encountered from December to March hovering for food between Haddiscoe and Belton. The main vantage points to look from are Haddiscoe Bridge or from the footpath that leads out through Waveney Forest to Fritton Marshes. Both sites are on the A143 between Beccles and Great Yarmouth.

❷Horsey Area - Norfolk
The Horsey area is often good for Rough-legged Buzzards and these can be seen from Winterton North Dunes or from the main coast road between Martham and Sea Palling. There often tends to be a late March/April peak to sightings, suggesting an exodus of wintering birds from elsewhere in East Anglia.

❷Warham Saltmarsh/Wells Lodge Marsh - Norfolk
In influx years, the extensive saltmarshes north of Wells-next-the-Sea can attract up to eight Rough-legged Buzzards, all hanging in the wind as they hover for prey. Although distant, they are best viewed from the main coast road (A149) just east of Wells, opposite the Wells-Walsingham railway station. This area is also excellent for wintering Hen Harrier. Alternatively watch from the car park at the end of the track to Warham Greens.

Golden Eagle *Aquila chrysaetos*

The Golden Eagle is a breeding resident in Britain in internationally important numbers, with a population of between 540 and 600 pairs. It is almost exclusively confined to the west coast islands and highlands of northern Scotland, with just roving immatures accounting for records to the south and east of the range.

Like most species of birds of prey, Golden Eagles are heavily persecuted, with poisoning, egg-collecting and continuing loss of habitat taking an annual toll on numbers. Once again, because of these dangers, it is not possible to publish the whereabouts of breeding pairs and I can only recommend those sites in which there is a high chance of connecting with a non-breeding wandering individual.

✱Haweswater, Lake District - Cumbria
For over thirty years, a pair of Golden Eagles has been resident in the steep fells at the south end of Haweswater. They are the only pair to regularly nest in England and are fully afforded protection by the RSPB. The birds are present throughout the year but are best observed between April and June, when the RSPB wardens have their summer camps trained on the location of the nest. A telescope is essential here as for the security of the eagles, there is no access into their nesting valley. Please note that the site suffers heavily from hill fog and the area may be shrouded in mist for long periods, particularly in late winter and early spring. Leave the M6 at junction 39 and drive north into Shap village. Turn left towards Rosgill and continue to Bampton village where a sign will direct you to Haweswater. Follow this road for some six miles to the very south tip of the reservoir and park sensibly at the end of the road. Follow the footpath around to the west side and then continue north into Riggindale Beck, the first valley on the left. Walk into the beck as far as the RSPB observation centre, manned 24-hours per day. This is a relatively hard walk across typical fell terrain and may take up to 25 minutes to reach the site. Peregrine, Raven, Ring Ouzel, Common Redstart, Pied Flycatcher, Grey Wagtail and Wood Warbler can also be seen in this area.

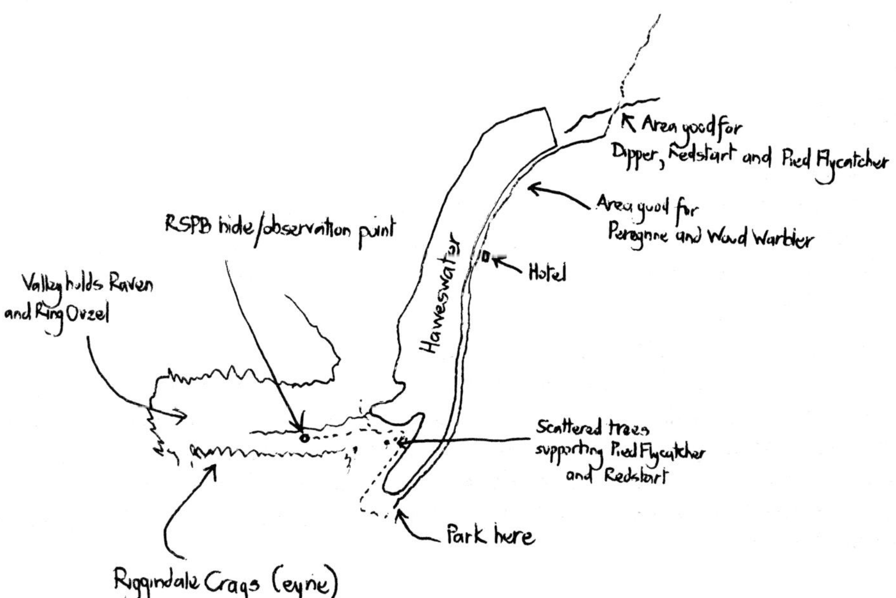

❶Findhorn Valley, Speyside - Highland Region

Away from breeding areas, Findhorn Valley is the best site in Scotland in which to see Golden Eagles. They are present all year and generally number up to 6-8 birds. Most are immatures, but occasionally adults wander in as well as first or second year White-tailed Eagles from the reintroduction programme on the west coast. The road up the valley is about twelve miles long and extremely treacherous in winter and is only single track throughout most of its length. There are several spots along the road which continuously produce sightings, but which require extreme caution when parking. It is essential to use Landranger Map 35 for this purpose, with recommended stops near Laggan (NH 749 227), opposite Creag Dhubh (NH 736 195), around Coignascallan (NH 730 185) and from Coignafearn Old Lodge (NH 709 178). Golden Eagles tend to soar or fly high over the valley, and particularly favour the rocky cliff-sides and the steep side-valleys on the east side of the river. The walk SW from Coignafearn Old Lodge to the new lodge at NH 680 155 will also produce sightings as well as the minor and gated road from Garbole to Farr.

This valley should also produce sightings of Goosander, Peregrine, Common Buzzard, Grey Wagtail, Common Sandpiper, Ring Ouzel and Red Grouse, as well as Red Deer and Mountain Hare. Goshawk is also resident.

The valley is reached from Tomatin (next to the A9, 15 miles south of Inverness and about 12 miles north of Aviemore. The single-track road begins at Findhorn Bridge.

❷Glen Tanar, Deeside - Grampian

The Glen Tanar National Nature Reserve is a huge area of Old Caledonian Forest and heather moorland situated south of the main A93 Banchory-Braemar road. Leave Aboyne south across the River Dee, then west on the B976. At Bridge o'Ess, take the minor road left to Braeloine, where there is a car park. From Glen Tanar House, there are a number of nature trails leading off through the surrounding woods. Two old high-level drove roads - the Mounth Road and the Firmounth Road - lead southwards along the Glen and eventually climb up into the hillsides. This area is very good for Golden Eagle although sightings are sporadic. The woodland walks are also productive, with Capercaillie, Black Grouse and Scottish Crossbill all being regularly recorded.

❷Glen Affric - Highland Region

Glen Affric is a relatively good site in which to see Golden Eagle, with birds being noted over hillsides on both sides of the road between Fasnakyle and Loch Affric. Leave the A831 west at Cannich and continue past Fasnakyle Power Station on the minor road towards the lochs. There are numerous stopping places along the road and the area is also very good for Capercaillie, Black Grouse, Scottish Crossbill and Crested Tit.

❷Glen Clova - Tayside

The steep corries and craggy hillsides of Glen Clova are an excellent site in which to see Golden Eagle, Peregrine, Raven and Ring Ouzel. Leave Kirriemuir northwards on the B955 and continue north as far as Glendoll Lodge, at the end of the road. The route up the glen should also produce Black Grouse, Short-eared Owl and Twite, the latter in the fields around Braedownie.

The west coast islands of Mull, Islay and North Uist are also very good for Golden Eagles, as well as Inverpolly National Nature Reserve, west of Knockan on the A835.

Osprey *Pandion haliaetus*

The Osprey is a rare breeding summer visitor and passage migrant to Britain, with a nesting population of at least 94 pairs in 1995. Although formerly quite common in England, it last nested there (in Gloucestershire) in 1840, and is now confined to the highlands of Scotland, where recolonization began in 1954, following immigration from Scandinavia.

Ospreys return to Scotland from wintering sites in Africa between late March and early June and remain in the highlands until late August through to late September. They favour Scots Pine forests in which to nest and feed on freshwater lochs, rivers, estuaries and fish-farms. The population in Scotland is largely centred around the Spey Valley and the RSPB has established a reserve centre, where the birds can be watched from the comfort of a hide throughout April-September.

Outside Scotland, Ospreys can be seen in increasing numbers as they head north (and south) to and from their breeding grounds, and can be encountered virtually anywhere in England during late March to May and from late July to October. Norfolk, Kent and Sussex have proved to be particularly attractive to this species in recent years, with some remaining at certain reservoirs for well over a month at passage times.

Breeding Ospreys

✱Loch Garten - Highland Region
Loch Garten is part of the RSPB's huge reserve at Abernethy Forest and has been home to breeding Ospreys since at least 1954. The eyrie is protected by closed circuit TV and is under surveillance 24-hours a day, but can be viewed without disturbance from the observation hide. This large, spacious hide is fully equipped with telescopes and binoculars, as well as having live video coverage of the birds, and is open from late April until August from 10.00am until 8.00pm.

Loch Garten can easily be reached from Aviemore and is situated midway between Nethybridge and Boat of Garten villages. It is very well signposted from Boat of Garten and lies east of the B970. The Osprey observation centre is on the opposite side of the road to the loch and the hide can be reached from the car park along a short track. This footpath is also very good for Crested Tit and Redstart, whilst the hide itself, can produce feeding Capercaillie. The Ospreys are most active when feeding young in July and August.

✱Inverdruie Fish Farm, Aviemore - Highland Region
The fish farm at Inverdruie is an excellent site at which to see fishing Ospreys between mid April and late August. There are two hides overlooking the ponds. The peak periods to visit are between 8 and 9.00am, 10-11.00am and from 6.30-9.00pm.

The farm is reached by taking the road from Aviemore towards Coylumbridge and the Cairngorm Ski-lift complex. After about half a mile east of Aviemore, the entrance is on the left at NH 898 117 (Sheet 36). Park sensibly at the site where there may be a nominal entrance charge. The fish pools are at the rear of the centre. The nearby bridge over the River Spey is an excellent site at which to see Dipper.

✱Loch of the Lowes - Tayside
The Loch of the Lowes has been a regular breeding site for Ospreys for over twenty years and birds are present there from late April until late August. There is a hide overlooking the eyrie and the Scottish Wildlife Trust has established a large reserve centre there.

The site can be reached from a minor road off the A923 Dunkeld-Blairgowrie road about 1.5 miles east of Dunkeld. The car park is situated at NO 050 440 (Sheet 52). The reserve is also very good for Northern Goshawk and Common Redstart, as well as breeding Great Crested Grebe and roosting Greylag Geese in autumn.

Ospreys on Passage

❷Bough Beech Reservoir - Kent
Ospreys are regularly present at Bough Beech Reservoir between April and May and from late July to September, and are best seen from the causeway at the northeast end.

The reservoir is situated five miles SW of Sevenoaks and is reached from the B2042, off the A25. Turn left at Cooper's Corner towards Winkhurst Green. The causeway is reached after half a mile. Mornings and evenings are best.

❶Weir Wood Reservoir - Sussex
Up to three Ospreys at a time have been present at Weir Wood Reservoir in recent summers, and are almost continuously present from late July until early September. The birds plunge for fish primarily at the eastern end and fly in from surrounding woodlands mainly in mid morning and in the evenings. They can be viewed from the car park at the western end, either by the hide or from the lane along the NW corner.

The reservoir is situated two miles south of East Grinstead and is reached by a series of minor roads. Leave East Grinstead on the B2110 and turn left after a mile towards Saint Hill. The road approaches the western end of the reservoir after a mile and a half.

❶Rutland Water - Leics
Ospreys have been increasingly reported from Rutland Water over the last five years, with 1-2 taking up residence from early August until September. The favoured fishing and roosting area appear to be Manton Creek and birds can be watched from the hide overlooking the bay or from Manton Bridge on the A6003. They can also be observed from the nearby Egleton Reserve and from the hides at the Lyndon Reserve along the southwest shore.

Merlin *Falco columbiarius*

The Merlin is a declining resident which breeds on moorland habitat in upland areas of northern Britain, Wales and Scotland. There are estimated to be about 550 pairs in Britain, but because of habitat loss and persecution, these numbers are believed to be in sharp decline. In autumn, many move south to winter in southern counties of Britain, where they inhabit farmland and marshes between November and February. They use traditional roost-sites and these are often shared with Hen Harriers.

The following are reliable sites in which to see Merlins in winter or in autumn.

❶Roydon Common - Norfolk
A limited number of Merlin still use the traditional site at Roydon, although disturbance from model aircraft and other leisure pursuits has seriously curtailed the attraction of this site. At the northern roundabout on the King's Lynn bypass, take the A148 east and after 50 yards, turn immediately right along the minor road to Roydon and then turn right across the common to view the bog from the raised hillock at TF 680 224.

❶Horsey Broad - Norfolk
The largest raptor roost in East Anglia is the extensive reedbed bordering the west side of Horsey Mere. This regularly holds up to ten Hen Harriers, eight Marsh Harriers, several Merlin and a resident flock of nine Common Cranes. The area can be viewed from two vantage points - from Horsey Mill on the east or Stubb's Mill on the west side. Leave Martham on the B1152 and head north towards Horsey on the B1159. Park at Horsey Mill and take the public footpath west along the north side of the mere to the disused mill on the raised bank after about a mile. Alternatively, leave the A149 eastwards to Hickling about one mile north of Potter Heigham. In Hickling Green, turn right along the minor road to the NWT reserve and after one mile, park in the car park at TG 420 220. The observation point at Stubb's Mill is at the eastern end of the road, about three quarters of a mile away. This is a sensitive area and it is imperative that one keeps to the public right of way.

❶Westwood Marshes, Walberswick - Suffolk
The vast reedbed at Westwood Marshes is a traditional site for roosting Merlins and regularly holds up to three birds. The area can be viewed from Westwood Lodge, reached via the minor road leading south-westwards from Walberswick village. Park in the open space at the end of the road and 'scope the extensive reedbed to the south.

➊Capel Fleet, Isle of Sheppey - Kent

The area around Capel Fleet and Leysdown and Harty Marshes are an excellent area for wintering geese, swans and wildfowl and for numerous birds of prey. From mid afternoon onwards between December and March, good numbers of Marsh and Hen Harriers and 1-3 Peregrine, Merlin and Short-eared Owl can be seen. From the B2231 west of Leysdown, turn right on a minor road signposted to 'Harty Ferry Inn'. This single track road crosses Capel Fleet after one mile and continues on through Harty Marshes. The extensive area of farmland and marshland either side of the road can be checked from numerous stopping places.

➋Stodmarsh - Kent

The extensive reedbeds at Stodmarsh have traditionally attracted Merlins in winter, with up to two being regularly recorded in the 1995/96 winter. The birds can be viewed from the Lampen Wall or from any of the stilt hides on the reserve, and flight in from 3.15pm onwards. Stodmarsh is a National Nature Reserve lying in the Stour valley five miles NE of Canterbury. Leave Canterbury on the A257 towards Sandwich and after one mile, turn left on the minor road opposite the Bekesbourne turning. This leads to Stodmarsh village after about four miles. In the village, turn left immediately past the Red Lion public house and follow the narrow and rather rough track to the reserve car park. From the car park continue on foot to the end of the lane, turning sharp right over a stile and onto the Lampen Wall. The path runs alongside woodland and scrub and then opens out into extensive reedbed and tracts of open water.

➊Bursdon Moor - Devon

Up to three birds roost at this site from November to February. The site lies west of the A39, about thirteen miles west of Bideford. Take the minor road right towards Lutsford and Tosberry and view the moor from the road after 500 yards. It is best to watch these birds from inside the car as noise and disturbance may unsettle the birds.

➋Tregonetha Downs - Cornwall

A very small roost of 1-3 birds survives at Tregonetha Downs, north of the A30. Barn Owl, Peregrine and Hen Harrier are also occasionally seen here, with December to February being best. Leave the A30 north on the B3274, four miles east of Indian Queens. Follow for just over two miles and over brow of hill, park carefully on the side of the road near the turning to Kernick.

✱St Mary's - Scilly

Merlins are regularly present on Scilly in the autumn and can be relatively easy to encounter. St Mary's is generally the favoured island and up to three birds traverse the island throughout October in search of small birds. They are most often seen in flight, as they dart from one end of the island to the other, but can be located roosting in pines when not hunting. Locations such as the airfield, Peninnis Head, Kitty Down and Telegraph are those where Merlins are most frequently observed.

Hobby *Falco subbuteo*

The Hobby is a scarce summer visitor to Britain with a breeding population of at least 800 pairs. They arrive from wintering grounds in Tropical Africa from late April to early June and depart from early August to late September and may be seen in small gatherings on migration. The British population is restricted to England, with nesting recorded north to Yorkshire and Derbyshire and west to Gwent, Powys and Shropshire, but with the main nucleus centred around the heath lands of Dorset to West Sussex. Hobbies breed in disused nests, particularly those of Carrion Crows, and are frequently in large, isolated deciduous trees or in pine clumps. They also nest regularly on tall electricity pylons, and normally lay 2-3 eggs in early to mid June.

Hobbies are best observed on migration, especially in April-May, and can be reasonably easily encountered at a number of localities in southern England. Parties of up to seven birds can often be seen in May.

✱Little Paxton GP - Cambridgeshire

Parties of up to seven Hobbies can be regularly seen hunting over the complex of gravel pits north and east of Little Paxton village. Birds are best seen from mid to late May and on clear, windless, sunny days and can be seen from the Heron nature trail, especially around Post 10 near the river. The birds hunt high above the trees and lakes for insects and have been joined by Red-footed Falcons on several occasions at this site.

Little Paxton is situated east of the A1 about six miles south of Huntingdon. The nature reserve is signposted immediately after leaving the A1 and the car park is on the left after about a mile. The complex of gravel pits lie to the north and can be reached by a number of public footpaths and nature trails.

✱Tring Reservoirs - Hertfordshire

Hobbies are present at Tring Reservoirs throughout the summer and are best observed from 25th April to 25th May and from 10th August to 22nd September. They often number up to five and can be watched hawking for insects over the reedbed and central causeway on Wilstone during spring and over the reedbed at Marsworth chasing roosting hirundines or Starlings in autumn. Exceptional close views are often obtained from the Drayton Bank hide on Wilstone, and birds favour the tall trees (normally containing large numbers of roosting Cormorants and nesting Grey Herons) on the central causeway in which to rest. Warm evenings are the most productive times in which to see these Hobbies.

Wilstone Reservoir lies east of Aston Clinton south of the B489. Park in the car park along the north side of the reservoir at SP 904 135 (Sheet 165) and follow the reservoir bank to the right to the Drayton Bank hide on the western shoreline. Marsworth and the other two reservoirs, Tringford and Startopsend, lie to the east and are best accessed from the Angler's Retreat car park adjacent to the B489 at SP 919 140. Startopsend and Marsworth are bisected by a causeway affording good views over the reedbed at the latter.

✱Stodmarsh - Kent

The Stour Valley is an excellent site in which to see numbers of migrant Hobbies in spring, and up to ten may be watched from the Lampen Wall or from the new Marsh hide at Stodmarsh throughout May. They fly high over the reedbeds in their search for insects and favour the river valley in which to spiral over.

Leave Canterbury on the A257 towards Sandwich and after 1.5 miles, turn left on a minor road towards Stodmarsh and Grove Ferry. This leads to the village on a narrow, meandering road after about four miles. In the village, turn left immediately after the Red Lion public house, along a narrow, rough track to the reserve car park on the right after 400 yards. The flood protection barrier, known as the Lampen Wall, leads from the car park and provides an excellent vantage point in which to view the reedbeds and marshes. The footpath reaches the River Stour after about a mile and continues along the south bank past a number of wet meadows, pastures and reedbeds on the right.

✱Cotswold Water Park - Wiltshire

The county of Wiltshire is one of the traditional strongholds of the Hobby population and the large complex of lakes of the Cotswold Water Park is one of the best sites in Britain in which to see this beautiful and charismatic falcon. The birds can be seen almost anywhere in the park but tend to congregate in spring along the river valley east of Waterhay Bridge, one mile east of Ashton Keynes (SU 061 933 - Sheet 163) and over Freeth's Wood and the complex of pits south of Somerford Keynes. It is not difficult to find up to fifteen Hobbies in this area from late April until mid May and from early August into September. Furthermore, a total of eight Red-footed Falcons have been identified amongst Hobbies at this site in May since 1988.

Cotswold Water Park lies south of Cirencester and northwest of Cricklade and can be reached via a number of minor roads from the A419.

✱Thursley Common - Berkshire

A number of Hobbies are present at Thursley Common throughout the summer and can be seen from a number of footpaths which criss-cross the heath. They are best seen in the afternoons and evenings and favour warm, sunny days. Park at the Moat car park at SU 899 416 (Sheet 186) and walk east along the boardwalks to view the area. Thursley Common is situated three miles south of Elstead.

✱Beaulieu Road Station - Hampshire

The area of heathland adjoining Denny Wood is an excellent area in which to observe feeding Hobbies between early May and early September. Evenings are best and birds can be best viewed from the footpath from the station leading down to the bridge over the railway or from the cumulus to the right of the track. This area is also excellent for Woodlark, Dartford Warbler, Common Buzzard and Lesser Spotted Woodpecker.

Beaulieu Road Station is best reached from Lyndhurst. Leave on the A35 and turn immediately southeast on the B3078 towards Beaulieu. After three miles, there is a designated car park on the right immediately prior to the railway bridge and the station hotel and public house. Park by this small patch of pines and walk south towards Bishop's Dyke and Denny Wood.

✱Acres Down, New Forest - Hampshire

Acres Down is another excellent site in the New Forest in which to observe Hobbies and other birds of prey. The birds can be seen hunting over the main valley at the site and are best seen on days of clear, sunny weather with light winds.

Leave the A31 westbound at Stoney Cross (note that there is no access from the eastbound carriageway) and drive south-eastwards on the minor road towards Emery Cross. After just under two miles, turn right to Acres Down picnic site and park at SU 268 097 (Sheet 195). Walk SW from the car park along an obvious ridge and watch from the raised heather-covered slope at about SU 266 090. This site is also good for Common and Honey Buzzard, Common Redstart and the occasional Hawfinch.

Peregrine *Falco peregrinus*

The Peregrine Falcon is a scarce resident in Britain with an internationally important breeding population of about 1,100 pairs, about 32% of the total breeding population in western Europe. Apart from breeding sites in mountainous areas of Scotland and in parts of northern England, the population is mainly sedentary, with many individuals wintering within the nesting area. The population is widely distributed throughout Britain, but shows a western and northern bias, with main strongholds in the uplands of Wales, Scotland and NW England and along rocky coastlines in SW England. There are many sites in Britain to watch this spectacular raptor, both in summer and in winter. They are widespread in Scotland and may be found relatively easily there in suitable habitat. It must be stressed that Peregrine is a Schedule One species and is afforded protection at all times.

Peregrines in summer

✱Craigellachie Cliffs, Aviemore - Highland

A pair of Peregrines has nested on the cliffs overlooking Aviemore for many years and can be watched from the large car park area adjacent to the artificial ski-slope and Osprey Hotel to the west of the town. The birds can be regularly seen performing sorties over the sheer cliff and may also be detected at the eyrie or roosting on the cliff face. They are generally present from mid March until early August.

➊Findhorn Valley - Highland

The Findhorn Valley is an excellent site at which to see Peregrine and it would be unlucky to drive its eleven miles without a sighting. The sheer rock face on the opposite side of the valley after about seven miles is the favoured area for sightings (and for Raven) and these birds may be watched soaring high over the valley or making sorties out from the crags.

The valley is reached from Tomatin, about twelve miles north of Aviemore on the A9. A single-track road follows the north bank of the River Findhorn for eleven miles to Coignafearn Lodge. This entire route is excellent for raptors, Golden Eagle being frequently observed over the ridges during the last four miles.

✱Haweswater - Cumbria

The steep crags surrounding Haweswater are excellent for Peregrines and birds can be seen regularly throughout the area. The 25 minute walk to Riggindale Crags for Golden Eagles will normally produce sightings between April and September, as well as the steep slopes adjacent to the hotel.

Leave the M6 at junction 39 and drive north into Shap village. Turn left towards Rosgill and continue to Bampton village where a sign will direct you to Haweswater. Follow this road for some six miles to the very south tip of the reservoir and park sensibly at the end of the road. Follow the footpath around to the west side and then continue north into Riggindale Beck, the first valley on the left. Walk into the beck as far as the RSPB observation centre, manned 24-hours per day. This is a relatively hard walk across typical fell terrain and may take up to 25 minutes to reach the site.

➊Leighton Moss/Silverdale Area - Lancashire

Peregrines are regularly noted at Leighton Moss RSPB reserve and nearby Morecambe Bay in summer and can be seen from numerous footpaths in the area. The Pilling Marsh wader roost is regularly visited by up to three birds throughout the winter.

Leighton Moss is accessed off the minor road between Silverdale and Yealand Redmayne and is open daily (except Tuesdays) from 9.00am until dusk. The Pilling wader roost is best seen from the Lane Ends car park, on the sea wall just north of the village.

❶South Stack RSPB Reserve, Anglesey - Gwynedd
Peregrines can be regularly encountered from the clifftop footpath which leads from the end of the road towards Holyhead Mountain. They can also be seen from the steep flight of steps down the cliff-face which leads to the lighthouse. This area is also excellent for Chough and in summer, large numbers of auks nesting, including Puffin.

The reserve is reached from Holyhead. Leave westwards on a minor road and after 1 mile, South Stack is signposted on the right. There is a car park at the reserve and an information centre and hide at Ellin's Tower overlooking the cliffs.

❶Ladybower Reservoir - Derbyshire
Peregrines can be regularly seen from the 'Goshawk observation point' at Windy Corner from late March to May and can be watched displaying over the adjacent moorland on bright, sunny days.

The best vantage point is from Windy Corner at SK 168 927 (Sheet 110). Park sensibly at the side of the road and 'scope the wooded valley to the north and the uplands to the east. Mornings are best at this site. Leave Sheffield west on the A57 to Ladybower Reservoir and then turn north up the minor road immediately as you cross the causeway. Follow this road north for about seven miles.

✱Willington Power Station - Derbyshire
In recent years, a pair of Peregrines has successfully reared young in an artificial nest site on one of the cooling towers at Willington. The birds are present between April and September, the power station being situated immediately north of the A5132 at the east end of Willington. The cooling towers are at the east of the complex and can be viewed from the road or from the car park just inside the security gate. Please act responsibly at this site and adhere to instructions.

✱Symonds Yat - Gloucestershire
A pair of Peregrines has nested at Symonds Yat for at least fifteen years and an information centre and viewpoint has been established by the RSPB so that the general public and birdwatchers can watch the birds at the eyrie. The site can be reached from the A40 at Old Forge. A minor road leads south from the B4229 to the observation point at Symonds Yat Rock (SO 564 160 - Sheet 162).

❶Portland Bill - Dorset
Peregrines are a regular sight on the Isle of Portland during the summer months and are best seen over the West Cliffs. Follow the footpath from the Bill car park to the Stopery and wander as far north as the A.R.E perimeter fence. The birds can often be seen 'hanging in the wind' above the sheer cliff-face.

✱Pendeen Watchpoint - Cornwall
The Peregrine is a relatively common bird in Cornwall and may be seen at a number of coastal localities. Pendeen Watchpoint, however, is one of the most reliable sites in which to see this magnificent raptor. The birds hunt along the coastline and may be seen from the lighthouse or the rocks below. Pendeen lies to the north-east of St Just.

Peregrines in winter

❶Caerlaverock WWT - Dumfries & Galloway
Several Peregrines hunt over the extensive mudflats and saltmarshes at Caerlaverock in winter and can be seen from a number of hides overlooking the area. The Eastpark Refuge is open daily in winter and is reached via the B725 from Dumfries, 1 mile south of Bankend.

❶Heysham Power Station - Lancashire

At least two birds can be found throughout the year roosting on the stage 2 building and are often seen in hot pursuit of Feral Pigeons about the structure.

In similar circumstances, a pair is normally present throughout the winter on the Blowick, Southport Gas Holder.

❶Martin Mere WWT - Lancashire

Peregrines are present at Martin Mere throughout the winter and can be regularly observed from the hides hunting over the reserve during afternoons and evenings. The refuge is situated on the minor road between the B5246 at Holmeswood and the A59 at Burscough Bridge. It is well signposted.

Marshside Marsh NR, north of Southport, is also an excellent site for Peregrine in winter.

❶Elmley Marshes/Harty Marshes - Kent

This extensive area of grazing farmland and marshland intersected by dykes and creeks is regularly visited by at least one Peregrine between September and March. Birds roam widely throughout the area but are often seen over the RSPB reserve and over the farmland at Capel Fleet and at Leysdown Marshes.

✸Slimbridge WWT - Gloucestershire

Up to three Peregrines winter on the Dumbles between October and March and can be seen from the Holden Tower at the refuge. They often spend long periods perched on strategically sited stumps and afford observers excellent opportunities for viewing. Slimbridge WWT is well signposted from the A38 south of junction 13 of the M5.

❷Langstone and Chichester Harbours - Hampshire/West Sussex

Up to three birds wander the various coastal localities between Thorney Island and Farlington Marsh during the winter and are most often sighted at the latter-mentioned site, hunting over the mudflats or rough grazing fields.

For Thorney Island, leave the A27 at Southbourne onto the minor road south to Prinsted. Park sensibly within the yacht club complex on the right and walk south along the sea wall to view the tidal mudflats and saltings.

For Farlington Marsh, leave the M27 at the intersection with the A2030 immediately north of Portsmouth. A metalled road leads off the roundabout and runs directly parallel to the motorway to the car park. A footpath along the sea wall gives ample views over the lagoon, grazing fields, marshes and saltings.

Red Grouse *Lagopus lagopus*

The Red Grouse is the only species endemic to Great Britain and is confined to Scotland, northern and south-west England, Wales and Ireland. The birds are sedentary with the highest densities occurring in northern England and eastern Scotland. They breed on open, treeless moorland dominated by heather or crowberry and nest between March and early June. There is a large number of sites in northern England and Scotland at which to see this species, a selection of the most reliable being detailed below.

✱Glenshee - Grampian
Large flocks of Red Grouse can be seen in winter along Glen Clunie, a steep-sided valley south of Braemar. The Old Military Road (A93) leads south to Spittal of Glenshee and the stretch from Auchallater southwards is highly productive. The area is also good for Peregrine and Arctic Hare and is one of the only places in Britain where Ptarmigan can be watched from the comfort of your car. Park in the ski-centre car park at NO 140 778 and view the high peak of Cairnwell to the west. The Red Grouse are present lower down the slope in winter and around the upper levels in summer.

✱Tomintoul Area - Highland
The A939 between Tomintoul and Grantown-on-Spey is an excellent road in which to see Red Grouse. The road passes through extensive areas of heather moorland west of Bridge of Brown and Red Grouse are relatively abundant in this area.

Other roads in the Scottish Highlands good for Red Grouse include the A9 between Blair Atholl and Newtonmore (particularly around Dalwhinnie) and from Carrbridge to Moy and the B9007 between Duthil and Lochindorb. They are also present in good numbers on the hilltops around Findhorn Valley and in the heather moorland on the gated Farr road nearby.

✱Pentland, Moorfoot and Lammermuir Hills - Lothian
Red Grouse are relatively abundant throughout the entire upland areas south of Edinburgh and can be seen from many of the roads passing through these areas. In the Lammermuirs, the minor road between Duns and Longformacus and its junction with the B6355 will produce sightings, up to 100 birds being noted on Kilpallet Heights in recent years. The B7007 south of the A7 east of North Middleton is productive in the Moorfoot Hills.

➊Northumberland
The extensive tracts of moorland in Northumberland hold large numbers of Red Grouse and birds may be encountered at widely scattered localities within the National Parks. Favoured localities are the B6341 between Elsdon and Hepple (over 200 birds survive in the Darden Lough area) and the A686 between Hayden Bridge and Alston, where many are present on Whitfield Moor.

➊Scaling Dam Reservoir - Cleveland
Red Grouse breed on the heather moors adjacent to the reservoir and can be seen and heard from the footpaths criss-crossing the area. The reservoir is situated on the A171, west of Whitby. Red Grouse are also reasonably common on the moorland of North and West Yorkshire and in County Durham.

✱Upper Derwentdale - South Yorkshire
The vast areas of heather moorland south and west of Langsett village hold large numbers of Red Grouse and are most abundant on Howden Moor. The minor road between Langsett and Low Bradfield should produce sightings, otherwise trekking out along the numerous public footpaths onto the moorland should suffice.

➊Horseshoe Pass - Clwyd
A small population of Red Grouse survives in the heather moorland bordering Llantysilio Mountain. Leave Llangollen northwards on the A542 and park at Horseshoe Pass at SJ 185 469. Explore the area of moorland to the west of the road. Early morning visits are essential.

Ptarmigan *Lagopus mutus*

The Ptarmigan is a bird of mountainous terrain and is restricted to the Scottish Highlands. They occur on all large areas of Arctic-alpine ground in Scotland and obtain cover from rocks and feed on scant alpine vegetation. They are beautifully camouflaged in both summer and winter and can be quite difficult to detect, allowing approach to within a few yards. The following are amongst some of the best sites in which to see them:-

✱Glenshee - Highland
The ski-ing complex at Glenshee is also an excellent site in which to record Ptarmigan and one of the only places in Britain where the species can be watched from the warmth and comfort of a vehicle. Up to 20 birds are present here during winter and can be seen between December and March feeding on the often snow-covered peak of Cairnwell. They can be 'scoped from the tourist car park at NO 140 778 and are to be found amongst the scree-slopes and rocks just below the skyline. During heavy snow, they often move down the hillside and can be seen at the same level as the car park.

Glenshee is located some seven miles south of Braemar on the A93.

✱Cairngorm Plateau - Highland
The Cairngorm plateau holds small numbers of Ptarmigan and these can be found within walking distance of the Ptarmigan Restaurant located near the peak of Cairn Gorm. They favour the rocky carns to the left of the peak and venture as far north as two miles from the restaurant. They can be extremely difficult to locate due to their camouflage and it is recommended to visit only when it is relatively clear and not windy. Low cloud, mist and rain will result in failure and is inadvisable. The walk south towards Ben Macdui can also produce sightings and is good for Dotterel and Snow Bunting.

Cairngorm is reached via Aviemore and is signposted from the town (Ski-lift centre). Drive the seven miles to the top car park and either walk or chair-lift to the peak. The lift costs £3.50 and operates from 9.00am until 5.00pm daily (weather permitting).

✱Carn Ban More - Highland
Carn Ban More lies to the south of Cairngorm and peaks lower at 3,443 metres above sea level. There is no chair-lift and birding here involves a gruelling uphill walk taking at least two hours. Ptarmigan are relatively easy to find at the top, as are Dotterel, and frequent the areas around the small carns for protection and cover.

Take the B970 south from Inverdruie and at Feshiebridge, turn south onto the Lagganlia road and follow to the end passing the airfield on the right. Park sensibly by the houses and follow the well worn trail through the heather to the start of the climb. This involves a long, difficult and gruelling ascent and is not recommended for birders with heart conditions or those with limited time. Do not attempt ascent in poor weather conditions.

Black Grouse *Tetrao tetrix*

The Black Grouse is a declining resident of Scotland and Wales, with much smaller numbers in northern England. They inhabit forest edges and heather moorland and groups of males indulge in lekking displays close to the breeding areas. Birds are generally secretive and are best seen early mornings, lekking males in spring being the most easily observed.

The males are sociable throughout the year but females are less gregarious and are usually seen in only ones and twos. Lekking usually commences in late February and continues until late May. Peak activity is at the end of April/early May and generally occurs during the first two hours of daylight. The lek is usually a flat area of short grass in a forest clearing, an open moorland or near a young conifer plantation.

Traditional Lekking Sites

A selection of the best sites are given below. All are in areas visible from the road - **please watch only from the car and do not trespass or disturb**.

✱Loch Ruthven - Highland

The Ruthven lek regularly contains up to twelve males and these can be seen from the RSPB car park at the east end of the loch during early morning. Scan the fields and moorland to the east of the road. The lek disperses early morning but birds can still be seen throughout the day feeding on the bilberry and birch scrub on the hillside.

Loch Ruthven is situated SW of Inverness and is reached via a minor road west of the B851 at East Croachy. Slavonian Grebes can also be seen at this site as well as Whinchat and Twite.

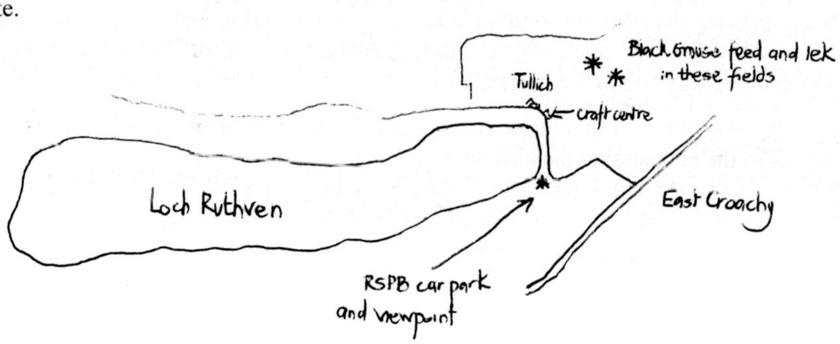

❷Duthil Burn - Highland
A small lek of up to five birds can be seen at dawn to the west of the B9007 about a mile north of the A938 east of Carrbridge. These birds are highly susceptible to disturbance and it is imperative not to leave the road. They favour the lush green areas at the edge of the moorland and fly in from the plantation on the opposite side of the road.

❷Kingussie Golf Course - Highland
A lek of up to eight birds can be seen during the first three hours of daylight at Kingussie. The golf course is situated at NH 764 015 and the eastern end is favourite.

❷Gairnshiel - Grampian
A lek of up to ten birds is regularly present to the east of the A939 at Gairnshiel. Leave Ballater northwards and after five miles the A939 crosses the River Gairn. The lek is in the fields to the north of the river.

❶Pitlochry - Tayside
A small population of up to eight birds survives on the moorland to the east of Moulin. Leave Pitlochry eastwards on the A924 and after three miles the road branches out into open moorland. The Black Grouse favour the Kinnaird Burn to the north of the road and may be located by parking sensibly at the side of the road at NN 965 597. The lek is operational between 4.30 and 7.00am in April and May.

❷Amulree Area - Tayside
Black Grouse are relatively abundant in the Amulree area and there are several sizeable leks along the A822 between Crieff and Dunkeld. The birds favour the lush green fields adjacent to the young conifer plantations in which to display, the ten mile stretch of road between Amulree and Trochny being the most productive.

❶Braco Moor - Perthshire
A sizeable population of Black Grouse survives in Strathallan, with a small roadside lek taking place during the first two hours of daylight three miles west of Braco. Take the B827 north of Braco west towards Comrie and after four miles, park in the roadside layby just east of Glenlichorn Farm. The Blackcocks lek in the field next to the road and can be seen exceptionally well from the car. Please do not leave the road and watch only from inside the car. Disturbance from birdwatchers at this site in recent years has seen a dramatic decline in the number of males attending the lek and this trend can only be reversed if birders behave responsibly. Away from the lek, birds can be seen at the roadside in the plantation to the north of the road, between the river crossing and the farm. This area is also good for Red Grouse, Short-eared Owl, Curlew and Whinchat.

✱Allenheads - Northumberland
Northumberland is one of the last strongholds of Black Grouse in Britain with large numbers surviving in the Otterburn Training Area and in the moor lands in the southwest of the county. The B6295 between Cowshill and Allendale Town passes through prime habitat and various leks can be seen from the road during the first two hours of daylight between February and May. Favoured areas are just south of Allenheads, on Allenhead Common at NY 835 464 (Sheet **87**) and at Coatenhill, near Spartylea (NY 843 480).

Plenmeller Common to the northwest is also excellent for Black Grouse. Drive south from Haltwhistle on a minor road to Boghead and several lekking fields can be seen from the road 1-2 miles south of Plenmeller.

❷Alston - Cumbria
A small population can be seen from the A686 NE of Alston village. View the edge of the conifers from the county boundary sign at NY 740 500 (Sheet **86**).

✸Langdon Beck - County Durham

Langdon Beck is a traditional site for lekking Black Grouse with up to thirty birds being counted in recent years. Early morning visits are essential and birds should not be disturbed. Leave the B6277 west of Langdon Beck on a minor road to St John's Chapel and after half a mile view from the first cattle grid. The grouse frequent the fields to the east of the road at approximately NY 852 326 (Sheet 92).

✸Arkengarthdale - North Yorkshire

A superb feeding site for Black Grouse is an area of moorland west of the minor road between Arkle Town and Thwaite. Leave the A66 south of Barnard Castle and drive south for about seven miles to Shaw Farm (NZ 004 053). The birds feed on the moorland surrounding the farm.

❷Grisdale - North Yorkshire

A small population exists near Garsdale Head and can be seen adequately from the road. Leave the A684 at the Moorcock Inn and drive north up the B6259 towards Kirkby Stephen. After 1 mile turn right towards the Blades Restaurant and scan the hillsides above.

✸Swallow Moss - Derbyshire

Swallow Moss is one of the best places in Britain in which to study Black Grouse and the RSPB has established a hide from which they can be watched without disturbance. The site is situated about eight miles NE of Leek and is best reached via the B50053 to Warslow. Leave Warslow on a minor road north-westwards and after two miles the road opens up into open moorland. Turn right towards Shawfield and Swallow Moss is immediately on the right after half a mile. Park sensibly at the side of the road or in the designated car park at SK 065 605 and follow the track eastwards to the hide. The Blackcock lek in the field in front of the hide and at least six are generally present. Please approach the hide quietly. Early mornings and evenings are best.

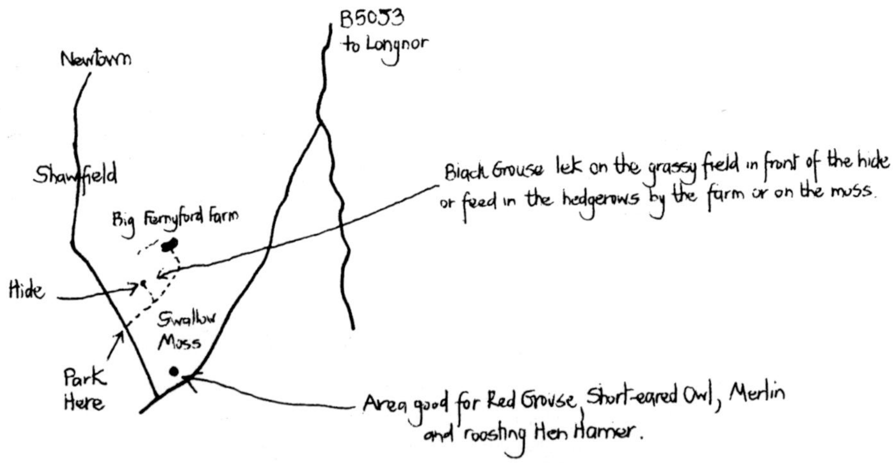

✸Pendinas - Clwyd

A population of 10-15 birds survives on Moel Garegog, an area of moorland adjacent to forestry commission land 1000' above sea level by the A525, seven miles west of Wrexham. Take the A525 west from Wrexham towards Ruthin and after seven miles park in the large lay-by at SJ 225 520 (Sheet 117). Scan the moorland opposite and also walk back up an unmade road on the south side of the A525 to a high point at SJ 235 523 and scan the edge of the forestry north of the road. It is essential to visit during the first two hours of daylight between February and March.

Capercaillie *Tetrao urogallus*

The Capercaillie is a localised Scottish breeder which is markedly decreasing in numbers, and is restricted to the Scots Pine forests of Tayside, Grampian and Highland. It is an incredibly difficult bird to locate and can take many hours of seemingly endless walking to find. Very occasionally a rogue male will take up territory near or on a public footpath and afford excellent views but more often than not, searching of suitable forest will prove fruitless.

Capercaillies frequent open mature pine woods on hills and valleys with an undergrowth of heather and bilberry and spend much of their time feeding on needles and shoots high in the treetops. The sound of them crashing through the canopy as one approaches is often all that is witnessed of them and it is a rare sight nowadays to see males lekking. It is probably one of the hardest birds on the British List to find and the felling of natural pine forest, as well as the hazard of deer-fencing, has had a devastating effect on their numbers.

Seeing Capercaillies requires patience and skill. It is essential to visit woodlands at dawn and to to be fully diligent. Keep to open tracks and footpaths and at the first sign of departure, scan the canopies and surrounding woodland as they fly through the trees.

❶Glen Tanar - Grampian

Glen Tanar holds one of the healthiest populations of Capercaillies in Britain and an early morning visit to this site should reap rewards. The forest is an important remnant of the old Caledonian Forest and contains many fine mature trees and an understorey of bilberry and heather.

From Aboyne, turn south off the A93 across the River Dee and head west along the B976. At Bridge o'Ess, take the minor road south to Braeloine and park at Glen Tanar House. A number of trails lead into the woods but for the best results, walk parallel with the east side of the river and walk upstream for up to two miles (either along the main footpath or the track that runs alongside the river). This area is also excellent for Capercaillie, Black Grouse, Dipper and Scottish Crossbill.

❶Grantown-on-Spey - Highland

The extensive tracts of forest to the east of the town hold Capercaillies, small numbers of which can be seen at dawn. Take the B9102 north-eastwards towards Lettoch and after half a mile, park sensibly by the road just past the golf course. A track leads southwards from the road and the whole area can be explored from a maze of footpaths. The Capercaillies favour the raised banks within the woodlands and are generally at least a mile in from the road. Again, it is more likely that all you will get is a quick flight view as one crashes out of the canopy.

In recent years, the area of woodland to the north of the Cromdale road has seen higher densities and it may well be worth checking this area. Please keep to designated footpaths.

Red Squirrel, Common Redstart, Tree Pipit, Crested Tit, Spotted Flycatcher and Scottish Crossbill are all frequently seen in these woodlands.

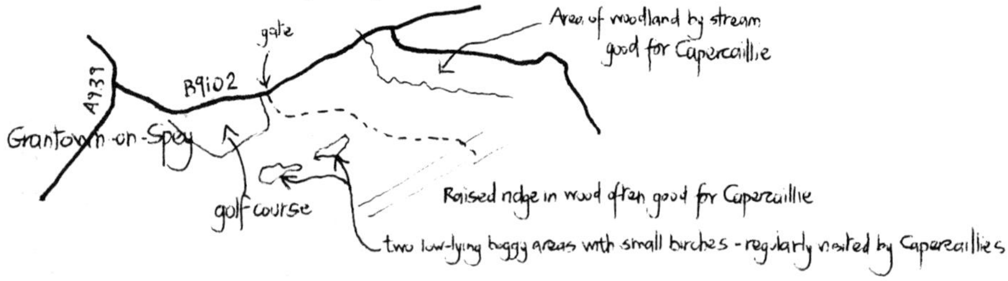

❶Loch Garten/Abernethy Forest - Highland

Quite often in April, Capercaillies take up residence in the woodland to the east of Loch Garten and can be seen from the 'Osprey' hide feeding on heather or small birch saplings. They are generally in the area to the right of the hide and can be seen up to mid morning. The wardens will be able to offer information on current sightings and whereabouts.

The extensive woodlands at nearby Forest Lodge hold up to eight Capercaillies but these are difficult to locate. An early morning search is essential and the tracks to the left of the main access track are the most reliable. Forest Lodge is reached from the Tulloch road at NH 998 167, an RSPB reserve car park being situated at NJ 019 161. The four mile walk to Dell Lodge should produce sightings during the first four hours of daylight.

❶Rothiemurchus - Highland

The Rothiemurchus is a huge area of Caledonian forest housing good numbers of Capercaillie and Black Grouse. Once again, it is essential to visit at dawn and birds may take many hours to locate. Leave Inverdruie on the Blackpark road and follow to the car park at the end at Whitewell (NH 915 086 - Sheet 36). Follow a series of public footpaths to the Cairngorm Club Footbridge at NH 926 078 and explore the triangle of woodland formed by the two rivers north of Carn Eilrig.

❶Inshriach Forest - Highland

Inshriach Forest is another huge tract of old Caledonian forest supporting reasonable numbers of Capercaillie. Park by the B970 at NH 860 060 and explore or follow footpaths from Feshiebridge or Lagganlia into the woodland. Early morning visits are recommended and birds may prove difficult to locate.

Quail *Coturnix coturnix*

The Quail is a rare summer visitor to Britain, with a small number arriving in May and the bulk of the population arriving from late June to July. Numbers fluctuate immensely from year to year, presumably due to climatic conditions, from as little as 150 males in 1991 to 570 in 1992 for example.

Quail inhabit cereal and hay fields and although easily heard, are almost impossible to see without flushing. Their far-carrying '*wet-my-lips*' call can be heard on warm, clear summer evenings but as it has ventriloqual powers, pin-pointing the bird's position can be extremely misleading.

Seeing Quail is a rather difficult pursuit and may take many hours of patiently watching over cereal crops for a brief flight view. It is an offence to tape-lure Quail and to tresspass in crops. A lost migrant on a coastal footpath or on an island is probably the only likely way that one will actually *see* Quail.

Because Britain is highly dependent on climatic conditions in its attraction to Quail, it is not possible to determine sites, as birds present one summer are not likely to return the next. However, there are a number of downland sites that do seem to attract Quail on almost an annual basis and can be recommended. The largest influx of Quail in Britain occurs in late summer, when many young birds disperse northwards out of Spain and France and adults are forced northwards in years of good populations in the Mediterranean.

The following sites are those visited by Quail on a regular basis but by no means annually:-

❶Royston Area - Hertfordshire

The Royston area is a traditional site for Quail with fluctuating numbers present from year to year. The whole area of farmland enclosed within the triangle formed by the A505 in the north, A10 in the east and A507 in the south is productive, but favoured spots are near Kelshall and Sandon. Leave the A505 southwards towards Kelshall and listen from vantage points south of Coombe Farm and just east of Stump Cross. The barley fields between Gannock Farm and Bury Barns north of Sandon are also regularly visited.

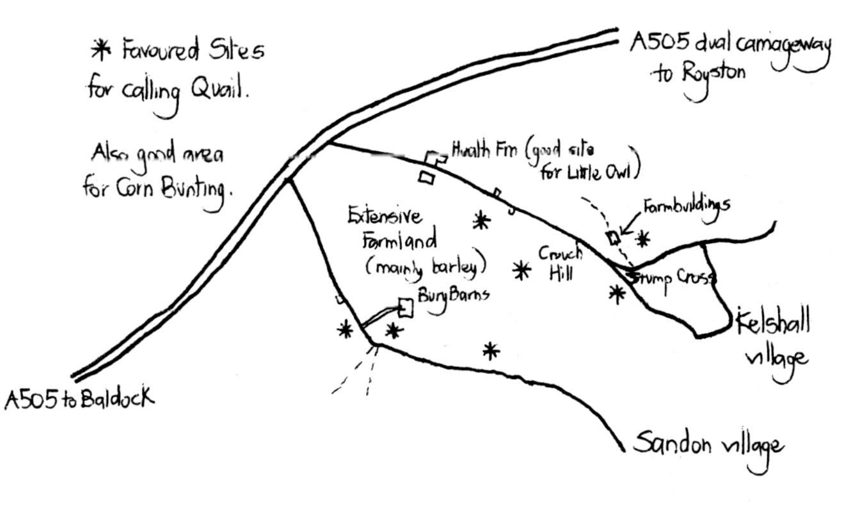

❶North Wessex Downs - Oxfordshire/Berkshire bdr

The Berkshire Downs attract Quail almost annually and up to 20 birds are generally present. They arrive from late April onwards and can be heard calling until mid July but are very unlikely to be seen. They favour tall cereal crops in which to nest and are best heard between 5.00 and 7.00am and from 6.00pm onwards (dry, warm and still nights are best).

The favoured areas are Compton and Blewbury Downs to the east of the A34 and East Ilsey and Bury, Cow, Sheep and Chilton Downs to the west of the A34. Leave the A417 at Blewbury on a minor road to Churn Farm and park at the end of the metalled road. Explore the immediate area of downland on a series of public footpaths and listen. Alternatively, leave the A34 at East Ilsey on a minor road to West Ilsey. Drive north from the village to the Ridgeway Path and park sensibly at the side of the road. The Ridgeway Path leads westwards across the downs and Quail should be heard within the first twenty-five minutes of walking.

❶Martin Down - Wiltshire

The extensive downland to the west of Martin village is a traditional site for calling Quail, although birds can be absent in some years. Apart from the odd calling bird in May, mid June to late July is the best period in which to see or hear them at this site.

A public footpath leads to the downs from the west end of the village and the whole area can be explored via a maze of tracks. The meadows at Martin Down proper are the favoured area, this also being an excellent site for Marbled White and other downland species of butterfly and Stone Curlews.

Martin village is situated south of the A354, nine miles SW of Salisbury.

The cereal fields at nearby Down Farm and Wyke Down are also excellent for Quail. Leave the A354 1 mile SW of the Handley roundabout on a minor road to Monkton Up Wimborne. The birds can be heard either side of the road within the first 600 yards.

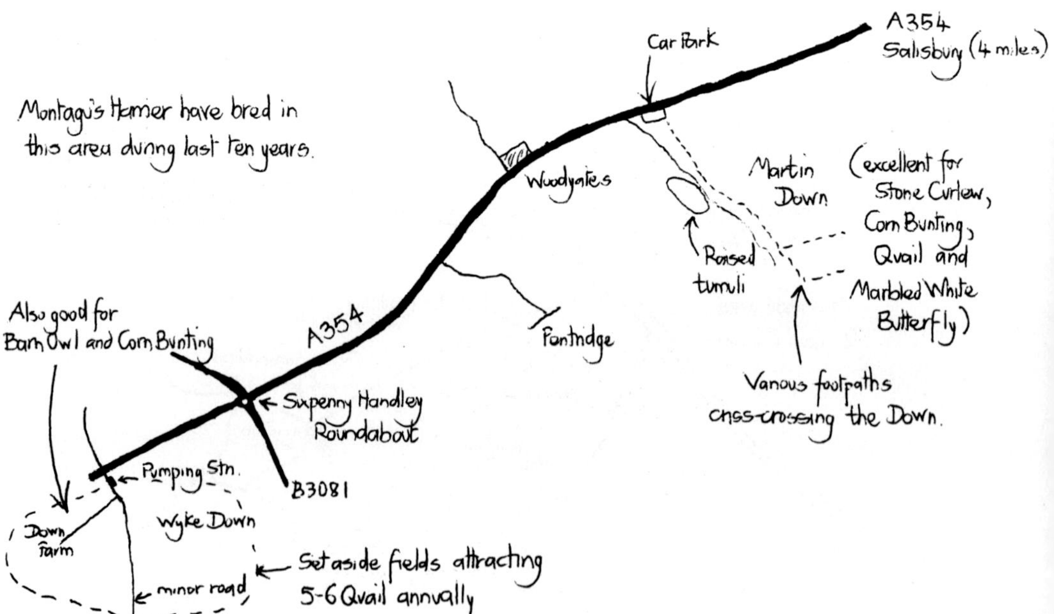

Golden Pheasant *Chrysolophus pictus*

The Golden Pheasant is a native of the foothills of China and was introduced in Britain in 1845, when small numbers were released in Norfolk. It quickly spread, and soon became established in the Norfolk/Suffolk brecklands, and was later introduced in SW Scotland, Anglesey, Northamptonshire, Sussex and Dorset.

Like Lady Amherst's Pheasant, it was admitted to the British List in 1971, and has since become a major attraction for avid year-listing or day-listing birders. Despite its gaudy plumage, it is an extremely difficult bird to locate, and skulks deep in thick cover throughout much of the day. It has a particular preference for beech woodlands, especially if they are mixed with fir, and is found almost exclusively in plantations of between ten and thirty years old. Like Lady Amherst's, Golden Pheasants are rarely seen in flight, and only leave the dense ground cover at dusk, when they roost in the firs. They are generally solitary creatures and are best observed at dawn or dusk, but in winter, may gather into groups of up to 20 birds, especially if a good feeding area in leaf litter is found.

The commonest sign of their presence is their distinctive call, a twice repeated loud, harsh and grating 'croak'.

The total population of self-supporting Golden Pheasants in Britain is estimated at 200 birds, with most occurring in the East Anglian breckland and in the woodlands along the South Downs. The best sites are in Norfolk and Suffolk, and West Sussex.

❷Dumfries & Galloway: at least 12 birds present at two sites.
A total of nine males and three females were recorded in 1993: four in the Palnure-Cairsmore area and five just outside Creetown. These birds are difficult and normally only encountered by chance.

❷Gwynedd: 1-2 pairs survive in the Pen-y-Parc woods, near Beaumaris, Anglesey.
These birds are extremely secretive and can be occasionally glimpsed from the gateway on the minor road towards the golf course at Beaumaris at SH 596755. The surrounding woods are strictly private.

❷Northamptonshire: a total of 3-5 pairs survive at one site.
Prime Site: Glapthorn Cow Pasture.
Glapthorn Cow Pasture is an isolated patch of deciduous woodland situated three miles NW of Oundle. It is best reached from minor roads out of Oundle and is north of the road about two miles west of Glapthorn village. Park at TL 006 903 (OS Sheet 141) and explore woodland via a series of footpaths. The pheasants are extremely elusive but can be seen with patience, especially at dawn. April visits are best.

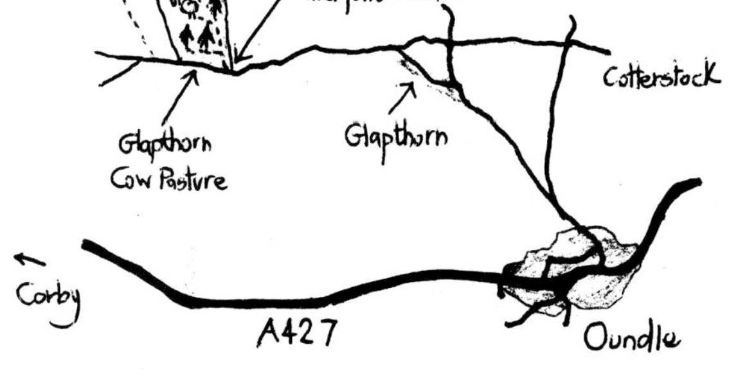

Norfolk: estimated population of 110-115 birds.

Brettenham Heath	7
Croxton Heath	6
Downham Market	1
East Wretham	13-18
Grime's Graves	15+
Hoe Rough	2
Lynford/Mundford area	25+
Swaffham	1
Thetford Forest	3
Wayland Wood	24-28
West Harling Heath	7
Wolferton/Sandringham	6

Favoured sites in Norfolk are the woodlands at Sandringham and Wolferton Triangle, where small numbers skulk in the dense rhododendron scrub adjacent to the A149, in the beech mast and leaf litter at Hockham's Belt and East Wretham Heath (adjacent to A1075, NE of Thetford) and at Croxton Woods (accessed from the A11 Thetford By-pass). A population of some 40 birds still exists in the woodlands north of Grime's Graves and to the east of Mundford.

✱Prime Site: Wayland Wood, Watton.

The best site by far in Norfolk is the isolated Wayland Wood reserve south of Watton. This NNT reserve holds a large population of birds which are relatively easy to see, especially in the mornings and just prior to dusk. It is a small deciduous wood situated to the east of the A1075, just south of Watton, with a trust car park at TL 923997, capable of holding eight cars. The woodland is criss-crossed by a number of damp and muddy footpaths, from which the pheasants can easily be encountered. In fact, a most favoured site is the first ride right, less than 50 yards from the car park.

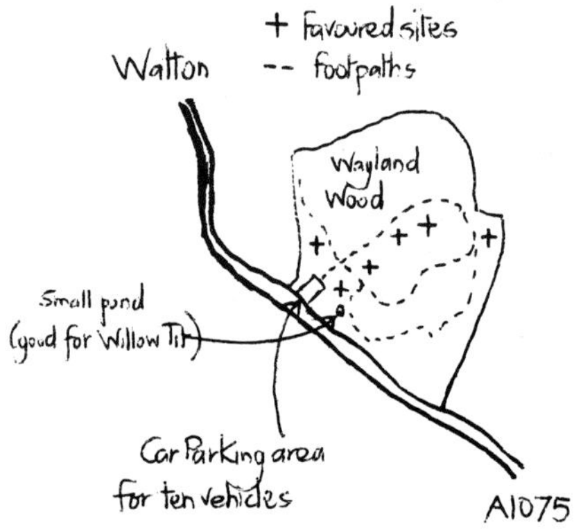

Suffolk: a dwindling population of at least 18 birds still survives in breckland.

King's Forest	6+
Wordwell	2
Mayday Farm/Brandon CP	8-10
Knettishall	2

❶Prime Site: Mayday Farm, Brandon

A population of up to 10 birds can still be found in the coniferous woodlands to the west of the B1106 at Mayday Farm. Park at TL 795835 (OS Sheet 144) and walk SW along the clearly marked *Goshawk Nature Trail.* Golden Pheasants favour the woodland on both sides of the track by the radar antennae, and can be seen (mostly in March-May) feeding on the tracks or in the leaf-litter from dawn until 7.00am. This area is also extremely good for Woodlark, Nightjar, Crossbill, Siskin, Sparrowhawk, Tree Pipit and migrant/wintering Great Grey Shrike. On fine, clear, sunny days in March and April, Goshawks from neighbouring territories can be seen displaying over the open heath (view from watchpoint at TL 786827), with sometimes up to five birds together (LGR Evans, 1990 and 1993).

Sussex: a population of 14 breeding pairs survives at two sites: Kingley Vale and West Dean Woods.

✱Prime Site: Kingley Vale NR

Kingley Vale NR lies six miles NW of Chichester (**West Sussex**) and is best reached by driving NW on the B2178 and turning right towards West Stoke village. Follow west through village and park in designated site at SU 825088 (OS Sheet 197). Follow metalled road northwards into steep valley and after about a mile turn left along the nature trail by the signs and information hut. Follow up steep valley until one reaches the mature yew forest (at about SU 819 103), between posts 19 and 21, and Golden Pheasants can be encountered on both sides of the track, especially in the last two hours of daylight.

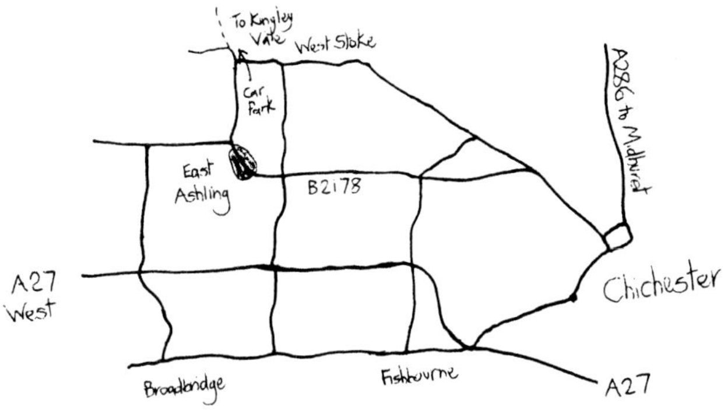

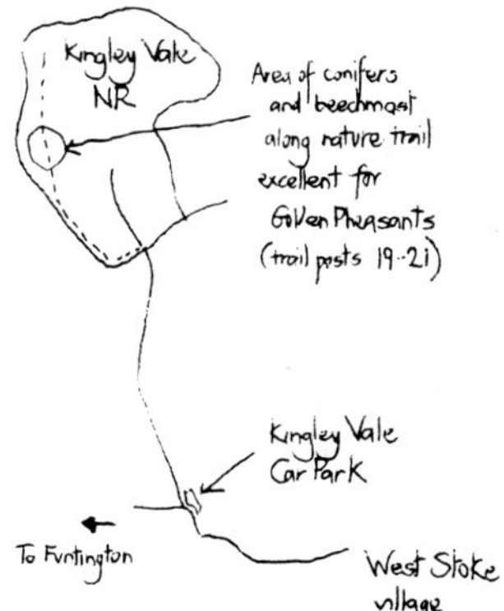

❷Hampshire: noted at two sites, with a population of 10-12 birds.

Up to three pairs remain in Head Down Plantation, in the Queen Elizabeth Country Park. These are believed to have spread from the neighbouring population in West Sussex.

At Exbury in the New Forest, an area of mixed conifer and deciduous woodland with dense rhododendron scrub cover, at least four birds still survive. These are the remnants of the 16 Golden Pheasants released there as recently as 1987.

❷Dorset: an isolated population in Poole Harbour consisting of at least 30 birds.

Up to 29 birds survive on Brownsea and Furzey Islands in Poole Harbour. A single female was also present on Green Island throughout 1993.

A feral population of at least six pairs also exists on Tresco (**Scilly**) and isolated birds, no doubt escapes from local collections, were recorded in at least five counties in 1995, including 1-3 birds in Dropmore Woods (Bucks) and 2-10 birds on the Mull of Galloway (Strathclyde).

Lady Amherst's Pheasant *Chrysolophus amherstiae*

The Lady Amherst's Pheasant is an extremely secretive and retiring species which inhabits the mixed woodland glades along the Greensand ridge in Bedfordshire and Buckinghamshire. It has a particular preference for beech and young conifer plantations, but relies heavily on woodlands with thick, dense cover in which to hide and forage.

It was first imported into England by Lady Sarah Amherst in July 1828, but did not successfully breed in captivity until 1871. It was initially introduced into Bedfordshire in 1890, when small numbers were released for ornamental purposes in Woburn Park. These gradually colonised neighbouring woodlands and by 1960, at least 200 were considered to be at large in the county. In 1971, it was officially admitted to Category C of the British List, and now maintains a precarious foothold as a British breeding bird. Naturalised it most certainly has become, confiding and easy to observe it definitely has not.

Its true origin is in the uplands of SW China, Tibet and Burma, where it inhabits the isolated mountain ridges covered in rhododendron and bamboo thickets and is now somewhat endangered.

Finding Lady Amherst's Pheasants at their known haunts is a difficult task, and requires patience and persistence. They are notoriously skulking and scamper away at the slightest hint of disturbance (branch cracking/voices/rustling/etc.). They are best observed during the winter months, particularly in snowy conditions, up to two hours after dawn and up to two hours prior to dusk. At these times they often frequent open woodland glades and tracks, and can sometimes be observed foraging amongst leaf litter under the canopy. In spring, they become increasingly active and can be heard uttering their diagnostic, rather mellow, disyllabic call from the woodland floor. April and May can be good months to see this species, especially at dawn, when amorous males may be encountered fighting for the attention of females.

The current population is estimated to be in the region of 72-90 birds, with the best sites at which to see them being detailed on the following pages. Up to five Lady Amherst's Pheasants survive at Halkyn Churchyard (**Clwyd**), in North Wales, but this population is considered to be not self-sustaining.

All currently known sites for Lady Amherst's Pheasant

Bedfordshire/Buckinghamshire border

Charle Wood and Brickhill Woods (4-6 pairs)
Lowes Wood (6 pairs)

Bedfordshire

Woburn Park	11-14
Washers and Daintry Woods -Eversholt	8-15
Moneypot/Breakheart Hill - Millbrook Area	8-10
Steppingley Wood	up to 6
Maulden Woods	3-5
Chicksands Wood	4
Flitwick Plantation	3
Warden Great Wood	3
Kingshoe Wood	3
Luton Hoo	2
Exeter Wood	2
Northill Home Wood	1
Total	54-68

Prime Sites

The following sites are the most reliable as at May 1996, but it should be noted that untoward disturbance to this species can result in its premature departure.

✱Maulden Woods, Clophill
This extensive woodland is situated 1 mile north of Clophill, to the west of the main trunk road, the A6. Access is from the large lay-by at TL 073394 (OS Map Sheet 153), where there is ample parking space for at least 70 cars (*Be careful as to not leave any valuables or optics on show here as many vehicles have been broken into, whilst birders are searching for pheasants). The Forestry Commission here have marked out a nature trail which takes you through all parts of the wood, and may produce Woodcock, Lesser Spotted Woodpecker, Hawfinch, Grasshopper Warbler, Marsh Tit and Willow Tit. The Lady Amherst's Pheasants however, favour the densely coppiced north and western sections of the wood, and can be watched feeding in the oak litter or on the wide tracks at dawn or in the evening. From the car park, walk west along the wide sandy track for about 400 yards, then explore the series of tracks on your right, or continue past the new plantation on the left and explore the tracks to the west. This is an excellent site in April and May, when the distinctive grating call of the male can be heard echoing through the wood. The site holds a population of between 3 and 5 males.

✱Lowes Wood, Brickhill
Up to ten Lady Amherst's Pheasants are known to survive in Lowes Wood, west of Woburn. Park by the county boundary sign at SP 924 327 (Sheet 165) and walk south into the wood opposite and explore the various footpaths that traverse the area. Two males can regularly be seen by the gate where you park at dawn in winter and in April and May, several males can be heard calling from the dense vegetation bordering the footpaths. Lady Amherst's Pheasants are frustratingly difficult to see and your only real chance is at dawn or dusk.

❶Charle Wood

Charle Wood is an extensive woodland, comprising mainly of oak, chestnut and conifer. It lies on the border between Bedfordshire and Buckinghamshire and is largely private, but is criss-crossed by a number of public footpaths. There is limited parking facilities on the county border at SP 925327 (OS Map Sheet 165) or on the roadside just west of there. Cross the stile, then follow the obvious footpath north until it curves around to the right and finishes at a crossroads. A long, private track leads NE, with another running almost parallel 100 yards to the south. These tracks are the best sites in which to see the pheasants, especially at dawn and dusk, particularly in the sloping hollows, where they forage in the oak litter. Also worth checking, is the rhododendron scrub running adjacent to the road.

*Note that this site is popular with dog walkers, etc., and is subject to disturbance, especially at weekends.

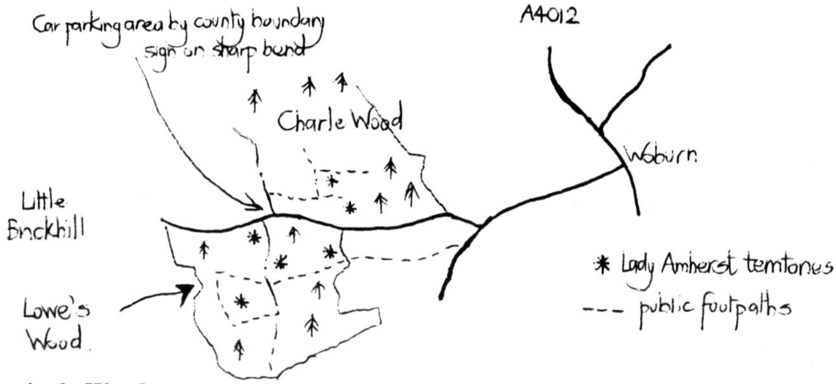

❷Steppingly Wood

This is one of the most reliable sites for this species, where up to six birds may be encountered in winter. It lies adjacent to the M1, two miles west of Flitwick, and is owned by the Crown Estates. Park at TL 000344 (OS Sheet 153) and watch from gate overlooking the long ride running SE. Flitwick Plantation, on the opposite side of the M1, also houses at least four birds, and these can be viewed from the public footpath at dawn or dusk.

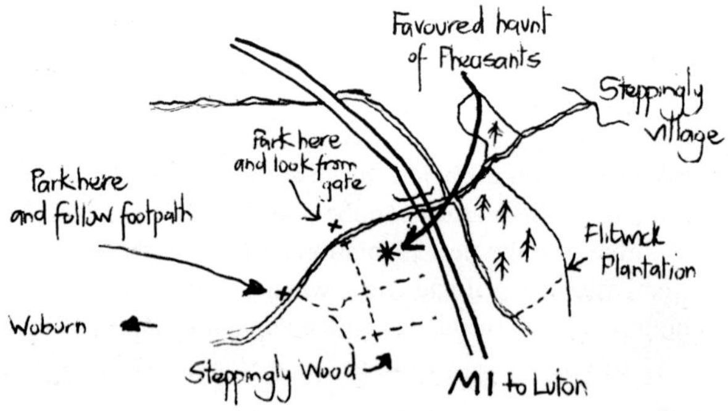

❷Moneypot and Breakheart Hill

This extensive, mainly coniferous woodland, lies south of the A507 at Millbrook, and is accessed from a public footpath running SW from the road. Park sensibly at the A507 lay-by and explore the greensand ridge woodland at TL 006375 (OS Sheet 153). At least six Lady Amherst's Pheasants survive in this wood and it has been a past site for breeding Nightjar.

Spotted Crake *Porzana porzana*

The Spotted Crake is a skulking wetland dwelling species which breeds in very small numbers in Britain. It is amongst one of the most difficult species to detect in summer and, unless the observer is extremely fortunate, hearing the characteristic call is the only chance of success. This consists of a sharp ascending whistle, somewhat reminiscent of a whip lashing, uttered in rapid succession on fine summer nights.

In autumn, Spotted Crakes on return passage from Scandinavia and northern Europe pass through Britain in varying numbers and are often easy to see. Summer droughts, as in 1995, will cause large numbers of Spotted Crakes to be seen, and these may be encountered feeding out in the open on mud-fringed reservoirs or gravel pits. As they are migrants, it is not possible to determine specific sites in which to see them as these depend on weather conditions and other aspects affecting their distribution at the time. They appear from mid August through to early October and are featured on *Birdline* and all the relevant regional birdlines. Favoured localities include Tophill Low Reservoir (Humbs), Cley NWT Reserve (Norfolk), Titchwell (Norfolk), Minsmere RSPB Reserve (Suffolk), Fowlmere RSPB reserve (Cambs), Titchfield Haven (Hants), Farlington Marshes (Hants), Radipole Lake NR (Dorset) and Chew Valley Lake (Avon).

Summer Sites

There are just two main sites in summer, where the birds may only be heard. There is no chance of seeing the birds.

❷Insh Marshes, Speyside - Highland

The Insh Marshes lie to the SW of Aviemore and are a huge area of shallow pools and marshy vegetation. Up to four Spotted Crakes are regularly present and these can be heard at dusk or at night from the B9152 between Loch Insh and Lynchat, especially from the area opposite the wildlife park. Birds are present from at least mid May to mid June.

❷Ouse Washes RSPB Reserve - Cambridgeshire

Up to three calling male Spotted Crakes can be heard from the hides overlooking the washes at Welches Dam in May and June. Visits at dusk are essential. The reserve is well signposted from Manea village and there is a car park and visitor centre adjacent to the Old Bedford River.

Autumn Sites

As stated above, it is not possible to give sites for migrant birds as they can and do turn up at virtually any site suitable to their liking or predicament at the time. However, there are two sites that warrant inclusion as Spotted Crakes occur there on an annual basis and there is a high degree of likelihood of connection during the passage period.

✱Marazion Marsh - Cornwall

A total of up to ten Spotted Crakes are known to visit the lush and heavily vegetated pools at Marazion between early August and late September and can be seen early morning or late evening from the main footpaths that criss-cross the reserve. They favour the shallow, well-vegetated pool immediately left of the main footpath and can be watched feeding between the clumps of reedmace. It is essential to be quiet and patient at this site.

The marsh is situated north of the minor coastal road between Penzance and Goldsithney west of Marazion village. There is a large car park immediately south of the reserve and a footpath leads to the marsh from the bridge.

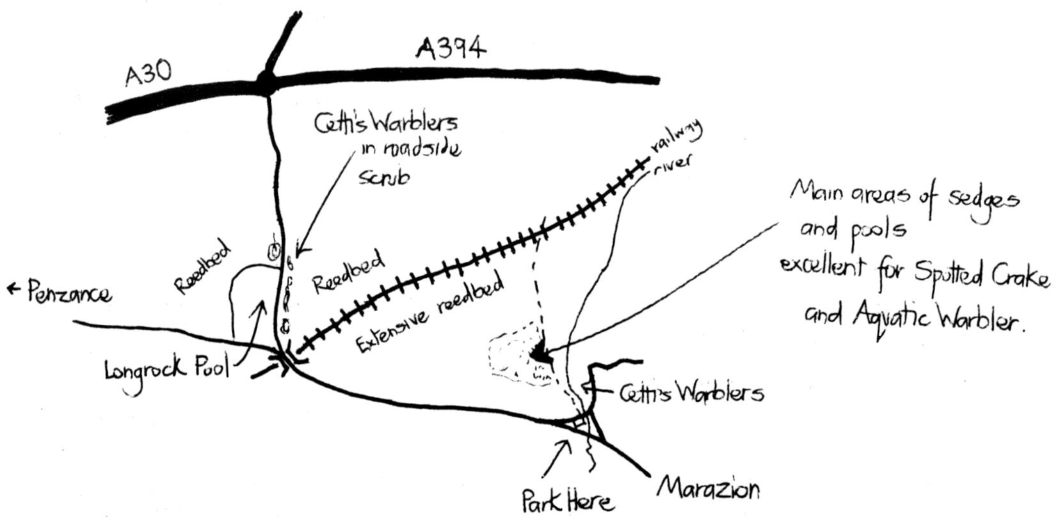

❶Porthellick Pool, St Mary's - Scilly

Porthellick Pool is guaranteed to host Spotted Crakes in autumn but sightings are dependent on the water level during the period. They favour the mud-fringed area at the north end and can be watched from either of the two hides along the west side. Numbers peak in late August but singles may be present until mid October.

Lower Moors can also be suitable at times and these birds can be watched from either of the two hides overlooking the scrapes.

Corncrake *Crex crex*

The Corncrake is a rapidly declining summer visitor that winters in south-east Africa, from southern Tanzania to northern South Africa. Birds depart wintering grounds in March and head through Morocco and the Iberian peninsular to breeding grounds in France, Ireland, NW Scotland, Belgium, the Netherlands, Poland, Sweden, Denmark, Norway and Finland. They arrive in Scotland between the third week of April and last week of May and remain through the summer until early August. Although formerly widespread throughout Britain, it is now confined almost entirely to north and west Scotland, where it is most numerous in the Outer and Inner Hebrides.

Like Quail and Spotted Crake, they are extremely difficult to see, but utter an unforgettable sound which announces their presence. Careful scrutiny of moving vegetation and patient watching should eventually reap rewards.

Because Corncrake is now so rare and is highly susceptible to disturbance, it is not possible to detail the whereabouts of certain sites, particularly those on mainland Britain and I have endeavoured to offer only those sites where respectable numbers of Corncrakes still continue to exist. Do not tape-lure these birds and do not trample vegetation in your search for them as this will not only cause untold disturbance to the birds but will also invoke criticism from landowners and the general public alike and may affect the continued presence of the species as a British breeding bird.

✱Tiree - Western Isles

A total of at least 111 calling males were counted on Tiree in June 1993 and the island remains one of the strongholds in Scotland for this declining summer visitor. The 640 hectares of herb-rich grasslands and cereal fields are vitally important for this species on Tiree and birds may be encountered virtually anywhere on the island. A drive from Scarinish to Crossapol and north to Kenovay will result in numbers of Corncrake heard, the males calling throughout the day and night during May and early June.

Tiree can be reached via Oban and involves a *Caledonian MacBrayne* ferry crossing. The service operates four days a week in summer and it is essential to book beforehand (☎ 01475 34531). There is no public transport available on the island, but cars or bicycles may be hired.

✱Benbecula, North and South Uist - Western Isles

A total of over 190 calling Corncrakes were recorded in the Uists in June 1993 accounting for a large proportion of the remaining Scottish population. They prefer the wet meadows and nettle clumps and are often found in concentrations of Iris beds. Finding Corncrakes on Uist involves driving around suitable roads and listening with the window wound down, but favoured areas are at Balranald and in west Benbecula. The entrance to Balranald is signposted 'RSPB', two miles NW of Bayhead, on the A865 Lochmaddy to Benbecula road. A mile after taking this turn, fork left to the reception centre at Goular. A warden is present from April-August inclusive and should provide any latest information.

The Uists themselves are reached from Skye (Uig Terminal) and sailing times and price can be obtained from Caledonian MacBrayne, Ferry Terminal, Gourock, PA19 1QP ☎ 01475 33755.

Corncrakes can also be found on Lewis, Harris, Barra, Vatersay, Canna, Coll, Mull, Iona, Colonsay and Islay.

Common Crane *Grus grus*

The Common Crane is a rare breeding resident and scarce passage migrant to Britain, with at least one pair attempting to breed in East Anglia since 1981. A population of up to nine birds now exists in Norfolk and these birds can be seen throughout the year. The main breeding population of Common Cranes in Europe is in the bogs and marshes of Scandinavia and western Russia but also in reed and sedge beds in Germany and Poland, so the isolated group in Norfolk is quite unique. However, Cranes formerly bred in East Anglia up until the seventeenth century but became extinct due to persecution. The Norfolk group is also unique in that it is resident - Common Cranes are essentially migratory in northern Europe, leaving nesting grounds in Scandinavia, the Baltic and the Russian steppes to winter in Iberia, France and the Middle East.

The Cranes in Norfolk are present throughout the year but are best seen between October and March as they flight into roost or feed in roadside fields to the north of Horsey.

✱Waxham, Horsey and Hickling Area - Norfolk

A small population of Cranes has been present in Norfolk broadland since 1981 with the flock fluctuating up to a peak of nine as recently as January 1996. The birds roost daily in winter in reedbeds on Brayden Marshes and are best observed from the footpath at Stubb Mill at TG 437 222 (Sheet 134). They generally fly in from the north about 15-35 minutes before dark and land out of view in the reeds to the north of Horsey Mere. Please park only in the NWT reserve car park at Hickling at TG 427 223 and walk the final 0.75 miles to the mill. This area is reached from minor roads in Hickling Heath village, the NWT reserve being clearly signposted from therein.

By day, the Cranes roam widely over coastal farmland between West Somerton village in the south to Waxham in the north and often split up into pairs or groups of four and five. Much favoured feeding areas are the Brograve Levels, best observed from the B1159 near Brograve Farm (TG 444 253), and in ploughed fields west of the road between Horsey Corner and Brograve Farm. From mid July to September, the coastal fields and meadows around the area known as Hundred Stream are often visited and this area can be viewed from the public footpaths leading east and southeast from the Nelson Head pub in Horsey (TG 460 228) or from the B1159 coast road.

The Cranes are notoriously difficult to see between April and July and visits during this time are to be discouraged. They are also difficult to find in windy conditions, as they shelter in concealed ditches and dykes.

Black-winged Stilt *Himantopus himantopus*

The Black-winged Stilt is an annual vagrant to Britain from the Mediterranean basin in southern Europe, with over 300 being recorded by the end of December 1995. Most are recorded as overshooting spring vagrants, between mid March and early July, with much smaller numbers in autumn, from late July to late September. They have also attempted to breed in Britain on at least six occasions, two of these attempts being successful (in Nottinghamshire in 1945 and in Norfolk in 1987).

Since 16th September 1993 and up to at least March 1996, a single male has been resident in north Norfolk and can be seen daily feeding and roosting on the lagoons at Titchwell RSPB reserve. It favours the tidal pool at the north end of the reserve and can be watched from the raised bank leading to the shore. It also visits either one of the two freshwater pools and associates with the local Oystercatcher flock throughout the summer, and has been known to visit nearby Brancaster Marsh and Thornham Harbour. For a most recent update on the presence of this vagrant, it is recommended that you check *Birdline* or *Birdline East Anglia.*

✱**Titchwell RSPB** reserve is situated on the A149 coast road between Wells-next-the-Sea and Hunstanton, half a mile west of Titchwell village. The reserve centre and car park are clearly signposted and the public footpath along the seawall leads immediately north from the information centre.

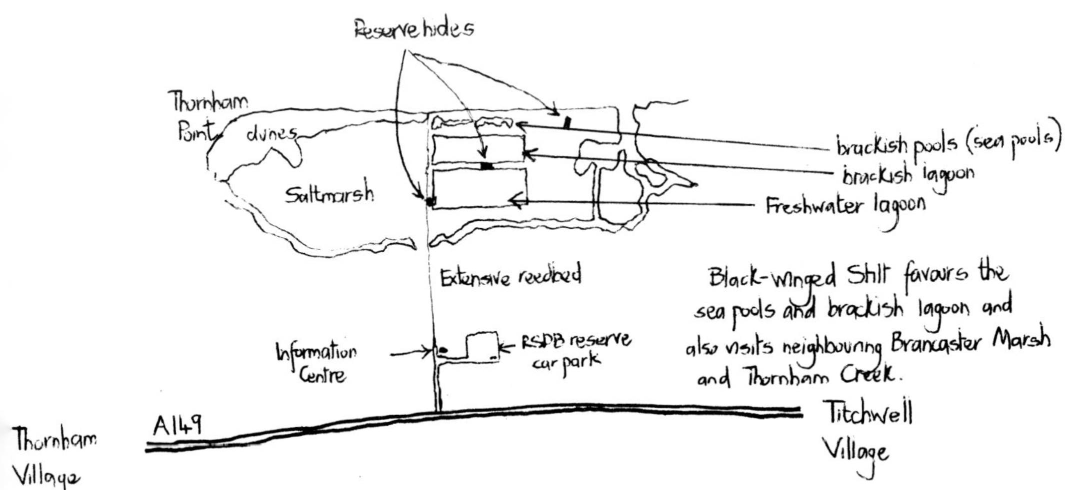

Avocet *Recurvirostra avosetta*

The Avocet is a scarce nesting bird in Britain with a breeding population of at least 600 pairs. Formerly a regular breeder in eastern England, it became extinct in 1842, but recolonised Suffolk in 1947 following isolated nesting occurrences in Norfolk in 1941 and in Essex in 1944 and has since flourished, spreading to neighbouring Norfolk and Essex and also, more recently, to Kent. It nests on banks or islets of shallow brackish lagoons and is present at breeding locations from late March until early September. Regular wintering in Britain also began in 1947, with birds being found in Devon and Cornwall, this phenomenon extended in recent years to include Dorset, Kent and Suffolk.

Summer Sites

✱Holme NOA/NWT Reserve - Norfolk

A small but increasing population of Avocets survives on Broad Water and the adjacent lagoons and these can be observed from a number of small hides overlooking the area.

At Holme next the Sea on the A149, three minor roads lead north towards the sea. Take the westernmost road and after just less than a mile, take the second right turn along the entrance track to the reserve, passing the holiday homes on the left. Continue along the rough track for about a mile until you reach the reserve car park by the pines. The reserve is open daily except Tuesdays and is by permit only.

✱Titchwell RSPB reserve - Norfolk

A total of up to eighty pairs nests annually on the freshwater scrapes at Titchwell reserve and excellent views of these birds can be obtained from either of the two hides overlooking the lagoons. Birds are present between late March and early September.

✱Cley NWT Reserve - Norfolk

A total of at least eighty pairs can be found on the numerous scrapes and meres on the reserve and can be amply observed from several of the reserve hides. Most nest on North Scrape, Simmonds Scrape and Pat's Pool and are a constant inhabitant between late February and early September.

✱Minsmere RSPB Reserve - Suffolk

A total of at least 70 pairs nest annually at Minsmere and these can be watched from a number of hides overlooking the scrape (especially East hide). Birds are present from about mid February until late August.

The reserve is open every day except Tuesday from 9.00am until dusk and can be reached from Westleton village or Eastbridge. Leave Westleton on the minor road eastwards towards Dunwich and after 0.75 miles turn right towards the reserve. This road eventually leads to the reserve car park and information centre, where a permit may be obtained.

A further population of equal provenance is found at the RSPB reserve at Havergate Island to the south of Minsmere.

✱Trimley Marshes NR - Suffolk

A small but increasing population of Avocets nests at Trimley Marshes NR and these birds can be observed from the series of hides overlooking the lagoons. The reserve can be reached from the village of Trimley St Mary, immediately west of Felixstowe, and involves a long walk from the designated parking area south of the railway station at TM 278 358 (Sheet 169) or from Fagbury Cliffs at TM 275 342. A footpath leads down to the information centre and hides.

✱Old Hall Marshes - Essex
A small but increasing number of Avocets breed on the scrapes of the RSPB reserve at Old Hall and can be seen from the sea wall east of the cottages and farmyard buildings. Observation involves a long walk from Tollesbury that may take over four hours. Leave Tiptree on the B1023 and continue east to Tollesbury village. Turn left on Woodrolfe Road and park at TL 964 106 (Sheet 168). Walk north along concrete track to sea wall and continue north to reserve at Old Hall Farm. The seawall then continues east from here towards Joyce's Head and Pennyhole Fleet.

✱Elmley RSPB Reserve, Isle of Sheppey - Kent
A small but thriving population of birds nests at Elmley Marshes and nearby Swale NR. These can be observed at both sites from a number of hides overlooking the scrapes. Birds are present between mid March and late July.

Elmley reserve is reached via the A249 north of Sittingbourne. About a mile north of the Kingsferry Bridge, turn right on to a rough track signposted to the RSPB reserve. There is a designated parking area at Kings Hill Farm, about two miles along the track, and the hides and wader pools are to the east after a further walk of at least 1.5 miles. The reserve is open daily except Tuesdays from 9.00am until dusk, permits being obtained from the farm building to the left of the track. Please note that observation will take a minimum of two hours visitation.

Wintering Sites

✱Breydon Water - Norfolk
In recent winters, up to 50 Avocets have chosen to visit Breydon in December and January before the general spring arrival of up to 100 birds takes place between mid February and early April. They favour the food-rich mudflats at the eastern end and may be observed at high tide from the small hide along the NE shore. Park at the SW corner of the Asda supermarket car park off the A47 immediately east of the main bridge and walk west along the northern sea wall for about 400 yards. At low tide, birds may congregate at the western end and may be better observed from Burgh Castle or from the seawall two miles west of the supermarket car park.

✱Alde/Ore Estuary - Suffolk
A regular wintering flock of at least 750 birds can be found on the Alde and Ore estuaries between September and March. They roam widely over the area dependent on tides but are best viewed from Iken Marshes, Orford village or from the minor road to Shingle Street. A car park at Ikencliff (TM 399 563 - Sheet 156) gives a wide panoramic view of the estuary and should provide sightings in the Long Reach.

An increasing number of birds is also wintering on the neighbouring Deben Estuary south of Woodbridge, these being best observed from the footpath east of Martlesham Creek or from Felixstowe Ferry.

✱Hamford Water - Essex
A wintering flock of up to 238 Avocets can regularly be found in the Hamford Water and Horsey Island complex of Walton on the Naze between October and March, and can be seen from various points along the seawall or from Stone Point (only accessible at low tide). Park in the designated Naze car park (by the cafe and tower) and walk north and west along the seawall.

✱East Tilbury - Essex
Up to 230 birds winter in the western reaches of the Thames Estuary and dependent on tides can be seen from either the Essex or Kent sides of the river. Large numbers can be seen feeding on the vast expanses of mud off Coalhouse Point, south of East Tilbury village. Park at Coalhouse Fort and view from the coastal footpath.

✱Thames Estuary (Medway and Swale) - Kent
A wintering population of at least 230 birds can be found at Higham Bight, an expanse of mudflats immediately west of Cliffe Marshes. Leave the B2000 in Cliffe village on a minor road and continue west between two quarries to West Court Farm. Park sensibly on the roadside at the entrance to the Alpha Pool and follow the public footpath to the west around the southern edge of the pool to view the Bight from the sea wall. Low or falling tide visits are essential.

A further wintering population of up to 350 birds may be found in the River Medway and these are best observed in Funton Creek. Leave the A249 at Iwade and drive west along a minor road towards Lower Halstow. After just over a mile, the road runs alongside the Barksore Marshes and the creek and the Avocets may be seen from any of the lay-bys along this stretch. A telescope is essential for viewing, the site also attracting up to 900 Pintail in winter and many other species of wildfowl and waders.

✱Poole Harbour - Dorset
A wintering population of up to 400 birds remains in Poole Harbour from October to February. The birds favour two main haunts - the freshwater lagoons at the east end of Brownsea Island and the mudflats off Arne NR. The Brownsea birds are best observed from the B3369 at Parkstone, where the road increases in height and allows one to 'scope the lagoons. The birds are extremely distant however and can only be identified on clear, sunny days. Views are better from Arne NR, the hide at Shipstal Point offering a good vantage point over the harbour. Wintering Spoonbill and Little Egret are also regularly seen from here. Arne RSPB reserve is reached from a minor road off the B3075 south of Wareham.

✱Exe Estuary - Devon
Up to 450 Avocets winter on the upper mudflats of the Exe Estuary between October and March and are best observed from the public footpath leading north along the west shore from Powderham Corner or from the canal footpath to Turf Lock on Exminster Marshes accessed from the Swan's Nest public house off the A379. They can also be viewed rather distantly from Riversmeet House, at the south end of Topsham, and occasionally at high tide on Bowling Green Marsh, Topsham.

✱Tamar Estuary - Cornwall
Up to 240 birds regularly winter on the upper reaches of the Tamar estuary and these can be seen well at low tide from Cargreen village on the NW shore. Leave Saltash north on the A388 and take the right turning to Landulph. Continue east on a maze of minor roads to the village and watch from various vantage points along the waterfront.

Stone Curlew *Burhinus oedicnemus*

The Stone Curlew is a rare breeding summer visitor to Britain with a dwindling population of less than 225 pairs. Its range has contracted considerably in recent years, and is now confined to East Anglia, Hertfordshire, Essex, Berkshire, Wiltshire, Hampshire and Dorset. It breeds on open stony ground with sparse vegetation and favours semi-natural grassland, chalk downland, arable farmland and Breckland grass-heaths. The birds return from wintering grounds in southern Spain, southwest France and northwest Africa from early March until mid May, and gather in large groups in autumn, prior to migration in September and October.

The majority of Stone Curlews in Britain nest on areas used by the Ministry of Defence and are therefore inaccessible. Many others nest on farmland or on heathland which are very susceptible to disturbance and because of the serious threat posed by egg-collectors, it is not possible to detail a large number of sites in this publication. However, the two sites below will provide opportunities to study this unique species in its natural surroundings.

✱ Weeting Heath - Norfolk

Stone Curlews return to Weeting Heath from 5th March onwards and by mid May, 3-5 pairs are generally in attendance. They are best seen during the first month of arrival, as the vegetation is still relatively cropped low and the birds are often in courtship display. Once nesting commences, the birds become difficult to locate and activity is largely restricted to early mornings and evenings. Furthermore, the heat haze on summer days affects viewing conditions at this site. The birds remain on site until at least late September and are often joined by many other breckland birds prior to southward migration. Flocks of up to 33 birds have been recorded in the last week of September.

The Weeting Heath NWT reserve lies to the north-west of Brandon and is reached by turning left off the A1065 towards Weeting on the B1106. In the village, turn left on to a minor road towards Hockwold and continue for 1.5 miles to the reserve car park and warden's caravan and shed on the left. This car park is somewhat hidden in the belt of roadside pine trees. Permits can be obtained from the wardens hut and they will be able to provide useful and up-to-date advice on the whereabouts and numbers of Stone Curlews present. At least three hides overlook the nesting heath. This reserve is also very good for Woodlark and at times Common Crossbill, and also for Lapwing, Red-legged Partridge, Northern Wheatear, Green Woodpecker and Little Owl.

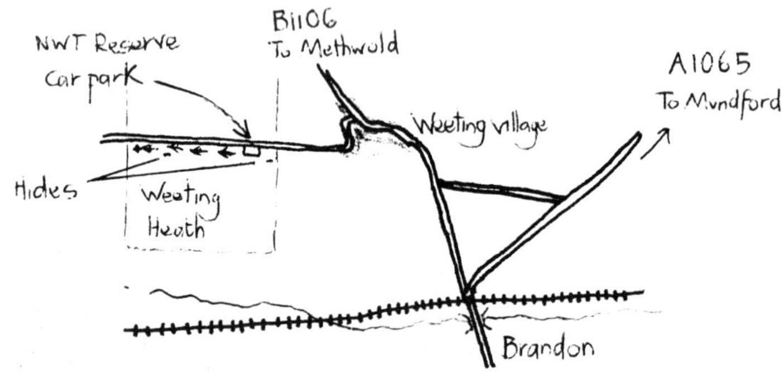

❶Foxhole Heath - Suffolk

Foxhole Heath is regularly used by Stone Curlews on spring and autumn passage and quite large gatherings can be encountered, especially in September. Leave Barton Mills roundabout north on the A1065 and after two miles turn right on to the B1112. Foxhole Heath is immediately on the left after 0.25 miles. Park on the triangle of grass at the entrance to Rakeheath Farm and scan. *Under no circumstances trespass on the heath.*

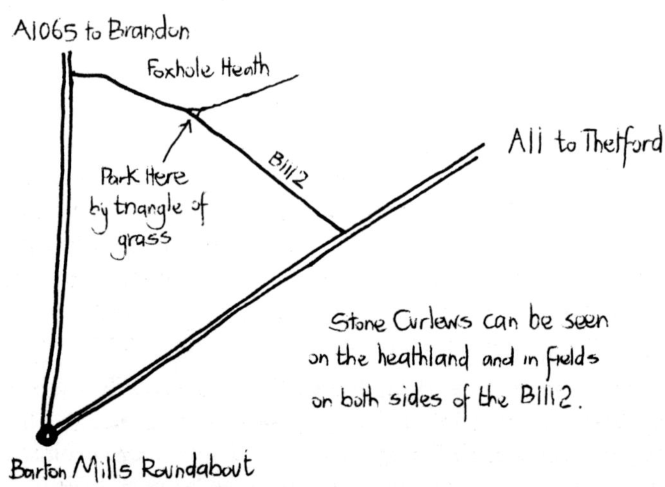

For security reasons, a number of other sites in the south of England cannot be published. *Stone Curlew is a Schedule One species and is afforded protection at all times. It is an offence to disturb this species.*

Kentish Plover *Charadrius alexandrinus*

The Kentish Plover is a scarce passage migrant to Britain with up to 63 birds recorded per year, primarily in spring. A few birds are also known to overwinter and these join Ringed Plovers at coastal locations. It is also an occasional breeder, with sporadic nesting in 1952 and 1979. Although there are no specific sites in which to guarantee seeing Kentish Plover in Britain, there are a number of sites visited regularly in April-May and September.

At the time of publication, only one Kentish Plover was known to be wintering in England - a female in Lancashire (detailed below), but in recent years, at least one has regularly returned to Ferrybridge, in Dorset.

❶Rossall Point, Fleetwood - Lancashire
A female has wintered with Ringed Plovers at Rossall Point since March 1992, and is generally present between late November and early March. It can be located at high tide on the shingle beach adjacent to the coastguard lookout. It is highly susceptible to disturbance from dog walkers and it may be necessary therefore to search the beach eastwards to the marine lake and beyond. On the very highest tides, all of the beach will be covered, so the bird will alternatively roost on the marine lake (if drained, which it often is) or on the grassy car park near the coastguard station or on the adjacent golf course, particularly at the south end.

Kentish Plovers are most likely to be encountered on the south coast between early April and mid May. A number of favoured localities include Dawlish Warren (Devon) (often roosting on the shingle beach with other waders at high tide), Ferrybridge (Dorset) (with other waders at high tide, viewable from the information centre), Stanpit Marsh, Christchurch (Dorset) (feeding on the mud on an incoming tide), Dungeness RSPB Reserve and ARC Pit (Kent), Pegwell Bay (Kent) (from the hide), Breydon Water (Norfolk) (mostly at the east end visible from the hide) and Cley NWT Reserve/Salthouse Pools (Norfolk). Small groups are also occasionally recorded at similar localities in September.

Dotterel *Charadrius morinellus*

The Dotterel is a scarce summer visitor to Britain with an estimated breeding population of about 840 pairs. Apart from isolated nesting attempts in Cumbria and North Wales, the breeding population is confined to the Scottish Highlands. Nesting takes place on high plateaux between mid May and early July but human disturbance and egg collecting on the high tops has a sustained impact on breeding success.

Dotterels also breed in Fenno-Scandia, northern Russia, Austria, Italy, Czechoslovakia and Romania, apart from Britain, and most winter in the Mediterranean area south to North Africa. They depart these wintering grounds in early April and pass through southern England on their way north between 22nd April and 13th June. Although many of these are on their way to nesting sites on Scottish peaks, a large proportion are heading for Scandinavia and Russia. A number of traditional staging-posts are used, particularly in Lancashire, Norfolk, Cambridgeshire, Essex and Hertfordshire, and 'trips' of up to 40 birds are regularly recorded. They seem to favour freshly-sown pea fields in which to rest on migration.

Return passage commences from mid August to October, with small numbers appearing in coastal localities between Norfolk and Scilly.

Summer Sites

❶Cairngorm Plateau - Highland

Small numbers of Dotterel nest on the high plateaux of Cairngorm and can be occasionally seen from the footpath leading to Ben Macdui. They favour the areas of short vegetation, interspersed between the scattered boulders, and are extremely well camouflaged and difficult to locate. The area beyond the Cairngorm peak and the weather station is best, particularly that within twenty-five minutes walk from the Ptarmigan Restaurant. Fortunately, Dotterels will sit tight and tolerate people passing very close to the nest, but disturbance caused by hill-walker's and birdwatchers can have a detrimental effect. Please take care not to disturb these rare birds and beware of damaging sensitive vegetation.

The Cairngorm Plateaux is within easy reach of Aviemore and the ski-lift centre is well signposted. A chair-lift operates throughout the summer and takes one to the Ptarmigan Restaurant near the summit. Alternatively, a well-trodden footpath leads one to the peak taking a minimum of an hour and a half to climb. It is advisable to visit only during dry, sunny conditions. Dotterels are present from mid May to August.

❶Carn Ban More - Highland
Dotterel, as well as Ptarmigan, nest on Carn Ban More in small numbers and can be seen, with patience, at the edge of the footpath over the peak. They are present from mid May to July and can often be detected by their far-carrying *'pwit pwit pwit'* call. Be careful not to wilfully disturb these rare breeding birds and do not linger near nests or near newly-hatched chicks.

The climb to Carn Ban More is hard and arduous but truly spectacular. It takes at least two and a quarter hours and should only be undertaken in clear and sunny conditions, not windy or wet. Take the B970 south from Inverdruie, one mile east of Aviemore, and at Feshiebridge, turn left on to the Lagganlia road and follow to the end at Achlean cottages. There is extensive car parking space here and the start of the well-worn fell track to the summit is obvious. The track becomes very steep on the final ascent and then gets easier as it heads towards the summit. The birds are generally within sight of the two or three cairns.

Dotterels on spring migration

Dotterels are well known for visiting traditional sites on migration and can be seen between 20th April and 30th May. Peak passage is during the period 12th-23rd May.

❶Pendle Hill - Lancashire
The classic site in NW England is the summit of Pendle Hill (557m), six miles east of Clitheroe in east Lancashire. Park sensibly alongside verge by the start of the track to Pendle House Farm (SD 815416). Walk past the farm and take the LEFT hand track across slope of the hill to reach the summit ridge half a mile south of the trig point. Search bare areas just beyond ridge crest and then walk north to search stony ground near summit. 5-15 birds are normally present daily during the first ten days of May.

❶Danby Beacon - North Yorkshire
Dotterels appear in the vicinity of Danby Beacon on an annual basis between 4th and 22nd May. They normally turn up in small groups of up to eight birds and generally favour the area around the beacon itself but because of disturbance from general public quickly move to adjoining areas. A track runs due east from Beacon Hill and the birds may be found in any of the bare or burnt heather areas.

Leave the A171 about nine miles east of Guisborough almost opposite the B1366 to Loftus. Turn left off the road to Danby and park sensibly by the beacon.

In years of peak Dotterel passage, birds may be located at a wide range of North York moorland sites, including Rosedale Moor.

❶Burbage Moor - South Yorkshire
A total of up to 20 Dotterel are recorded at Burbage Moor almost annually, primarily between 20th April and 17th May. Leave Sheffield south-westwards on the A625 and after two miles turn right towards Ringinglow. Turn left at the village crossroads and park by Lady Canning's Plantation at SK 290 835 (Sheet 110). A public footpath leads up over the moor, the Dotterel favouring the bare areas of heather near the peak.

❶Totley Moor - South Yorkshire
Totley Moor is another regular site for Dotterel and lies about ten miles southwest of Sheffield. Leave the city on the A621 and continue south-westwards to New Totley village. Continue for a further two miles and then turn right on to the B6054. A footpath leads up over Flask Edge to a high point from SK 277 782 (Sheet 119) and the areas of cleared heather moorland near the peak are favourite.

❶Swinefleet Area - Humberside
Large 'trips' of Dotterel are regularly recorded in the pea fields to the south of Swinefleet. Explore the maze of roads around Mount Pleasant and Swinefleet Moor Farms. Swinefleet is situated two miles east of Goole on the A161.

❶Deeping St Nicholas Area - Lincolnshire
The large pea fields to the north of the A16 are extremely attractive to migrating Dotterel in spring and are recorded in varying numbers between 28th April and 18th May. The fields either side of Deeping St Nicholas are the traditional sites.

❶Gringley Carr - Nottinghamshire
Gringley Carr is a traditional site for Dotterel and regularly attracts up to 40 birds in May. It is part of the Idle Valley and is an area of large open fields (cultivated and flat) with limited access and few public footpaths. Please view only from the road. Leave the A631 at Gringley-on-the-Hill on a minor road at the west end of the village and head northwest over the Chesterfield Canal into an area of extensive flat farmland. Gringley Carr and associated farmland lies immediately east of the road north of the canal.

❷Morridge - Staffordshire
The former glider field at Morridge has proved extremely attractive to migrating Dotterel since 1989. They tend to appear in the first two weeks of May. Morridge lies to the north of the A523 two miles east of Leek.

❶Bredon Hill - Worcestershire
Trips of up to 20 Dotterel have been recorded during the first week of May at Bredon Hill, the southern slopes being the favoured area. Bredon Hill is an outlier of the Cotswolds and lies between the A435/A438 Evesham to Tewkesbury, B4080 Tewkesbury to Pershore and A44 Pershore to Evesham roads. The southern slopes are best approached from Kemerton or Bredon's Norton. There are numerous public footpaths in the area but parking is rather restricted in the narrow village streets.

❶Happisburgh/Eccles/Winterton/Hemsby area - Norfolk
Two distinct areas of east Norfolk are traditional resting grounds for small parties of Dotterel. They favour the fields of rolled beet and peas and these are rotated from year to year. The coastal fields between Happisburgh and Eccles-on-Sea are the most regularly visited and these can be checked from the B1159 or from minor roads in the area. The peak period is between 28th April and 20th May. Further south, the fields immediately inland of the B1159 between Winterton-on-Sea and Hemsby are also often visited if in suitable condition.

❶Cottenham Area - Cambridgeshire
A traditional resting area for migrating Dotterel in early May are the fenland fields between Cottenham and Wilburton. The B1049 runs between both villages giving ample views over the farmland. Trips of up to 40 birds have been recorded in the past.

Elsewhere in Cambridgeshire and depending on the field crop, previously used sites worth checking are the B1381 between Sutton and Earith, the A141 between Warboys and Chatteris, the B1040 between Ramsey and Warboys, the B660 between Holme and Ramsey St Mary's and the minor roads between Holme and Yaxley and Witcham and Downham Hythe. Remember, pea and rolled beet fields are favoured.

❶Baldock and Royston Areas - Hertfordshire

The Baldock area has long been a traditional site for passage Dotterel and trips of up to thirty birds may be encountered. The crop rotation varies from year to year and so locating the birds involves driving around searching for suitable resting sites. They favour the freshly sown pea fields in this area and peak between the 5th-16th May, particularly during easterly conditions. The favoured fields are between Newnham and Ashwell, but in some years, Guilden Morden is visited, the road between Baldock and Wallington and the road between Bygrave and Ashwell.

Another traditional area is that between the A505 and the B1039, although fields in recent years have not been suitable. The road between the A505 and Chrishall Grange is good as is the farmland west of the B1368 north of Flint Cross.

❶Balsdean - East Sussex

Balsdean is a traditional area for Dotterel in both spring and autumn, with trips of up to ten birds regularly recorded. It is an area of the South Downs NW of Newhaven and involves a lengthy walk of at least 40 minutes from Iford or Swanborough Manor on the minor road between Newhaven and Lewes. The Dotterels favour the south-facing slopes and are particularly fond of the area of bare fields at Bird Brow. Newmarket Hill, north of Woodingdean, can also be visited if field conditions are to their liking. Peak passage periods are 5th-18th May and 3rd-22nd September.

Another area nearby worth checking is Gardener's Hill, immediately east of Denton, which in May 1995, held a trip of Dotterel for over two weeks. A public footpath leads out of the east end of the village onto the downs.

❶Mynydd Du (Black Mountain) - Carmarthenshire

Trips of up to twenty birds are regularly recorded on the Black Mountain between 25th April and 10th May. The Tair carn Ridge is the best site and the stony area around the lower three cairns is where they are most likely to be found (Tair Carn Isaf and Tair Carn Uchaf). Park near the crossroads at SN 669 162 and walk north-east up the slope. Leave Bryamman on the minor road towards Trapp and Llandeilo. This area is also good for Red Kite, Golden Plover and Raven.

❷West Cornwall

A surprisingly large number of Dotterels are recorded in Cornwall each year with small parties of up to 12 birds in spring and up to eight birds in autumn. In early May, Land's End Airfield, south of St Just, is well worth a look, and from late August to the end of September, birds are regularly encountered on the moorland above Porthgwarra village, in the fields around Sennen and on the Land's End airfield and the disused runway at Davidstow.

❶St Mary's - Scilly

It is very unusual for an autumn to pass without the presence of at least one Dotterel on the islands. They turn up from the last week of August and continue to appear until early October. Prime locations are the airfield and golf course on St Mary's and Wingletang Down on St Agnes.

Purple Sandpiper *Calidris maritima*

The Purple Sandpiper is a passage migrant and fairly common winter visitor to Britain and has, in recent years, nested in very small numbers in the Scottish Highlands. The chosen habitat is rocky coastlines and birds arrive from breeding grounds in Iceland, Scandinavia and Arctic Russia from mid July onwards. Many remain well into April and even into May, and the total wintering population is somewhere in the region of 16,000 birds. The main population is concentrated in NE Britain and the species is common throughout Shetland and Orkney southwards down the east coast of Scotland and into Northumberland, Durham and Cleveland down to Yorkshire. South of Bridlington, birds become scarce as the coastline becomes predominantly sand or shingle, and elsewhere, small groups are found in Norfolk, Suffolk, Essex, Kent, Sussex, Dorset, Devon and Cornwall. Purple Sandpipers are inevitably joined by numbers of scavenging Turnstones.

✱Burghead - Grampian
Burghead Harbour and the rocky coastline eastwards to Hopeman regularly holds a wintering population of up to 86 birds and these can easily be seen from the coastal footpath north of the town.

✱Lossiemouth - Grampian
The largest wintering numbers in Moray & Nairn are found in Lossiemouth and its environs. Check the rocky coastline to the north of the town.

✱Buckie - Grampian
Another large concentration of up to 120 birds winters on the rocky coastline between Spey Bay and Portknockie.

✱Fraserburgh - Grampian
The rocky coastlines around Kinnaird Head and Cairnbulg Point hold large populations of wintering Purple Sandpipers and these can be easily encountered between July and April.

✱Peterhead - Grampian
Counts of up to 130 Purple Sandpipers are commonplace in Peterhead in winter, with birds frequenting the rocky coastlines surrounding the bay.

✱Girdleness, Aberdeen - Grampian
A small wintering population of up to 32 birds frequents the rocky beach near the outfall and lighthouse at Girdle Ness. Leave the A956 on the south side of the river and follow a minor road through Torry to the Ness.

✱St Andrews - Fife
A wintering population of up to 40 birds can be found on the rocky beaches below St Andrews castle and cathedral. Numbers tend to build up in spring, between 19th March and 18th April.

✱Fife Ness and environs - Fife
A population of at least 200 birds winters in the Fife Ness area and can be seen around the headland or at Tullybothy Craigs just to the north.

✱Ruddons Point, Largo Bay - Fife
Up to 70 birds frequent the rocky coastline around the point between October and May and can be seen from the coastal footpath adjacent to the caravan site. Ruddon's Point can be reached from the A917 between Lower Largo and Elie. About a mile north of Elie, turn right towards Shell Bay caravan park and follow to end. Between 9.00am and 4.00pm, the gate is open and one can drive to within 600 yards of the point, but at other times, a walk of one mile must be made from the gate.

✱**Dalgety Bay - Fife**
The rocky coastline at Braefoot Point and in Dalgety Bay holds a regular wintering population of at least 80 birds. Dalgety Bay lies two miles east of Inverkeithing south of the A921.

✱**Port Seton - Lothian**
A small wintering population of 40 birds regularly frequents the stretch of coastline between Prestongrange and Port Seton. They roost in Port Seton Swimming Pond at high tide.

Elsewhere in Lothian, the rocky coastline between North Berwick and Peffer Sands holds at least 50 birds.

✱**Saltcoats Harbour - Strathclyde**
A wintering population of up to eighty birds can be found in Saltcoats Harbour and an equal number between Ardrossan North Bay and Seamill. Check suitable beaches from the A78.

✱**Ayr Harbour - Strathclyde**
Ayr Harbour holds a wintering population of at least 80 birds and these can be easily seen between late November and April.

Smaller numbers can be encountered off Turnberry Point and in Girvan and Troon Harbours.

✱**Southerness Point - Dumfries & Galloway**
The most regular site in D & G is the area of rocks at the southern tip of Southerness Point, 14 miles south of Dumfries, off the A710. The wintering population numbers between 25-40 birds.

✱**St Abb's Head - Borders**
The rocky coastline to the north of the lighthouse supports a wintering population of up to 50 Purple Sandpipers. From Coldingham, take the B6438 to St Abb's Nature Reserve and park at Northfield Farm and walk north along the coastal footpath or drive further north to the lighthouse.

✱**Workington Harbour -Cumbria**
Up to 38 birds remain from October to April in Workington Harbour and can be seen from the promenade. A further group of up to ten birds can regularly be found at Harrington, three miles to the south.

Elsewhere in Cumbria, a regular flock of over 30 birds can be found at Mill Scar on North Walney.

✱**Heysham Harbour - Lancashire**
A small wintering group of up to ten birds may be found amongst Turnstones at the north end of Heysham Harbour. There is straight-forward access off the A589, turning west at the traffic lights towards the heliport. Skirt round the helipad and follow the seawall to the end at the harbour entrance. The birds roost with Turnstones on the walls at high tide.

➊**Blackpool North Pier - Lancashire**
At low tide, up to seven birds may be found feeding amongst rocks below the North Pier between mid November and late March.

✱**Hilbre Island - Cheshire**
A wintering flock of up to 25-50 birds feed on the intertidal zones, rocks and seaweed around Hilbre Island. At high tide, depending on the wind direction, the flock will be roosting on Lion Rock, in a sheltered cove on the east side, or on sandstone ledges on the cliff-face near the North End.

Hilbre Island is accessed on foot by a two mile walk from the Dee Lane slipway in West Kirby, setting off three hours before high tide and returning three hours after. The birds are present from the last week of October to the first week of May, peaking in April.

✱Rhos Point - Clwyd
A small wintering flock of up to ten birds may be seen at high tide roosting with Turnstones along the promenade. Rhos-on-Sea lies NW of Colwyn Bay, the point and rocky escarpment being obvious from the coast road.

✱Treaddur Bay, Anglesey - Gwynedd
Up to ten birds frequent the rocky islet at high tide and can be seen from the hotel overlooking the bay. Leave the A5 at Dyffryn (Valley) on the B4545 and after three miles, turn west in Treaddur village along the northern shore of the bay.

❶Aberystwyth Harbour - Dyfed
A wintering population of up to twenty birds can regularly be found on the College Rocks or in the harbour at Aberystwyth between October and May.

✱Seahouses Area - Northumberland
A large wintering population of up to 280 birds can be found along the rocky stretches of coastline between Budle Point (NW of Bamburgh) and Seahouses village. Check suitable rocky islets from various parking bays along the B1340, particularly Harkess Rocks and Creenhill Rocks. Seahouses harbour and the rocky beach to the west is also very good.

Elsewhere in Northumberland, gatherings of over fifty birds are regularly seen at Seaton Sluice, Tynemouth, Newbiggin (Newbiggin and Beacon Points) and Cullercoats. There is also a regular high tide roosting flock of over 150 birds in Blyth Harbour between October and April.

✱Seaham Harbour - County Durham
A wintering population in excess of 125 birds is regularly present in Seaham Harbour in winter and at high tide can be seen roosting on breakwaters, groynes and harbour walls. At low tide, many birds forage the coastline north to Ryhope. Seaham is situated six miles south of Sunderland.

✱Hartlepool Harbour - Cleveland
A large concentration of Purple Sandpipers winter in Hartlepool Bay between October and April and are to be found at high tide roosting on the walls and rocks in the West Harbour in Hartlepool. Leave the A179 on a minor road to view the harbour.

To the south of Tees Bay, large numbers can be found roosting on South Gare breakwater in late April.

✱Whitby Harbour - North Yorkshire
A population of up to 70 birds is normally present in Whitby Harbour and its environs in winter. They can be seen from either the East or West Piers at high tide or at Saltwick Nab to the east at low tide.

✱Scalby Mills, Scarborough - North Yorkshire
A large concentration of up to 110 birds can regularly be seen on Scalby Ness Rocks. Leave the A165 north of Scarborough and follow a minor road to the beach in Scalby Mills. Park by the amusement arcades and observe from the promenade or from the coastal footpath that runs north along the cliffs.

✱Filey Brigg - North Yorkshire
Filey Brigg is one of the few sites in Britain where Purple Sandpiper can be seen virtually all year round. A population of between 40 and 150 birds can be seen foraging amongst the rocks on the Brigg and can be seen from the hide or from the footpath that leads down from the car park to the Brigg. Leave the A165 at the northern Filey roundabout and head east along the A1039. Filey North Cliff Country Park and caravan park is well signposted from here. Park at the north-eastern end of the car park.

✱Flamborough Head - Humberside
A small wintering population of up to sixty birds can be found roosting at high tide on the South Rocks. Park at the designated pay & display car park at the Head and follow the footpath down to the foghorn and then south along the clifftop for about 500 yards to where the hedgerow borders the field.

✱Heacham North Beach - Norfolk
A small flock of up to eight birds can be found foraging amongst the groynes and breakwaters on Heacham North Beach between December and February. Drive north on the A149 to the southern roundabout in Hunstanton. Turn left along the B1161 and at the next roundabout and follow the road towards the beach. The sandpipers frequent the beaches immediately south of the obvious ski-ramp along the promenade but may wander south as far as the main beach car park at Heacham or round to Old Hunstanton.

Elsewhere in Norfolk, 1-4 birds can occasionally be found on the breakwaters at Walcott, six miles east of North Walsham on the B1159. A few also visit the harbour mouth at Great Yarmouth from time to time.

✱Ness Point, Lowestoft - Suffolk
This traditional site for wintering Purple Sandpiper regularly attracts up to 20 birds and these can be seen roosting on the breakwaters at the point or on the large rocks between there and the harbour wall and coastguard lookout. Lowestoft Ness is situated north of the river and is accessed from the A12. Turn east off a roundabout immediately north of the harbour and park by the raised concrete sea wall at the end.

❶Holland Haven - Essex
A small flock of up to eight birds can be found on the seaweed-covered groynes just south of Frinton-on-Sea. They are present between November and March and are best seen at high tide. Park at the south end of the promenade at Frinton and walk south along the beach checking the breakwaters and groynes. This area is also excellent for Sanderling and Ringed Plover.

✱Foreness Point - Kent
A wintering population of up to 60 birds frequents the rocky platforms at the base of the cliffs between November and early May. They are best observed at high tide when they gather to roost on the narrow strip of sand above the high water mark in one of the bays. However, this area is popular with dog-walkers and the birds quite often roost in nearby Margate Harbour. From the B2052 north of Broadstairs, a public footpath leads north from the Captain Digby pub. Follow north-westwards for about a mile, passing White Ness and Botany Bay.

❶Glyne Gap - East Sussex
The seaweed covered breakwaters at Glyne Gap regularly attract a wintering flock of up to ten birds and are present from November until late April. The site is situated immediately east of Bexhill-on-Sea. Park at the side of the A259 or on the minor road at TV 775 085 (Sheet 199), cross the railway and walk west along the beach for up to 500 yards.

A similar number of birds can also be found in nearby Newhaven Harbour, where at high tide the birds roost on the east pier.

❶Brighton Marina - West Sussex
Up to five birds can be found feeding on the seaweed covered rocks at the entrance to the marina between December and April. They can be viewed from either the West or East Arm but are difficult to find during rough sea conditions. The marina is situated at the east end of Brighton promenade and provides ample parking facilities. This area is also good for wintering Red-throated Diver, rare grebes, Shags and auks.

➊ Southsea Castle, Portsmouth - Hampshire
Up to five Purple Sandpipers have wintered in recent years on the stretch of coastline bordering Southsea Castle. Birds are present between December and March but are highly erratic in their appearances. Follow the A288 to the southern tip of Portsmouth and turn on to a minor road skirting the edge of Southsea Common. Park sensibly at SZ 636 985 (Sheet 196) and follow the coastal footpath around the seaward side of the castle.

✱ Foreland - Isle of Wight
A fluctuating wintering population of up to thirty birds can be found on the rocky islets at the eastern end of the island. Park at the end of the road in Bembridge and follow the public footpath around the headland from the entrance to the lifeboat station. Black Rock Ledge is their most favoured haunt.

✱ Portland Bill - Dorset
Up to thirty birds regularly winter on the rocks by the Bill and are generally present from November to mid May. Depending on the state of the tide, they can be found either on the rocks immediately south of the obelisk or along the East Cliffs. Park in the pay & display car park at the southern tip of the Bill and walk south to the end.

➊ Lyme Regis - Dorset
Lyme Regis is a small and very picturesque fishing and tourist resort at the western end of the Dorset coast and the spit of land known as the Cob is a regular wintering site for up to 15 Purple Sandpipers. These birds are present from mid November until the first week of May, access to the Cob being straightforward.

➊ Hope's Nose, Torquay - Devon
Hope's Nose is an obvious premonitory of land to the east of Torquay consisting of sheer cliff ledges, a small disused quarry and an area of bracken, bushes and scrub. A sewer outfall is situated near the tip of the point and Lead Stone and the rocky beach to the right of the sewer regularly hold up to 12 wintering Purple Sandpiper. From Torquay Harbour, head inland up Torwood Street and turn right at the first set of traffic lights towards Meadfoot Beach. Follow the beach and turn sharp right on to Ilsham Marine Drive and park at the highest point of the road. A public footpath leads down the Nose to the tip of the headland.

Small numbers may also be seen in some years on the rocks off Prawle Point, the southernmost projection in Devon. Up to nine also regularly winter in Plymouth Hoe.

✱ Godrevy Point - Cornwall
A flock of up to 60 birds can be found on the seaweed-covered reef around Godrevy Point between September and May. Leave Hayle northwards on the B3301 and about half a mile north of Gwithian, take a left hand turning to the Point and car park.

Small numbers can also be seen nearby in St Ives harbour and on rocks around the Island.

✱ Mount's Bay, Penzance - Cornwall
Up to 55 birds are regularly present between Penzance Harbour and Newlyn outfall and at high tide can be found roosting on the rocks in front of the outdoor swimming pool. They spend the day foraging the seaweed-covered rocks along the beach and are often seen along Newlyn beach. Drive west along the A30 into Penzance and then turn left by the bus station towards the main harbour. This road then traverses the promenade, passing the swimming pool on the left just beyond the harbour entrance. The birds can also be seen feeding on the rocks behind the harbour wall.

Jack Snipe *Lymnocryptes minimus*

The Jack Snipe is a relatively scarce winter visitor to Britain which inhabits shallow ditches, sewage farms, cressbeds and marshy field edges. It feeds in dense grass and low cover and can be told from Common Snipe by its diagnostic up and down rhythmical bouncing feeding action, smaller size, much shorter bill and brighter golden-yellow mantle streaking. It is also less susceptible to flushing, rising silently from within a few feet and re-landing almost immediately or within 50 yards.

Due to its secretive nature and its often inaccessible habitat preference, the true wintering population in Britain is unknown, but in periods of severe weather, large numbers may be forced out in the open to feed in brooks and streams and become much easier to see. However, Jack Snipe are never easy to find and, apart from brief flight views gained after flushing, the species is rarely encountered on the ground. The most reliable sites in Britain for obtaining good views are Porthellick Pool and Lower Moors on Scilly.

✱Porthellick Pool/Lower Moors, St Mary's - Scilly
Superb views of up to eight feeding Jack Snipe are often obtained on St Mary's in autumn. The first birds start appearing on the island in the last week of September and throughout October can be seen daily feeding along the muddy fringes of Porthellick Pool and Lower Moors. Conveniently positioned hides provide ample viewing conditions, 'bouncing' Jack Snipes often feeding only yards away.

Away from Scilly, Jack Snipes can be recorded at a vast number of sites in autumn and in winter, but these will generally only involve flight views as birds are flushed. The following is a brief resume of all the main wintering sites in Britain but numbers largely depend on the condition of the fields or marshes or the severity of the weather conditions. For details of access, please consult the relevant regional site guides as detailed on page 8. It is not practical or necessary to give specific instructions to the finding of Jack Snipe because of the nature of the species. It is purely a matter of luck whether or not one is flushed from suitable habitat.

Large numbers pass through the Northern Isles in September and October and can be seen almost daily on Fair Isle, Stronsay and North Ronaldsay and at many sites in Shetland.

Regular sites in **Grampian** include Inverugie, Netherly, Rattray Head, Meikle Moss and Logie Buchan Bridge, whilst in **Fife**, sites such as the Eden Estuary, Shell Bay and Cullaloe Reservoir are often visited. In **Lothian,** birds can be regularly seen at Belhaven Bay (by the Beil Burn mouth), Aberlady, Cobbinshaw, Stoneyburn and Tailend Moss.

Favoured areas in **Cumbria** include Biggar and Wylock Marshes at Walney, Cavendish Dock Marsh and Kingsmoor, Carlisle, whilst in **Lancashire**, Barden Lane, Burnley and Roundwood Marsh regularly attract 1-5 birds. It is a regular winter visitor to **Cheshire** and can be found at Ashton's Flash, Astmoor, Gayton Sands RSPB Reserve, Hockenhull Platts, Inner Marsh Farm, Rostherne Mere NNR, Sandbach Flashes and Fiddler's Ferry. Small numbers are also regularly seen from October to March in the vicinity of the Old Baths car park at the north end of Parkgate Parade whenever the marsh is inundated by a tide exceeding 10.0 metres. View from the car park itself at SJ 273 790. Nearby the waterlogged inner salting between Neston Old Quay and Marshlands Road, Little Neston, is a reliable site for birders prepared to put on wellingtons and skirt around the creeks and pools. The edge of the *Scirpus* clumps are favoured and the site can hold up to eight birds amongst large numbers of Common Snipe (SJ 288 764). In **Merseyside**, 1-2 birds have shown very well in recent years on the causeway at Seaforth NR in mid-October - looking out to your left from Hide A.

Most favoured sites in **Northumberland** are Druridge and Newton Pools, Cresswell Pond, Corbridge and in the Coquet Estuary, whilst in neighbouring **Co. Durham**, traditional sites include Chilton Moor, Shibdon Pond and Hetton House Wood.

In **Yorkshire** and **Humberside**, numbers can be seen at Spurn Point between mid September and late October and up to six are normally present at Gallows Hill Pond, Knaresborough. In the Doncaster district, Highfields Quarry is a good site, as is Broomhill Ings, Thorpe Marsh and Bentley Tilts, whilst around Sheffield, Treeton Dyke, Carr Vale, Poolsbrook Marsh and the tip at Bolehill Flash are all regularly visited in winter.

In **Lincolnshire**, coastal localities such as Gibraltar Point and Saltfleetby regularly attract birds in October, whilst Marston Sewage Farm is one of the better inland sites.

A number of sites are used in **Leicestershire**, including Priory Water, Kirby Lakes, Trent Valley Pit, Tinwell Meadows, Dishley Pool, Lubbesthorpe, Quorn Borrow Pit and Ullesthorpe Water Meadows, but the most reliable are the flooded meadows around Rutland Water and Frisby GP.

In **Northamptonshire**, regular wintering sites include Stanford Reservoir and Earls Barton GP, whilst in **Bedfordshire**, the boggy marshland by the inlet stream in Houghton Regis Quarry is best.

In the West Midlands region, Dosthill GP (**Warks**), Castlemorton Common (**Worcs**), Upton Warren (**Worcs**) and Sandwell Valley RSPB Reserve (**West Mids**) are the most reliable sites

The flooded meadows at Oakgrove, Milton Keynes, and at the Blue Lagoon, Bletchley, are the best sites in **Buckinghamshire**.

In East Anglia, they can be seen at a wide variety of coastal sites on autumn passage in **Norfolk** but only reliably on the scrapes and lagoons at Cley and Titchwell reserves. The same can be said for **Suffolk**, where Martlesham Creek and Lackford WR often retain wintering birds. There are several wetland sites in **Cambridgeshire**, particularly Block Fen near Mepal, the Nene Washes, Barleycraft GP and St Ives GP.

Regular wintering sites in **Essex** include Colchester Hythe and Dagenham Chase, whilst in **Hertfordshire**, Amwell GP, Purwell Ninesprings, Lemsford Springs NR, Rye Meads sewage farm and Tyttenhanger GP are regularly visited.

In **Kent**, the areas of flooded fields and meadows at Aldington Floods, Sandwich Bay, Cliffe Pools and Sevenoaks Reserve are all regularly visited in winter, whilst in **Sussex**, Thorney Deeps and Pulborough Brooks are best.

The most reliable sites in **Berkshire** are Slough Sewage Farm and Dinton Pastures CP, whilst in **Oxfordshire**, the main wintering site is Shipton-on-Cherwell Quarry. Otmoor is also a regular site. The flooded meadows at Littlecote have recently become a regular wintering site in **Wiltshire**, in addition to the Cotswold Water Park.

In **Hampshire**, small numbers may be encountered at Titchfield Haven, Farlington Marsh, Eling Great Marsh, Pennington Marsh, Keyhaven Marsh and Winchester SF, whilst in neighbouring **Dorset**, the cress beds at Winterbourne Abbas, Lodmoor NR and Stanpit Marsh are the most regularly visited.

The wetlands of the **Somerset** Levels are an excellent site for Jack Snipe with large numbers wintering in the county. The better sites are Tealham Moor, Chilton Moor, West Sedgemoor, Berrow, Shapwick Heath and Burtle Moor.

There are two main sites in **Devon**: Bicton Common and Venn Ottery Common, whilst in **Cornwall**, birds may be found wintering in the marshy fields adjacent to the Hayle Estuary and in the area of pools at the eastern end of Copperhouse Creek, at St Columb Porth and at Maer Lake, Bude.

Red-necked Phalarope *Phalaropus lobatus*

The Red-necked Phalarope is a very rare breeding summer visitor to Britain and a regular passage migrant in spring and autumn. It breeds at sites with open water and wet and dry mire and builds its nest amongst grass tussocks and other vegetation. Open water is used for feeding and courtship purposes and will be visited throughout the nesting period. Birds arrive from wintering grounds in the south Atlantic off West Africa between the last week of May and third week in June and after a short nesting period, adapted to the brief northern summer of the high Arctic, depart the breeding grounds between the last week of July and the last week of August. The present British population of 22-25 pairs is almost exclusively confined to Fetlar, on the Shetland Isles, where these birds are afforded full protection by the RSPB.

✱Loch of Funzie, Fetlar - Shetland

In June 1994, a total of 28 males were recorded on Fetlar and many of these (and females) could be regularly observed feeding, bathing and displaying on Loch of Funzie, at the eastern end of the island. This site can easily be viewed from the road and if possible, it is best to watch the birds from the comfort and shelter of your vehicle. A hide has also been erected overlooking the Mires of Funzie. Do not wander from the road as this may cause untold disturbance to the phalaropes and other rare breeding waders on the island. The phalaropes are protected by wardens and are under 24-hour surveillance. Whimbrel, Arctic and Great Skuas and Arctic Terns should also be encountered.

Fetlar is one of the most northerly of the Shetland Islands and is reached via a regular vehicle ferry from Yell (at Gutcher) or Unst (Belmont). Please check local ferry timetables for details and bookings - ☎Burravoe 259/268. The RSPB employ a summer warden from April through to September and during this time the reserve area is accessible only by prior arrangement with the warden. Visitors should report at Bealance Croft, 2.5 miles from the Oddsta ferry terminal (HU 604 916 - Sheet 2) or ☎01957 83246. Please respect the property of farmers and crofters on the island and do not disturb breeding birds. Loch of Funzie is situated at the east end of the island south of the B9088, six miles from the ferry.

Migrant Red-necked Phalaropes may be encountered at a number of coastal localities, peak periods of occurrence being from 23rd May to 18th June in spring and between 10th August and 17th September in autumn. Failed or non-breeding adults may also appear in July. Most regularly visited localities are Pat's Pool and North scrape on Cley NWT Reserve (**Norfolk**), the freshwater lagoons at Titchwell RSPB Reserve (**Norfolk**), the main scrape at Minsmere RSPB reserve (**Suffolk**), the various scrapes at Elmley Marshes (**Kent**), the complex of freshwater pools at Cliffe (**Kent**), Sidlesham Ferry Pool (**West Sussex**), the various freshwater pools on the Ouse Washes Purls Bridge RSPB reserve (**Cambs**),the series of pools at Welney WWT Refuge (**Norfolk**) and the scrapes at Blacktoft Sands RSPB reserve (**Humbs**). A number of inland gravel pits, reservoirs and beet factory pools are also regularly visited.

Grey Phalarope *Phalaropus fulicarius*

The Grey Phalarope is a scarce passage migrant to Britain, most often recorded following westerly gales in September and October. The largest numbers are recorded in the south-west, but birds can be encountered virtually anywhere around the coast, the North Sea often producing a spate of records following north-west or northerly gales. Such conditions will often result in 5-10 birds (generally juveniles) being storm-blown inland. The following is a guide to the better sites for seeing Grey Phalaropes, but views will often be distant and are by no means guaranteed. Sightings are dependent on weather conditions.

❶St Ives Island - Cornwall

This is the best site in Britain to see Grey Phalarope and under suitable conditions, birds can be seen close inshore sheltering from Atlantic gales. They are recorded annually between 20th August and 12th November but peak during 5th September to 17th October and are generally *only* seen immediately following or during westerly/north-westerly gales. Winds of force 5 and above will suffice, the birds favouring the swell around the sewage outfall to the right of the island. They can be watched gradually drifting west and are often very close inshore, but normally fly westwards once they drift round the headland.

For substantial Grey Phalarope and other seabird movement to be seen at St Ives Island, winds should ideally be between west-north-west and north. A deep Atlantic depression from the southwest will have recently passed through, preferably with its centre not much higher than Northern Ireland. As it passes through Cornwall and further upcountry, the wind should veer from southwest to northwest and at this point, seabird passage should begin. The stronger, more vigorous and faster the depression is, the greater the seabird passage will be. Also, less fortunate for the birder, depressions coinciding with heavy rain will force more birds inshore.

St Ives Island is situated to the north of the town. Leave the A3074 towards the harbour and follow the signposts to the island car park. This involves negotiating a series of very narrow town streets and some very tight turnings, but will eventually lead to a car park immediately below the steeply rising headland. This car park is relatively small and regularly full (particularly in August and September) and involves payment by the hour or on a daily basis. A tarmac footpath from the car park leads up the slope to the coastguard lookout and the seawatching site. One can watch from either the outer wall of the lookout or from the wall to the right.

The same weather conditions that cause Grey Phalaropes to drift into St Ives Bay, also result in birds moving west past Pendeen Watchpoint. The views here, however, are not so good and powerful telescopes are essential. Small groups of phalaropes may congregate in the surf breaking over the three pronged reef immediately north of the watchpoint prior to their entrance to the full fury of an Atlantic gale in the open sea to the west. Newquay Headland will also reap results in these conditions, as will Godrevy Point, but neither site offers much in the way of protection from the elements.

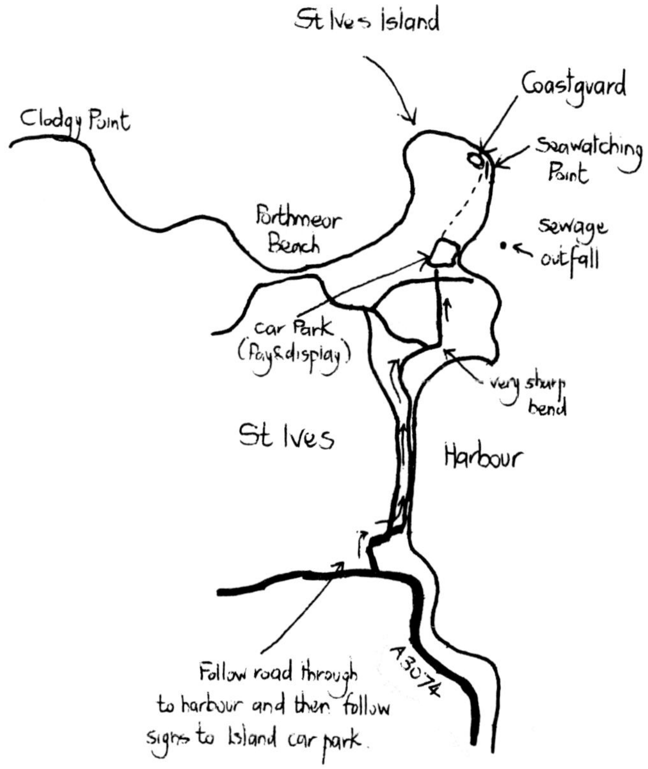

Grey Phalaropes can also be seen from Gwennap Head, Porthgwarra, but are always distant, a powerful 'scope being essential for viewing. The wind must be in a southwest or southerly direction and be a minimum of force 6 and preferably follow a westerly gale. The period late August to early October is best.

❷Strumble Head - Dyfed

A strong southwest/west/northwest gale between late August and late October will result in moderate numbers of Grey Phalaropes and other seabirds being seen off Strumble Head. The head rises to about 100 feet, mainly as steep grass and heather-covered slopes with rocky outcrops, and is at the western end of the Pencaer Peninsular. It can be reached from Fishguard Harbour via a well-signposted minor road and there is ample car-parking space at SM 895 414 (Sheet 51). The Pembrokeshire Coast National Park Authority building just below the car park offers protection from the wind and rain.

On the East coast, birds can be seen under suitable conditions (strong NW/northerly winds during late October to December) from most prominent headlands and recognised seawatching sites, but from in particularly Seaton Sluice (Northumberland), Whitburn (Tyne & Wear), Hartlepool Headland (Cleveland), Filey Brigg (North Yorks), Flamborough Head (Humbs), Spurn Point (Humbs), Gibraltar Point (Lincs), Cley Coastguards (Norfolk) and Sheringham (Norfolk).

Pomarine Skua *Stercorarius pomarinus*

The Pomarine Skua is an annual passage migrant to Britain that breeds on tundra in Arctic Russia, Alaska and Canada. Nesting depends largely on the availability of lemmings, the main food source, and in years of poor supply, birds will fail to breed. The majority of Pomarine Skuas that pass through Britain in spring will have wintered in the warmer upwelling waters off South Africa and depart this region between the first and third weeks of April. They reach the breeding grounds in early June, but small numbers of immature birds remain in the North Atlantic all summer and may be encountered at any time within this period. There are two distinct migration routes through Britain, one involving birds entering the English Channel and passing out into the North Sea past Kent and another involving birds migrating north up the Atlantic and passing the Western Isles and Shetland and some heading inland over the Southern Uplands. The birds migrate in large flocks of up to 100 birds and are a truly magnificent sight. Many in spring are in full breeding attire, the distinctive 'spoons' on the tail being readily apparent.

Pomarine Skuas in Spring

❶ Dungeness - Kent

Dungeness is the premier site in Britain for seeing Pomarine Skua, with up to 250 birds passing offshore between 26th April and 30th May. The peak period for passage is between 5th and 20th May, ideal conditions being light to moderate SE/E winds with mist or light drizzle. The birds will also pass in clear visibility and on light southerlies but in smaller numbers. For optimum results, be sure to be present from dawn until 11.00am or from 4.00 to 8.00pm, the largest groups often passing within the first three hours of daylight. The birds can be seen from anywhere along the shingle ridge in front of the power station perimeter wall, with limited seating available in the two seawatching hides provided by the Observatory. From New Romney, turn left on to the B2075 towards Lydd and at the roundabout turn left towards Dungeness. After about two miles and just before Greatstone-on-Sea, turn right towards the beach and lighthouse and park near the Britannia pub or the entrance to the concrete track leading to the beach and around the power station perimeter fence. The area known as 'The Patch' (a disturbed area of water caused by the power station outflow) regularly attracts large numbers of gulls and terns, often including Mediterranean, Glaucous and Little Gulls and Black and Arctic Terns.

❷Selsey Bill - West Sussex
A total of about 180 birds are recorded off Selsey Bill in an average spring, with movement noted between mid April and late June and peaking between 28th April and 22nd May. Conditions, like all sites along the south coast, should ideally be clear, misty or drizzly with light to moderate south-easterly winds. Selsey lies eight miles south of Chichester and is reached via the B2145. Continue to the southernmost point of the road and park sensibly in nearby roads or in the car park at the end. The car can be used as a hide in inclement weather conditions or one can seawatch from the car park or from the bill itself, 200 yards to the left. Pomarine Skuas and other seabird passage is from west to east in spring.

❶Seaford - East Sussex
An average spring at Seaford Head sees between 150 and 200 Pomarine Skuas flying east and these are best seen in the first four hours after dawn or in the evening. Conditions must ideally be light to moderate south-easterly or southerly with light mist and it is not unusual to see flocks of up to twenty birds fly past. The best place to seawatch from is Splash Point at the far southeast end of Seaford promenade, just west of the golf course. Seaford lies nine miles west of Eastbourne on the A259.

A similar number of birds can be seen under suitable conditions off nearby Beachy Head, with the sheltered cove at Birling Gap Hotel and Lifeboat Station offering the best view.

❷Brighton Marina/Worthing - West Sussex
The stretch of coastline between West Worthing and Brighton Marina will produce Pomarine Skuas under suitable conditions. Fresh south-easterlies and clear, hazy or misty conditions will see small groups pass close inshore between 26th April and 26th May. Watch from various parking areas adjacent to the A259 or from the end of Brighton Palace Pier or the end of the west arm of Brighton Marina.

❶Portland Bill/Chesil Cove - Dorset
Under suitable conditions, moderate numbers of Pomarine Skuas can be seen flying east past Portland Bill on their way up the English Channel. The winds must be south-east to south, birds moving between 25th April and 21st May. Mornings are best, particularly the first four hours of daylight, with some passing again in the evening if conditions prevail. It has become apparent in recent years that a proportion of the skuas take a short cut across the Fleet and have been seen heading east over Chesil Beach at Ferrybridge and out across Portland Harbour. This site is particularly good in the evening where birds are often known to linger offshore, but it is essential to have misty conditions combined with light to moderate south-easterly winds. This site is also good for wader passage, as well as for good numbers of terns. The peak period is between 28th April and 12th May.

Elsewhere in Dorset, birds may be observed under suitable conditions off St Alban's Head, Durlston Head and Hengistbury Head, but numbers are predominantly small.

❶St Catherine's Point - Isle of Wight
An average total of 50 birds is recorded off St Catherine's Point in spring, primarily between the 20th April and 20th May. South-easterly winds with mist is essential for passage. The Point is reached along a minor road west of Niton village.

❷Berry Head/Hope's Nose - Devon
The first three weeks of May generally sees the appearance of small parties of Pomarine Skuas flying northeast past the exposed headlands of Berry Head and Hope's Nose, Torquay. Passage is normally recorded during fresh to moderate south-easterly winds. Early mornings and evenings are best.

Berry Head is situated at the eastern extremity of Brixham and is accessed via a number of narrow, minor roads. Likewise, Hope's Nose lies at the eastern end of Torquay and is best reached from Babbacombe.

❷Porthgwarra - Cornwall

A moderate passage of Pomarine Skuas takes place off Gwennap Head, Porthgwarra, between late April and the end of May and generally peaks during 6th-16th May. The birds fly east up the English Channel and are best seen during light to moderate south-easterly winds in the first three hours of daylight or in the evening. Leave Penzance west on the A30 and turn left on to the B33283 at Catchall. Continue west to Polgigga and turn south off the B3315 at the sharp left hand bend to Porthgwarra. A narrow meandering lane eventually descends into Porthgwarra village, a pay & display car park being located at the end. Walk west along the coastal clifftop footpath for about 500 yards to the headland, using a large granite rock as shelter.

❶Bowness-on-Solway - Cumbria

Spring migration can be observed at Bowness-on-Solway Viaduct on the Inner Solway. Take the B5307 west out of Carlisle for ten miles to Bowness village and the viaduct is situated half a mile beyond here. Park by the roadside verge and walk 300 yards to the end of the disused viaduct which protrudes into the estuary at NY 213 628 (Sheet 85). Large numbers of Pomarine Skuas (as well as other species of skua) can occur from mid April to early June with the peak movement falling between 4th-16th May. The best conditions are rather unpredictable but overcast skies, an early morning high tide and southwest breeze may be favourable. Parties of over fifty birds together can be encountered gliding low over the water at close range past the viaduct before gaining height to head off purposefully north-east and over land.

❶The Uists - Western Isles

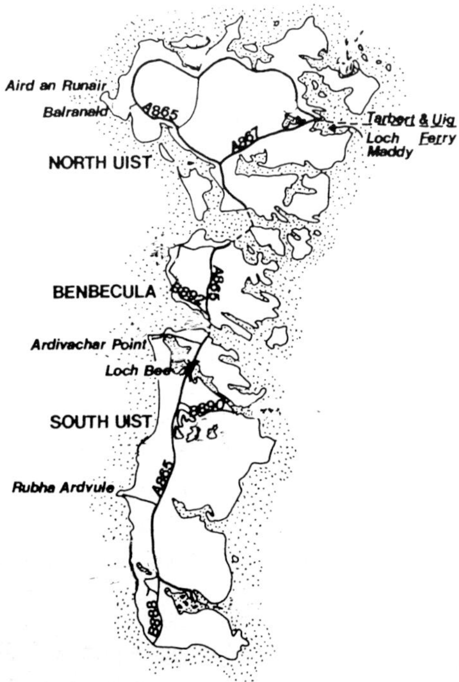

A large proportion of the west Siberian population of Pomarine Skuas migrates northeast over the Western Isles in spring, resulting in counts of up to 2,000 birds during the period. Flock sizes sometimes exceed 100 birds and these can be observed from a number of strategically placed sites along the west coast. The main passage period is between 6th May and 10th June, with a notable peak during 11th-26th May. However, movement is determined by weather conditions, an easterly element in the wind curtailing passage. The best site to watch from is Aird an Runair, on North Uist (at NF 686 706 - Sheet 18). Leave Lochmaddy on the A867 and head NW along the A865 towards Tigharry. After about five miles, turn left to Hougharry and then follow a minor track towards the headland and walk the rest. There is limited shelter at this site from the elements.

Another good vantage point is Ardivachar Point, on South Uist (NF 737 456). As you leave Benbecula across the Causeway, turn first right towards Clachan and Ardivachar and drive to the end. Walk north to the exposed headland and view. Smaller numbers can be seen from Rubha Ardvule, ten miles to the south (NF 709 300).

The ferry crossing between Skye (Uig) and Lochmaddy (North Uist) can also produce large movements of skuas, especially during a warm south-westerly airstream. There are two car ferries operating to the Uists: a service connecting Lochmaddy (North Uist), Tarbert (Harris) and Uig (Skye) and another from Lochboisdale (South Uist) to Oban. Ferry bookings and timetables can be obtained from Caledonian MacBrayne, Ferry Terminal, Gourock, PA19 1QP or by ☎ 01475 33755.

xEsha Ness/Wats Ness - Shetland North Mainland
The premier sites in Shetland for watching both Pomarine and Long-tailed Skuas in spring are the west facing headlands of Esha Ness and Wats Ness. In suitable weather conditions (light to moderate north-westerly winds), large flocks of both species can be encountered between mid May and early June.

Wats Ness lies to the south of Sandness and is best accessed from Bridge of Walls. Follow the minor road west from Walls to Swinister at the end of the road at HU 178 508 (Sheet 3) and walk west to view the coast.

Esha Ness is further north and lies at the western end of the B9078, ten miles west of Hillswick. Watch from the lighthouse car park at Calder's Geo at HU 205 784.

Pomarine Skuas in Autumn

The identification of Pomarine Skuas in autumn can be difficult, especially in immature plumage, the most useful reference material available being Lars Jonsson, *Birds of Europe*, pages 257-261, and Harris, Tucker & Vinicombe, *The Macmillan Field Guide to Bird Identification*, pages 99-103.

In autumn, Pomarine Skuas can be encountered in almost any coastal location, but the well-known seawatching locations generally produce the largest numbers. The entire east coast between the Black Isle (Highland) and Dungeness (Kent) will produce sightings, as will the west coast between Machrihanish SBO (Strathclyde) and Pendeen Watchpoint (Cornwall). Adults will begin drifting south into the North Sea from mid August onwards whereas juveniles should not be expected much before mid October. Peak passage is often during November, when strong northwest/northerly winds will produce large numbers.

Long-tailed Skua *Stercorarius longicaudus*

The Long-tailed Skua is a scarce passage migrant that visits Britain in varying numbers in spring and autumn. Apart from a trickle of sightings along the south coast in May, the bulk of the passage in spring passes the northwest coast of Scotland with flocks of over 100 birds being occasionally encountered. The return passage of adults depends largely on the success of the breeding season and in years of poor lemming populations can begin as early as late July. Most appear in the southern North Sea off the east coast of Britain from mid August onwards and continue to be recorded until early October. Juveniles begin appearing from late August, with peak passage occurring throughout the first three weeks of September. Identification of immature Long-tailed Skuas is problematical and is best discussed by Harris, Tucker & Vinicombe 1989, *The Macmillan Field Guide to Bird Identification*, pages 103-104. Remember, however, that skuas in juvenile plumage are highly variable and appear in dark ands pale morphs.

Long-tailed Skuas in spring

Each spring, the odd adult Long-tailed Skua is recorded at south coast seawatching locations by observers searching for Pomarine Skuas. The peak period of observation is from 26th May to 5th June and localities such as Portland Bill, Splash Point, Seaford and Dungeness are the most likely to produce one. The main passage, however, is in the Northern Isles, where large numbers migrate north between 6th May and 10th June.

❶The Uists - Western Isles
Totals of over 1,400 birds have been recorded in recent years, with the peak passage period occurring between 12th and 25th May. There are three main points of observation: Aird an Runair, North Uist; Ardivachar Point, South Uist and Rubha Ardvule, South Uist, birds passing these sites on light to moderate south-westerly winds (see site directions under Pomarine Skua).

❶Wats Ness/Esha Ness - Shetland North Mainland
As with Pomarine Skua, large numbers of Long-tailed Skuas can be witnessed moving north past these sites in late spring. Observations are highly dependent on weather conditions. For site directions, see under Pomarine Skua.

Long-tailed Skuas in Autumn

Like all seabirds, citing sites in which to see Long-tailed Skuas in autumn is purely conjectural. If it has been a good lemming year, then chances are that there will be exceptional numbers of birds entering south into the North Sea, but if it has been a poor lemming year, there may be lots of early returning adults in the North Sea but few records of juveniles or of adults after the second week of September. Furthermore, the weather conditions play a major role in their appearance. Strong north-easterly winds in early autumn will see many birds enter the North Sea from late July onwards but a persistent westerly will result in very few birds recorded and a somewhat west coast bias to sightings. In recent years, a pattern has emerged with Long-tailed Skua occurrences. From late July to mid August, a scattering of adults have appeared in NE England, from Northumberland south to Cleveland, and then dispersed south to Humberside and Lincolnshire. Many more adults have arrived into the southern North Sea by late August and these numbers are supplemented by the first juveniles of the autumn, and have appeared regularly off East Anglia and Kent as well as further north. September then sees the bulk of the passage and in several recent years, unprecedented numbers have been recorded passing Humberside and Norfolk. These movements can continue well into October and often involve 1-10 birds penetrating inland on gravel pits and reservoirs. Although the east coast attracts the largest numbers, a small number can be recorded at various west coast seawatching localities, particularly the sites around the Wirral, Strumble Head (Dyfed) and St Ives Island/Pendeen Watchpoint (Cornwall).

➊Flamborough Head - Humberside

Flamborough Head is probably the single most reliable site for Long-tailed Skua in Britain, with regular sightings (sometimes daily) between 27th July and 14th October, and peaking between 20th August and 12th September. Birds are often distant and require skill to identify, but generally pass during periods of NW to NE winds or in slack conditions. Park in the pay & display car park at the Head and walk east towards the foghorn. The seawatching site is half way down the cliffs below the foghorn.

➊Sheringham - Norfolk

The sheltered buildings along the promenade at Sheringham have recently become the premier seawatching localities in Norfolk. One of the major attractions has been the large numbers of Long-tailed Skuas being identified offshore, with up to 40 birds being reported on some days. Sheringham offers observers relatively good views of birds, especially at high tide, and also affords protection from the elements. Long-tailed Skuas are recorded from mid August onwards and peak in number in September. Passage depends on wind direction, but is eastwards during an easterly and westwards during a westerly. Mornings are primarily best.

Birds can also be identified during suitable conditions at any prominent east coast location, especially Chanonry Point, Black Isle (Highland), Girdleness (Grampian), Fife Ness (Fife), Hound Point (Lothian), St Abb's Head (Borders), Eyemouth (Borders), Newbiggin (Northumberland), Seaton Sluice (Tyne & Wear), Whitburn (Tyne & Wear), Hartlepool Headland (Cleveland), Filey Brigg (North Yorks), Hornsea (Humbs), Spurn Point (Humbs), Gibraltar Point (Lincs), Witham Mouth (Lincs), Cley Coastguards (Norfolk), Covehithe (Suffolk), Canvey Point (Essex), Leysdown-on-Sea, Sheppey (Kent), Reculver (Kent) and Foreness Point (Kent).

Mediterranean Gull *Larus melanocephalus*

The Mediterranean Gull is a regular winter visitor, a passage visitor and a rare breeding bird in Britain, which is largely distributed in eastern Europe and in the Russian States. It was formerly a rare vagrant to Britain, but after the first breeding attempt in 1968, numbers have gradually increased and by 1995, at least twelve pairs were attempting to nest within colonies of Black-headed Gulls in Southeast and Southern England.

Mediterranean Gulls can be recorded at a large number of coastal and inland sites throughout the year but the following are a selection of the most reliable:

✱Copt Point, Folkestone - Kent

Copt Point is far and away the best site in Britain to see Mediterranean Gull. The theory is that large numbers from Boulogne and Calais followed the channel ferries over and gradually infiltrated Folkestone. The birds favour the outfall and rocks at Copt Point and in recent years have peaked at a phenomenal 112 birds. They are present throughout the year but peak in winter and spring, with only a handful of birds remaining in late June to July. The vast majority are adults and second-years and at least four regular birds afford exceptional views on the playing fields immediately inland of the headland. The remainder can be found roosting on the rocks below the clifftop at low tide or feeding on the outfall offshore.

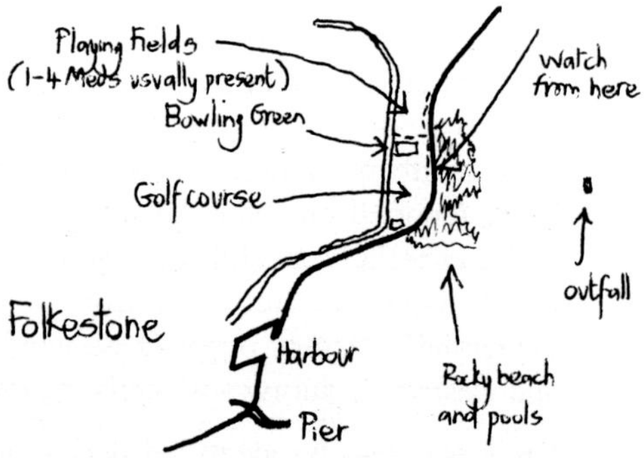

Leave the M20 at Junction 13 and follow the A20 eastwards. Continue east at the A260 roundabout but at the next turn right. At the next junction, continue towards the seafront and after about a mile, a number of playing fields appear on the left. Park sensibly along the road at about TR 240 367 (Sheet 189) and walk to the cliff edge along the track between the bowling alley and the playing fields. Observe the sewage outfall and rock falls from the clifftop at TR 243 365.

✱Dungeness Area - Kent
The Dungeness area has long attracted numbers of Mediterranean Gulls, especially in spring and in early autumn, with 1-5 pairs regularly attempting to breed on the RSPB reserve. They can often be seen feeding with other gulls on 'The Patch', the hot water outflow from the power station, or roosting on the shingle beach adjacent. Birds are present daily from mid March to late August and peak in June and July, and are made up of mainly adult and first-year birds. They can also be found amongst the Black-headed Gull colonies on the shingle islands at Dungeness Reserve but can be extremely difficult to locate due to the number of birds present and the viewing distance involved. However, many gulls and terns choose to roost and bathe on the near bank and Mediterranean Gulls can often be picked out from the main information centre or from the hide to the right. Several individuals (primarily adults) can also be found amongst the roosting gulls by the toilet block at Greatstone-on-Sea, about a mile north of the Pilot pub.

✱Radipole Lake RSPB Reserve/Lodmoor/Weymouth Bay - Dorset
Good numbers of Mediterranean Gulls are recorded in the Weymouth area throughout the year, with peak numbers appearing in winter and at passage periods. The best place to see them is at Radipole Lake, where they can be identified amongst roosting Black-headed Gulls in the evening and at high tide periods. The birds roost and congregate on the small islands and shallow water adjacent to the information centre.

The nearby reserve at Lodmoor is also very good and birds can be found roosting on the numerous islands and marshy grasslands at the beach end of the reserve. The roost in Weymouth Bay is also worth checking.

Radipole Lake lies close to Weymouth bus and railway stations and is situated north of the Westham Bridge in the centre of town. A large pay & display car park is available at the south end of the lake and the RSPB shop and information centre lies adjacent. Lodmoor CP is situated about a mile to the east immediately inland of the A353. Access to the reserve is via the public car park to the left of the minor road to the refuse dump, immediately east of the town. The reserve is open at all times and three hides are provided off the main footpaths.

Christchurch Harbour also attracts numbers of passage Mediterranean Gulls, particularly in spring, and these are best observed from Stanpit East Marsh.

✱Seaforth NR - Merseyside
This Lancashire Wildlife Trust reserve is situated on the A565 Liverpool to Southport road and is accessed from the Freeport entrance on Crosby Road South. Mediterranean Gulls visit the reserve throughout the year and are best seen from early to mid afternoon when they regularly appear on the freshwater pool in front of Hide A and bathe before roosting on the causeway. Adult birds peak in March, first-summers in May and July/August brings a mixture of moulting adults and first-summers (e.g., up to 16 birds were present on the reserve in mid July 1995).

✱Blackpill - West Glamorgan
The mudflats at Blackpill are an excellent place for gull-watching, with large numbers of birds and a good variety of species present throughout the year. Scanning through the large flocks of Common and Black-headed Gulls will inevitably turn up a Mediterranean Gull and occasionally, Ring-billed Gull. The Mediterraneans peak in spring, when up to twenty different birds may be recorded between 15th March and 15th June. They are present at other times of year but in much smaller numbers.

The A4067 follows the shoreline of Swansea Bay west between Swansea and The Mumbles, with Blackpill Beach lying immediately east of its junction with the B4436. The gulls tend to congregate in the area around the inlet channels.

✱Plymouth Area - Devon

The Plymouth area with its fish quays, sewage outfalls and riverside rubbish tips is hugely attractive to large numbers of gulls and Mediterranean Gulls are no exception. Birds can be seen throughout the year but peak in winter and early spring. The most reliable site for sightings is the sewage outfall off West Hoe, where they can be identified amongst Black-headed Gulls scavenging for edible waste. Most birds are immatures and are also known to wander to the nearby Plym Estuary at low or rising tides.

The Plym Estuary is best checked at low tide, preferably from the footpath at the northern end or from the east side, adjacent to Chelson Meadow Tip. The sewage outfall at West Hoe is accessed from Millbay and can be overlooked from a public walkway at the western end.

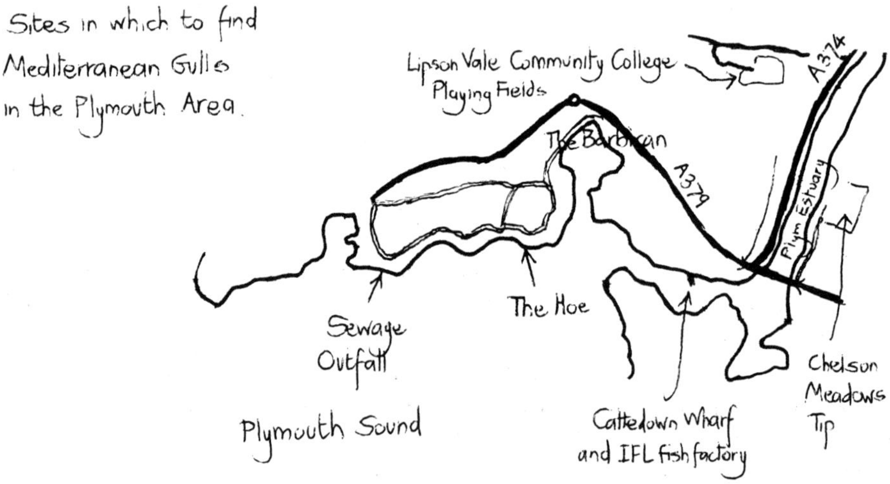

✱St Helens/Bembridge - Isle of Wight

St Helens Duver and Bembridge Harbour are both sites of ornithological interest situated at the eastern end of the Isle of Wight. The northern part of Bembridge Harbour is partially enclosed by the Duver, a low-lying spit of sand dunes colonised by coastal scrub and short grass. The seaward shore consists of muddy shingle, whilst on the opposite side of the harbour entrance, low tide reveals the vast expanse of Bembridge Sands. This entire area is good for roosting gulls and Mediterranean Gulls can be identified in varying numbers throughout the year. Although one or two birds winter at St Helens and can be regularly seen roosting on the rocky pools and on the shingle ridge offshore, the main influx is in July and August, when up to 30 birds may be recorded. The flock normally consists of mainly adult birds, and these numbers remain until at least mid November. Between December and early March, the birds move to nearby Foreland and up to twenty may be located roosting on the flat rocky ledges offshore or at various points around the headland.

Both St Helens Duver and the entrance to Bembridge Harbour are reached via the B3330 eastwards from Ryde. At a sharp left hand bend at the north end of St Helens, take Duver Road east to either the shore by St Helens Church or to the Duver car park. There is general access to the shore, the harbour entrance and the Duver, the gulls roosting at low tide on the vast expanse of mud flats or at high tide, on the rocky shore and shingle ridge just offshore. There is also a footpath from the Duver to St Helens Mill, traversing the seawall and separating the Old Mill Ponds from the main harbour.

Up to three adult Mediterranean Gulls are also present at nearby Ryde between November and March. These can be viewed from various vantage points along the seafront between Ryde Pier (SZ 594 929 - Sheet 196) and Nettlestone Point (SZ 629 918). The Canoe Lake is also worth checking, an adult Ring-billed Gull often being in attendance here.

✱Great Yarmouth/Breydon Water - Norfolk

A population of up to six birds (mainly adults) can be found in the Yarmouth area in winter, primarily in the coastal grassy areas along the South Beach or at the river mouth. They associate with the large wintering flocks of Black-headed Gulls and can be seen feeding on scraps in the car parking areas scattered along the promenade, roosting in grassy areas within the industrial complex directly inland of the promenade and on the groynes and breakwaters at the harbour entrance and river mouth.

In spring, large numbers of first-winter Mediterranean Gulls can be identified roosting amongst Common and Black-headed Gulls at the eastern end of Breydon Water. It is not unusual to see up to 15 birds roosting on the saltmarsh from the first hide along the north sea wall during April and May, with birds being recorded between early March and mid June and from late July to early October.

Elsewhere in **Norfolk**, Mediterranean Gulls can also be regularly seen at Mundesley (winter), Overstrand outfall (winter), Salthouse coastal marshes and fields to the west of Beach Road (mainly passage), Cley NWT reserve Pat's Pool and North Scrape (April-September) and on Blakeney Point (May-July).

❶Minsmere scrape/Sizewell Outflow - Suffolk

Mediterranean Gulls can be regularly recorded in the Minsmere area throughout the year, with 1-3 adults noted during October to April and up to ten birds between early May and late September. They commute between roosting islands on the scrape (best observed from West and East hides) and the outfalls off Sizewell Beach.

One to two birds are also regularly present at Ness Point, Lowestoft, in the winter, commuting between the harbour and the outfall offshore of the Point.

❶Farlington Marsh Lagoon - Hampshire

Up to four birds are regularly present in Langstone Harbour between October and March, with further passage birds recorded in April-May and August. They can be found roosting in the large Black-headed Gull flock on the northern shore of Farlington Marsh Lagoon or on nearby playing fields. Evenings are best. Leave the A27 westwards at the A2030 junction and turn immediately left on the roundabout into the Farlington Marshes car park. Drive eastwards parallel to the A27 and park. The lagoon can be viewed from the seawall, 400 yards south of the car park.

Elsewhere in Hampshire, Mediterranean Gulls can be seen at a number of locations, primarily in winter. These include Titchfield Haven NR, Pennington and Keyhaven Marshes, Southsea Canoe Lake, Weston Shore, Needs Ore Point, the IBM Lake at Cosham and Alresford Pond. Until recently, a regular adult wintered on the concrete slipway at HMS *Daedalus* in Lee-on-Solent.

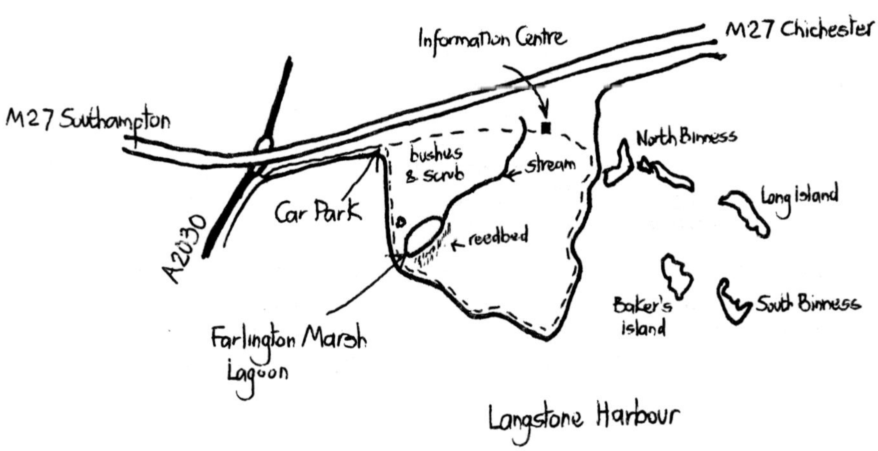

➊Chew Valley Lake - Avon

Numbers of Mediterranean Gulls visiting Chew Valley Lake have fluctuated from nine to 21 birds in the last twelve years, with most being recorded between December and March. The best place in which to view these birds is from the hide situated on the end of Nunnery Point, on the western side of the lake. The birds come in to roost on the reservoir towards dusk but may be difficult to locate amongst the thousands of roosting Black-headed and Common Gulls. The hide is for use by permit holders only, these being obtainable from Bristol Water, Woodford Lodge, Chew Stoke, Bristol, BS18 8XH.

Chew Valley Lake is situated about ten miles south of Bristol, north of the A368. Access to the Nunnery Point hide is east off the B3114 north of Compton Martin.

➊Fishguard Harbour - Dyfed

1-3 birds are regularly present in Fishguard Harbour between November and March and can be found roosting amongst the large numbers of Black-headed Gulls there. Check the harbour area around the ferry terminal.

✱Hayle Estuary - Cornwall

Mediterranean Gulls can be seen on the Hayle Estuary throughout the year, with up to five in winter and further birds on spring and autumn passage. They are best seen at low tide, when large numbers of gulls congregate on the saltings to roost and preen, and can be watched from the car park (or hide) at the Quay House Inn, at the extreme SW end of the estuary. In winter, birds are also regularly seen on Copperhouse Creek, particularly from the Library car park in Hayle.

Elsewhere in West Cornwall, birds can be found in winter in Mounts Bay, off Marazion Beach, and at Newlyn Beach and outfall. Small numbers can also regularly be seen during pre-roost gatherings on Drift Reservoir, two miles west of Penzance.

Swanpool Beach, Falmouth, and the Camel Estuary are also likely to provide sightings if visited between December and March.

Other locations in Britain which may produce 1-2 regular wintering birds are Craster Harbour (Northumberland), Sunderland Docks (Co. Durham), Hartlepool Bay and South Gare (Cleveland), Hornsea Promenade (Humbs) and Canvey Seafront (Essex).

A number of inland reservoirs also regularly attract 1-3 Mediterranean Gulls between October and March including Ogston Reservoir (Derbyshire), Blithfield Reservoir (Staffs), Stewartby Lake/Rookery Pit/Brogborough Lake (Beds), Draycote Water (Warks), Chasewater (Staffs) and Cheddar Reservoir (Somerset). These birds fly in to roost up to an hour before dark.

Little Gull *Larus minutus*

The Little Gull breeds in lowland freshwater lakes and pools, river valleys and coastlines in Siberia and from north-west Russia to the Baltic region and in isolated scattered colonies in the Netherlands, Denmark, Sweden, Finland and Poland. It has occasionally bred in Britain in the past but is primarily a passage migrant or winter visitor. Substantial numbers of birds occur in autumn as juveniles disperse from breeding grounds further north and adults move south to moult. One favoured moulting area is the Firth of Forth, with birds remaining in that area until mid September. The heaviest passage down both coasts is from late September to mid November, with birds eventually departing westwards to winter in the Mid Atlantic, off the west coasts of Iberia and North Africa. The size of the wintering population in Britain is variable, but averaging at 150 in the Irish Sea and up to 350 in the North Sea. The following are the best sites in Britain in which to see this small, somewhat fragile-appearing and dainty gull. It can, however, be seen in small numbers at virtually any coastal location in autumn. Furthermore, following easterly winds in spring, many individuals are displaced inland, and from late April until the third week of May, can be recorded in flocks of up to 80 birds at inland reservoirs and gravel pits in Southern England.

✱Hornsea Mere/Flamborough Head - Humberside

The first small parties of Little Gulls generally appear at Hornsea Mere in spring, with up to twelve birds remaining throughout May and possibly 1-3 birds remaining through the summer. The moulting flock of adults and dispersing juveniles start arriving in early August and by 18th, it is not unusual to see as many as 140 birds. Numbers continue to increase until a peak of about 180 birds is reached in early September and then numbers gradually decrease to only a handful of birds in mid October.

The largest numbers of birds are recorded further north in Bridlington Bay, where from the third week of July, the first returning adults can be seen from Flamborough Head. These numbers are rapidly swelled in August and by the end of the month 200 birds are regularly recorded offshore. The first week of September sees the peak arrival and up to 810 birds can be seen in one great raft in Bridlington Bay or moving in small feeding parties off Flamborough. Numbers quickly decrease from then on and by mid October, only a trickle of passage birds are noted.

Hornsea Mere lies some 15 miles south of Bridlington west of the B1242. In Hornsea village, follow signs to 'The Mere' out to Kirkholme Point. The Little Gulls can easily be watched from here.

Further south in Humberside, large numbers of Little Gulls can also be seen during easterly conditions moving north or south off Spurn Point. Peak movements are between late August and early November.

✱Fife Ness - Fife

From late July to early October, Little Gulls can be recorded off Fife Ness on virtually a daily basis, with numbers increasing from 1-10 in early August to 150+ in mid to late August and up to 300 birds throughout September. Peak counts of up to 526 birds have been made between 29th September and 10th October.

In winter, Little Gulls are regularly recorded in small numbers off Tayport, on the south side of the Firth of Tay. Look north from the toilet block by the caravan site along the southern shore. The once regularly used moulting site at Kilconquhar Loch seems largely to have been abandoned, but in recent years, large numbers have been recorded in spring at Roscobie Loch, in Tayside. This site lies four miles east of Forfar adjacent to the A932.

✱Seaforth NR/Crosby Marine Lake - Merseyside

The peak arrival of Little Gulls at both these neighbouring sites is determined by the hatching of the *Chironomid* midges on which they feed on over the freshwater pools. In recent years, these have hatched in mid April and up to 300 Little Gulls have been present. They disperse in early May and sightings throughout the rest of the year are rather sporadic.

Seaforth NR is accessed via the Freeport entrance on the A565 Liverpool-Southport Road (Crosby Road South), whilst Crosby Marina is first left at the traffic lights north of the Freeport entrance and at the end of the road.

❶Sheringham - Norfolk

During periods of strong to gale force northerly or north-easterly winds, large numbers of Little Gulls can be recorded off the north and east Norfolk coasts. Sheringham is one of the better vantage points as its shelters offer protection from the wind and rain. Depending on the wind direction, passage can either be from the east or west and can involve some impressive movements of birds as they are forced down into the southern North Sea. The same conditions are also conducive to many other sea-birds, with shearwaters, petrels, skuas, terns, auks and sea-duck also being regularly seen. The peak period for Little Gull passage off Norfolk is from early September to early November.

Elsewhere, Hunstanton Cliffs, Holme Dunes, Titchwell Beach, Cley Coastguards and Mundesley all offer similar opportunities for sea-watching, but there is less chance of sheltering from inclement weather conditions.

In neighbouring **Suffolk**, large numbers can also be recorded in easterly or northerly winds off Southwold and Covehithe and in smaller numbers off Ness Point, Lowestoft, and Minsmere Beach. More reliably, however, 1-20 birds feed around the rigs and outflows off Sizewell Power Station between late August and early October.

❶Staines Reservoirs - Surrey

Following a moderate spring passage of up to 30 birds between late March and late May, recent summers have seen a preponderance of adults moulting at the site and remaining present from mid July to mid August. These numbers are then swelled by an arrival of up to 15 birds in late August and these may then remain until at least late September. They seem particularly attracted to the South Basin and can be easily seen from the main causeway separating the two reservoirs. Leave the M25 at Junction 13 and drive east along the A30. At the first set of traffic lights, turn left on to the A3044 and after half a mile park sensibly at the side of the road, making sure not to obstruct the cycle path. Staines Reservoirs are on the eastern side of the road and there is a clearly marked entrance gate and stile allowing access to the causeway. It can also be accessed from the east side along the B378 in Stanwell.

❶Dungeness - Kent

Large numbers of Little Gulls can be recorded off Dungeness during spring and autumn, with peak movements coinciding with easterly oriented winds. Spring passage generally begins in mid March and continues until mid May, with the peak numbers occurring throughout April, whilst in autumn, birds can be present on 'The Patch' from mid July onwards but with the heaviest passage between the 29th September and 15th October. Lingering birds are best observed from the hide and shingle ridge overlooking the hot water outflow and passage birds from the hide at the end of the power station perimeter wall.

In neighbouring Sussex, similar numbers can also be recorded between late March and mid May from the regular seawatching locations of Selsey Bill, Splash Point, Seaford, and Worthing Promenade.

Sabine's Gull *Larus sabini*

The Sabine's Gull is a scarce passage migrant to Britain which occurs annually between mid August and early November. It is a highly pelagic seabird and is normally only encountered in periods of strong winds. The majority of birds are present in the Atlantic but smaller numbers are found in the North Sea and a trickle of birds end up in the English Channel. Cornwall is by far and away the best county in which to see this species, with Humberside the runner-up. However, under suitable conditions, birds may be seen at virtually any coastal site on the east coast between Girdle Ness (Grampian) and Dungeness (Kent) and from Islay (Strathclyde) to Hartland Point (Devon) on the west coast. Identification is relatively straightforward but best discussed and illustrated by Harris, Tucker & Vinicombe 1989 in *The Macmillan Field Guide to Bird Identification* on pages 108-111.

❶St Ives Island - Cornwall

The best place in Britain in which to get good views of Sabine's Gull is in St Ives Bay, where, after a deep depression has passed through Cornwall in autumn, strong north-westerly winds force small numbers of this beautiful and rather dainty gull close inshore. They tend to congregate about the outfall immediately right of the headland and are usually identified amongst Kittiwake flocks. If it has been a very deep depression and originated from far out in the Atlantic, it is not unusual to see up to ten Sabine's Gulls fly west during the day and on one occasion in September 1982, the author and many others recorded an exceptional flock of 100 summer-plumaged adults feeding and sheltering in the bay by the car park throughout the day. There is no set time for the passage of Sabine's Gulls past this point and birds can appear at all times of day providing the weather conditions persist. In fact, even on a force 4 westerly wind following a deep depression, Sabine's Gulls will be recorded. The vast majority of Sabine's Gulls at this site pass remarkably close to the headland and often afford tremendous views and at times of constant seabird passage, it can be very easy to overlook them as they fly west with large reorienting flocks of Kittiwakes.

St Ives Island is situated to the north of the town. Leave the A3074 towards the harbour and follow the signposts to the island car park. This involves negotiating a series of very narrow town streets and some very tight turnings, but will eventually lead to a car park immediately below the steeply rising headland. This car park is relatively small and regularly full (particularly in August and September) and involves payment by the hour or on a daily basis. A tarmac footpath from the car park leads up the slope to the coastguard lookout and the seawatching site. One can watch from either the outer wall of the lookout or from the wall to the right.

❶Pendeen Watchpoint - Cornwall

Pendeen Watchpoint is probably the best site in Britain for Sabine's Gull but the views are never as good as those at St Ives. The viewing circumstances at this site however, allow for a wider variety and larger number of seabirds to be recorded, and it is probable that many birds are just too far out from St Ives Bay to be seen or identified from there but relatively closer at Pendeen. In strong west to northwest winds (force 5 and above), Sabine's Gull will be recorded from here between mid August and the first week of November, with peak numbers occurring between 30th August and 21st September. If one is lucky, birds may pass with flocks of Kittiwakes between the viewing point and the three-pronged reef but more often within the outer third field of view to the horizon. A telescope is essential at this site as it is with all seawatching. There is ample shelter at the site, with birders sitting close to the lighthouse building perimeter wall. The lighthouse itself is reached from the B3306 at the east end of Pendeen village and is well signposted.

➊Strumble Head - Dyfed

Up to twenty Sabine's Gulls are recorded annually from Strumble Head and like Cornwall, are seen during strong SW/W/NW winds. Passage commences in late August and continues until mid October. A telescope is essential at this site and birds pass south-westwards with flocks of Kittiwakes. A limited amount of sheltered space is available by the Trust buildings below the car park at the head.

➋Flamborough Head - Humberside

In recent years, a total of up to twelve Sabine's Gulls have been seen off of the headland annually, with movement recorded between 26th August and 22nd October. Suitable conditions range from moderate north-westerly winds around to light south-east, but peak during northerlies. They associate with large flocks of Kittiwakes and can be difficult to locate or identify at long range. Occasionally birds linger in Bridlington Bay and may be seen several times flying to and fro with the Kittiwake groups.

Smaller numbers can also be seen from other prominent vantage points along the Humberside and Yorkshire coast, including Filey Brigg, Hornsea Promenade and Spurn Point.

Sabine's Gulls are also regularly recorded from other seawatching localities in suitable conditions including Girdle Ness (Grampian), Fife Ness (Fife), Uisead Point, Kintyre (Strathclyde), St Mary's Island and Newbiggin (Northumberland), Seaton Sluice, Tynemouth and Whitburn (Tyne & Wear), Hartlepool Headland and South Gare (Cleveland), Heysham Harbour and Blackpool Promenade (Lancs), Formby Point (Lancs), Hilbre Red Rocks (Cheshire), Point of Ayr (Clwyd), South Stack and Point Lynas, Anglesey (Gwynedd), Gibraltar Point (Lincs), Cley Coastguards and Blakeney Point (Norfolk), Sheringham (Norfolk), Ness Point, Lowestoft, Southwold and Covehithe (Suffolk), Canvey Island (Essex), Shellness Point, Sheppey (Kent), Foreness Point (Kent), Dungeness (Kent), Hengistbury Head and Portland Bill (Dorset), Hope's Nose, Torquay (Devon), Newquay Headland (Cornwall) and Porthgwarra (Cornwall).

Ring-billed Gull *Larus delawarensis*

The Ring-billed Gull is an annual winter visitor and passage migrant to Britain from North America, with an annual incidence of up to 110 birds. The majority of birds turn up in western Britain and there is a preponderance of individuals to remain at a site for very long periods, reappearing each winter for up to ten years. Identification is problematical, with separation from Common Gull being the biggest pitfall, and is best discussed by P.J. Grant in *Gulls: a guide to identification*, on pages 63-67 and photographs 120-145.

This guide details all long-staying and regularly returning individuals up to March 1996 and a number of sites where migrant Ring-billed Gulls regularly appear in spring.

Wintering Individuals

✱Copperhouse Creek, Hayle - Cornwall

At least one adult has appeared in winter at Copperhouse Creek since 1981 and at times, up to three have been present. The current wintering adult has been in winter residence since 9th November 1992 and can be easily seen from the Library car park in Hayle town. The bird regularly visits the car park for scraps and roosts with other gulls on the creek adjacent to the car park. It is generally present from mid October until early April and can wander as far as the main estuary, viewable from the Quay House Inn car park at the extreme western end.

The estuary is also a good site to see additional Ring-billed Gulls in winter and spring, with individuals being identified amongst the large numbers of roosting gulls present at low tide. They can also be found in nearby Penzance, with birds occasionally appearing in Newlyn Bay, Marazion Beach and Eastern Green Beach outfall east of Penzance railway station.

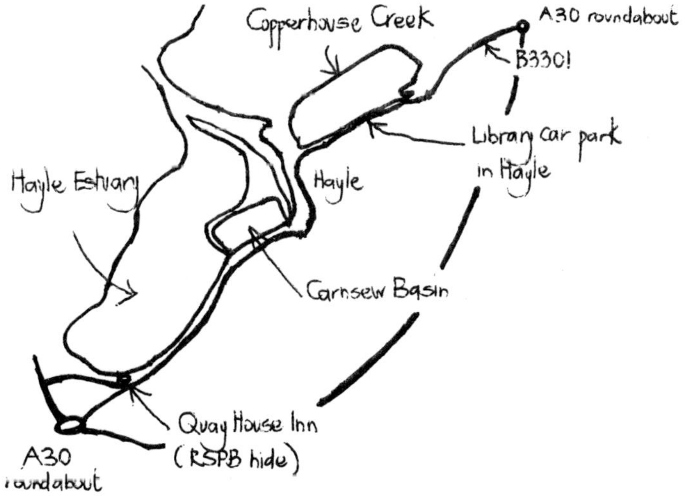

➊Bideford Quay/Barnstaple Leisure Centre - Devon

An adult Ring-billed Gull has wintered in the Barnstaple/Bideford area since December 1988 and is generally present from mid August until late April. It favours the lorry park and Leisure Centre car park on the south side of the river in Barnstaple. Leave the town centre on the A39 and immediately after crossing the river, turn left into the industrial estate and follow signs to the Leisure Centre. It also visits Bideford Quay, especially between August and November, and may be found with other gulls scavenging for scraps along the main riverside walk way up to half a mile north of the A388 road bridge.

➊Par Beach Pool - Cornwall
At least one adult has wintered at Par Beach Pool since at least 1987, with, at times, three individuals being present. They are often very confiding and are easily attracted to food and are present from mid November until mid April.

The pool is easily reached from the A3082. Take the minor road signposted to Par Sands Holiday Park, immediately west of the Ship Inn. There are speed ramps on this road and care must be taken to avoid dog-walkers and children. The pool is on the right after 200 yards and is best viewed from the western end. The gulls congregate on this pool to bathe, rest and feed. The adjacent beach is also worth checking for grebes, divers and seaduck in winter.

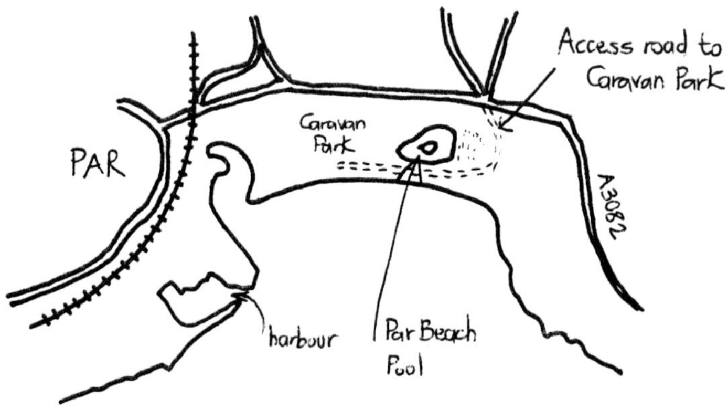

➊Lipson Vale Community College Recreation Fields/Plym Estuary - Devon
A regular adult can be found amongst large numbers of Black-headed Gulls and a few Common Gulls feeding on the playing fields around the Lipson Vale Community College. They are best seen early mornings and are highly susceptible to disturbance, with students, children and dog-walkers being generally present. 1-2 Ring-billed Gulls are present at the site from late November until late March and are also regularly seen at low tide at the north end of the River Plym. Between late February and late April, further passage Ring-billed Gulls may be identified at this latter site.

The playing fields are located to the west of the A374 and can be reached by turning west on the B3214 towards Lipson. Continue for about a mile and take second left after the railway bridge towards the college at Lipson Vale. The fields are on the left after 400 yards and extend to beyond the college buildings. Please respect college privacy and do not enter during college working hours. Weekend visits are preferred.

The River Plym can be easily checked from the A379 or from the eastern side at Chelson Meadows. A low or rising tide is best.

➊Ryde Canoe Lake - Isle of Wight
An adult has regularly wintered in the Ryde area since 1989 and is generally present from mid December until early April. It favours the Canoe Lake at SZ 602 926 (Sheet 196) but does wander west as far as the Pier.

➊Radipole Lake NR - Dorset
Although not present annually, Ring-billed Gulls are very regular visitors to the winter gull roost at Radipole (an adult remained from December 1995 to March 1996). The birds can be found from mid afternoon onwards roosting with large numbers of Common and Black-headed Gulls in the car park or from the main lake by the RSPB information centre and shop in Weymouth. It is also regular for them to be found in spring, when sub-adults can be identified between early March and mid April. Birds may also be found at nearby Lodmoor CP, about a mile to the east.

Passage Ring-billed Gulls

Ring-billed Gulls may also be found regularly at a number of coastal localities between late February and mid May. These are generally first or second years.

❶Seaforth NR - Merseyside
Ring-billed Gulls are regularly recorded at Seaforth NR between mid April and late May, with up to six birds being occasionally seen. The majority of birds are immature, mainly second-summer, and roost on the causeway visible from Hide A. Access to the reserve is via the Freeport entrance on Crosby Road South.

❶Teign Estuary - Devon
Up to six Ring-billed Gulls may be recorded on the Teign Estuary in spring, with occurrences peaking between mid March and mid April. They arrive with large groups of northbound Common and Black-headed Gulls and are best looked for from Passage House Inn, at the western end of the estuary, at low tide. Leave the A380 at Kingsteignton and at the first small roundabout turn left towards the Inn. This winding road eventually passes back under the A380 and ceases at the Passage House Inn car park. Check the roosting gulls from here, this site also being an excellent one for Little Egret.

❶Blackpill Beach - West Glamorgan
The first Ring-billed Gull to be identified in Britain was discovered at Blackpill in March 1973. Since that time, the species has been recorded virtually annually at the site, with a peak of up to nine birds recorded in March 1984. They can be found on the large expanse of sand in Swansea bay from mid February to late April and are best observed from Blackpill Beach, about three miles SW of Swansea, from the A4067.

❷Ogmore Estuary - South Glamorgan
Like Blackpill, a marked passage of Ring-billed Gulls can be encountered on the Ogmore Estuary between late March and mid May, with some first-year birds lingering on into July. The estuary is situated SW of Bridgend and is best viewed from the B4524 towards the river entrance or from the bridge just west of the Castle. Low tide is best.

❷Farlington Marsh Lagoon - Hampshire
The large gull roost at Farlington Marsh Lagoon occasionally attracts the odd Ring-billed Gull in winter or on spring passage and is easily observed from the seawall running south from the car park immediately south of the A2030 Southsea junction of the A27.

❷Chew Valley Lake - Avon
The Ring-billed Gull is an almost annual visitor to Chew Valley Lake, with occurrences peaking in March and early April. They can be identified amongst up to 18,000 Common Gulls as they come into roost at the reservoir from mid afternoon onwards. They are best searched for from the hide at the end of Nunnery Point. Furthermore, in past years, an adult remained for long periods in the Herriott's Bridge area and others have been recorded coming to bread in Heron's Green Bay.

Yellow-legged Gull *Larus cachinnans*

Yellow-legged Gull is considered by Evans & Turner (1996) and Vinicombe (1995) to be a separate species from Herring and Lesser Black-backed Gull, the identification and taxonomy of such being best discussed by Vinicombe *et al* in *The Macmillan Field Guide to Bird Identification* on pages 120-123. It is a regular summer visitor and passage migrant to Britain and appears in large numbers in south-east England in July-August. Birds can be found amongst flocks of Lesser Black-backed Gulls at numerous sites along the south coast but are much less recorded in northern Britain.

✱Thames Estuary - Essex and Kent

A large population of up to 500 birds can regularly be found on the River Thames from late summer onwards. They begin arriving from the last week of June and peak in numbers at the end of July and the first week of August.

On the Essex side of the river, they are best observed from Coalhouse Fort, at East Tilbury, where at low tide large numbers can be observed roosting or feeding on the mudflats. The foreshore at Rainham Marsh is also very productive, as well as the landfill site at Aveley.

In Kent, the best sites for observation are from Stone Marshes, Dartford, or from Higham Bight at Cliffe foreshore. Large numbers can also be seen from the foreshore in Greenhithe.

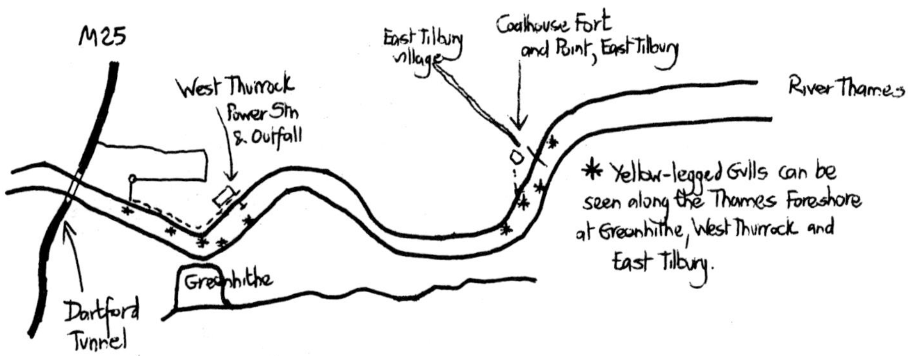

✱Pagham Harbour - West Sussex

A gathering of up to 300 Yellow-legged Gulls is regularly noted in Pagham Harbour between the last week of July and early September. They can be viewed from the seawall at the north end of the harbour (accessed via Church Lane in Pagham village) or from Church Norton. The gull roost can also be seen from the seawall running between Sidlesham Ferry and Church Norton. Low tide is best.

✱Dungeness - Kent

Dungeness is the third most productive site for Yellow-legged Gull in Britain, with numbers peaking to about 30 birds in late July. Most are aged between third-summer and adult and frequent the power station outfall and shingle beach or the RSPB reserve shingle islands nearby. They are generally present from mid July to mid September.

✱Chew Valley Lake - Avon

The largest number of Yellow-legged Gulls occur at Chew Valley Lake from July through to October, with a peak usually in September. Up to a dozen can sometimes be found in the evening gull roost; these are mostly adults. Reduced numbers remain through the winter to mid March. From April until June, only residual numbers of first-summer birds (1-3) are likely to be encountered.

They use the lake principally as a nocturnal roost, arriving usually from the south in the late afternoon and evening. Small numbers often stop to bathe on Herriott's Pool at the south end of the lake and these can be viewed from Herriott's Bridge. The largest numbers can be seen from Stratford Hide (south-west corner of the lake), particularly in late summer/early autumn when large pre-roosting flocks of Lesser Black-backed Gulls gather at the south end. The best way to find them is to scan through these flocks.

In recent years, a large adult male Yellow-legged Gull has taken up residence in Villice Bay or Herons Green Bay on the west side of the lake. He can often be seen during the day sitting on the fish pontoons in the mouth of either bay. A second, smaller adult may also be present. Both birds have become regular features of the lake from mid July to mid March.

Yellow-legged Gulls can be seen at a large number of other sites north to Yorkshire and Merseyside, especially in winter and in July-August. Recommended sites are Seaforth NR (Merseyside), Broomhill Ings (South Yorks), Southfield Reservoir (Humbs), Draycote Water (Warks), Westwood Pool and Throckmorton Landfill Site (Worcs), Chasewater, Belvide Reservoir and Blithfield Reservoir (Staffs), Rookery Pit and Stewartby Lake (Beds), Minsmere RSPB scrape (Suffolk), Holes Bay, Poole (Dorset), Radipole Lake NR (Dorset) and the River Plym (Devon). At one site in NW England, an adult has reappeared for nine successive years at Arnside Promenade, on the Kent Estuary (Lancs). It reappears in mid July and stays until mid November and favours the beach below the promenade at low tide and roosts near the viaduct at high tide.

Iceland Gull *Larus glaucoides*

The Iceland Gull is a rare winter visitor and spring passage migrant to Britain which breeds on the west and east coasts of Greenland. The winter distribution has a markedly west coast bias, with the largest numbers being recorded in Scotland. Wintering birds arrive in Britain from late October to early December, and remain until early April, but in recent years, there have been no regular returning individuals. The average incidence is of between 30 and 50 wintering individuals but in severe winters, numbers may be swollen with further arrivals from Arctic waters and can increase to as high as 400 individuals. A marked increase in sightings is normally apparent in spring, between late March and late April, as wintering birds further south reorientate northwards.

The largest numbers of Iceland Gulls recorded in an average year are in the Northern Isles, where between three and ten birds may be present in Lerwick Harbour (Shetland), several in Orkney, five or more in Stornoway Harbour, Lewis (Western Isles) and up to ten on St Kilda (Western Isles). The best mainland sites are in Northern Scotland, where up to 30 may be wintering at various localities in Highland Region. Influx years sees much larger numbers arriving and these penetrate further south, producing regular sightings at most Midlands reservoir gull roost-sites.

✱Ullapool - Highland

Large numbers of gulls, including Iceland and Glaucous, are attracted to Ullapool by the constant arrival of fishing trawlers and factory ships throughout the winter. The birds can be found anywhere around the harbour and fish quays or scavenging offshore about the factory ships and trawlers and are generally present between December and April. It is possible to find up to 30 Glaucous and 15 Iceland Gulls present during an influx year, but only 1-2 of both species in a poor year.

Further north, Kinlochbervie Harbour, 19 miles SW of Durness, can also produce large numbers of Iceland Gulls in influx years. Up to three birds are also regularly present in poor years, attracted in by the fishing fleet operating from there between November and March.

❷Wick Harbour - Highland

In years past, Wick Harbour was an excellent place in which to observe Iceland Gulls (mainly first-winters) scavenging about the harbour walls and quayside fish factories. However, recent years have seen a serious decline in numbers and it is now difficult to find a single bird. December-April is the peak period for occurrences. The harbour at Scrabster is also a regular site for Iceland Gulls, with up to four birds recorded wintering in recent years.

❷Longman Sewage Outfall, Inverness - Highland

The large wintering flocks of gulls which frequent the sewage outfall at Longman, immediately west of the Kessock Bridge, regularly attract up to four Iceland Gulls. The outfall can be easily checked from the minor road which runs along the southern shore and is accessed from the Longman Industrial Estate. Low tide is best.

❶Fraserburgh Harbour - Grampian

Large numbers of wintering gulls may be found in Fraserburgh Harbour and around Kinnaird Head between November and April and in most winters, 1-4 Iceland and up to five Glaucous Gulls may be found amongst them. They favour the fish quays and fish factories in the harbour but at low tide, may be found roosting with other large gulls in Fraserburgh Bay and off Cairnbulg Point and on the rocks off Kinnaird Head and about the outfall there. Furthermore, the large fish factory on the B9031, just west of the town, produces regular sightings.

➊Oban Harbour - Strathclyde
Oban Harbour regularly attracts up to six birds between December and April and these can be easily observed from the harbour walls and ferry terminal.

➊North Shields Fish Quay - Tyne & Wear
Up to four individuals can regularly be found in the vicinity of the fish quays in North and South Shields. They generally appear in December and remain through to March and can be seen scavenging through fish bins or roosting on groynes on the River Tyne.

North Shields is situated on the north side of the River Tyne and the fish quays and river can be reached via a minor road south of the A187. Birds commute from here to the fish quays on the south shore, at South Shields.

Further south, Sunderland Harbour and fish quay also regularly attracts 1-2 Iceland Gulls during the winter period.

➊Plymouth Area - Devon
Iceland Gulls are regularly recorded in the Plymouth area between December and April. They favour the mudflats of the River Plym at low tide and feed regularly on the rubbish tip at Chelson Meadows. Up to three birds may be present and in spring, further individuals appear for a few days. The Cattedown Wharf IFL fish factory outlet is also an excellent site in which to find birds, as well as the sewage outfalls at West Hoe and the fish quays near the Barbican.

➊Penzance Area - Cornwall
Like Plymouth, commercial fishing activities in Penzance attract huge numbers of gulls in winter and it is not unusual to find small numbers of Iceland and Glaucous Gulls amongst them. These can be found at all of the traditional gull-watching haunts, including Marazion Beach, Newlyn Harbour, Eastern Green Beach and at Drift Reservoir.

St Ives and Carbis Bay also regularly attract 1-3 birds, the majority of sightings involving first or second-year birds.

Iceland Gulls can also be recorded at a number of inland reservoirs and rubbish tips in winter and regularly roost throughout the winter at traditional sites. These include Broomhill Ings (South Yorks), Wintersett Reservoir/Angler's Country Park, Wakefield (West Yorks), Fairburn Ings (West Yorks), Southfield Reservoir (Humbs), Newsam Green Tip/Swillington Ings (West Yorks), Eccup Reservoir, Leeds (West Yorks), Huthwaite Tip (Notts), Ogston Reservoir (Derbyshire), Eye Brook Reservoir (Leics), Draycote Water (Warks), Cannock Tip, Chasewater and Blithfield Reservoir (Staffs), Stewartby Lake, Rookery Pit and Brogborough Lake (Beds), Hanningfield Reservoir (Essex) and at William Girling Reservoir (Essex/London bdr).

They are also recorded infrequently at a number of coastal localities in winter, including Flamborough Head (Humbs), Great Yarmouth (Norfolk), Lowestoft Harbour and Ness Point (Suffolk), Felixstowe Beach (Suffolk) and Dungeness Beach (Kent). Until recently, a long-staying adult regularly appeared in Weymouth Harbour and roosted at Radipole Lake (Dorset).

In NW England, no regular individuals can be relied upon, but inland gull roosts at Fishmoor Reservoir, Blackburn (Lancs), Audenshaw Reservoirs (Gtr Manchester) and Lower Rivington Reservoir (Lancs) have frequent winter/early spring birds, as also do the refuse tips at Richmond Bank, on the River Mersey, near Warrington (Cheshire), Maw Green, near Crewe (Cheshire) and Whinney Hill, near Accrington (Lancs). Seaforth NR (Merseyside) has up to six birds per annum, but these are largely one-day birds.

Glaucous Gull *Larus hyperboreus*

The Glaucous Gull is a regular winter visitor to Britain from the high-Arctic and sub-Arctic coasts of Svalbard and eastern Greenland, with up to 250 birds normally present between late August and mid April. Occasional influxes of up to 600 birds are noted, and these are usually associated with northerly or easterly gales in mid winter. The majority of birds wintering in Britain occur in the Northern Isles and Scotland, with the rest being found sparsely about the coasts and roosting inland on large reservoirs. At the time of publication (March 1996), only 25 birds were known to be regularly returning birds.

Apart from regular Northern Isles sites such as Lerwick Harbour (Shetland), Burrafirth (Shetland), Stromness Harbour (Orkney) and Stornoway Harbour, Lewis (Western Isles), the following are the most regularly visited sites in Britain.

➊Thurso Harbour - Highland
Thurso Harbour is a regular site in which to see up to eight, mostly first-winter, Glaucous Gulls between December and April. They can be found in the concentrations of gulls that congregate at the river mouth and outfall and also around the fish factories in the harbour at Scrabster.

Wick Harbour, 21 miles to the south-west of Thurso, also regularly attracts Glaucous Gulls, but in much smaller numbers. As with Iceland Gulls at this site, the majority of birds are first-winters.

✱Ullapool - Highland
Up to thirty Glaucous Gulls are regularly recorded in Loch Broom, Ullapool, and these may be located either scavenging the fish quays in the harbour or following the factory ships and trawlers offshore. The majority of birds are first-winters, with very few adults ever being noted, and are present from December until mid April. The winter of 1995/1996 saw a dearth of sightings.

✱Fraserburgh Harbour - Grampian
The fish quays, factories and harbour at Fraserburgh are naturally attractive to many thousands of gulls in winter, and inevitably 1-9 Glaucous Gulls are identified amongst them. The best points of observation are the rocky foreshore and sewage outfall off Kinnaird Head, the gull roost in Fraserburgh Bay and the Birds Eye factory along the B9031, immediately west of the town.

1-3 birds can also be found at nearby Peterhead Harbour and in Aberdeen Harbour and the Girdle Ness area.

➊Arbroath Harbour - Tayside
Arbroath Harbour regularly attracts up to three Glaucous Gulls during January and February, but with a change in the fishing fleet in recent years, birds may soon cease to appear. The refuse tips at Dundee and Loanhead, Forfar, may well yield better results, particularly as both sites regularly host over 5,000 Herring Gulls.

➊Longannet Power Station Lagoons - Fife
In recent years, this site has regularly hosted 1-3 Glaucous Gulls and up to ten Iceland Gulls between early January and mid April. The lagoons are situated at NS 955 853, about two miles east of Kincardine Bridge, and can be viewed from the B9037 to the north.

➊Oban Harbour - Strathclyde
Up to three birds can generally be located amongst wintering gulls in the harbour between December and March.

❶Saltcoats Harbour - Strathclyde
Up to three birds are regularly present in Saltcoats Harbour between mid November and late April and can be seen from the harbour walls.

❶Ayr Harbour/Doonfoot
Glaucous Gulls, mostly first-winters, are present in the harbour area at Ayr from late November until mid April and are also regularly seen at the mouth of the River Doon, on the south side of the town. Large numbers of gulls congregate at low tide at this site, the area being readily viewable from a number of car parks along the southern shore. Turn right off the A719 immediately after crossing Doonfoot Bridge and follow to the shore.

Troon Harbour, six miles to the north, also regularly attracts at least one Glaucous Gull throughout the January-March period.

❷Edinburgh Area - Lothian
In most years, 1-2 Glaucous Gulls are seen in the Edinburgh area between January and April. They favour the fish quays and sewage outflows and are most regularly recorded at Cramond, Seafield and at Musselburgh Lagoons.

❷Seaforth NR - Merseyside
Seaforth NR is the best site in NW England for Glaucous Gulls, with individuals being recorded there in virtually every month of the year. Although only 1-2 birds are generally present, a marked influx of individuals occurs in spring, with up to four birds visiting the freshwater pools in March and April. December also sees a further peak in occurrences. Access is via the Freeport entrance on Crosby Road South. View from Hide A.

Elsewhere in **Merseyside**, the flooded fields at Marshside Marsh RSPB reserve, north of Stockport, can hold up to three birds over the mid winter and early spring period, but these require extreme patience and expertise as they occur amongst 30,000 other large gulls roosting in the area.

❷River Tyne - Tyne & Wear
Up to six birds are regularly recorded in North Shields Fish Quay between mid November and April. They frequent the offal skips and fish factory roofs, as well as the groynes and breakwaters on the river. Leave the A187 in North Shields and drive south towards the river. The road eventually leads to the fish factories and quay.

Further north, in **Northumberland**, Glaucous Gulls can regularly be encountered between Holy Island and Seahouses Harbour and in the Druridge Bay and Widdrington area. There are no regular wintering individuals however.
Further south, the harbours at Hartlepool, Seaham and Sunderland also attract individuals during the winter period.

❷Scarborough Harbour - North Yorkshire
In most winter periods, 1-4 Glaucous Gulls are recorded in Scarborough Harbour. They are most reliably seen in January and February and frequent the main harbour or roost amongst large numbers of gulls at low tide in South Bay.

Birds are also often seen off Flamborough Head (Humbs), in Filey Bay (North Yorks), off Scalby Mills (North Yorks) and in Whitby Harbour (North Yorks), but sightings are sporadic. In recent years, returning adults have been seen in the Scarborough area and regularly at the mouth of the harbour at Whitby.

❷Broomhill Ings - South Yorkshire
The large number of wintering gulls at Broomhill Ings inevitably attracts up to three Glaucous Gulls between January and late March and these are best searched for from late morning onwards as they come in to bathe and preen on the flash. They spend much of the day at the nearby Stairfoot Refuse Tip and generally roost at Wintersett Reservoir.

Broomhill Ings is situated east of Wombwell. Leave the A633 on a minor road to Broom Hill and the flash is on the left in less than half a mile. Wintersett Reservoir is NW of Ryhill and can be viewed from the dam on the minor road between Ryhill and Wintersett.

Pugney's Country Park, near Wakefield, also regularly attracts Glaucous Gulls in January to March.

❷Swillington Ings/Newsam Green Tip - West Yorkshire
The Swillington Ings and Newsam Green Tip area attracts huge numbers of gulls in winter and inevitably small numbers of white-winged gulls are found amongst them. Glaucous, mostly first-winters, are generally discovered in late winter and can be found with patience between mid January and early April.

Swillington Ings are situated in the Aire river valley immediately east of Woodlesford. Access is via Astley village. Newsam Green Tip is to the NW of Woodlesford and is clearly signposted.

❷Ogston Reservoir - Derbyshire
Although sporadic in their occurrences, Glaucous Gulls are annual winter visitors to the large gull roost at Ogston Reservoir, with 1-4 recorded between January and March. The reservoir lies west of the A61 about five miles east of Matlock. The gull roost is best observed from the west side, either from the sailing club car park or from the minor road which runs parallel with the west shore.

❶Cannock Tip/Chasewater - Staffordshire
Each winter, Glaucous Gulls are attracted in small numbers to the Cannock area and can be seen during the day scavenging at the large refuse site east of Cannock or roosting on Chasewater reservoir at dusk. The reservoir lies north of Brownhills and is accessed from the A5. It can only be reached from the eastbound carriageway, so if travelling from the west, proceed past the entrance and on to the roundabout at the junction with the A452. From the carriageway, turn left in to Pool Road and left again into the amusement park. The gull roost can be watched from the southern shore here.

Blithfield Reservoir to the north can also be good for white-winged gulls in January-March and these can be looked for from the main causeway on the B5013, west of Abbot's Bromley.

❶Eye Brook Reservoir - Leicestershire
The large gull roost at Eye Brook Reservoir inevitably attracts 1-2 Glaucous Gulls in January-February and these can be seen from the western shore from 3.15pm onwards. The gulls roost off the promontory on the west shore. If the reservoir is partially frozen, many gulls visit the reservoir during the day and Glaucous Gulls can be searched for standing on the ice. The reservoir lies west of the A6003, about five miles north of Corby.

❷Pitsford Reservoir - Northamptonshire
The gull roost at the dam end of Pitsford Reservoir often holds at least one Glaucous Gull in January and February and these may also be found in the fields between Brixworth Tip and the reservoir. The dam is reached via a minor road leading east from the A508 roundabout, 1.5 miles south of Brixworth.

❶Draycote Water - Warwickshire
Draycote Water attracts vast numbers of roosting gulls in winter and these are often joined by Glaucous Gulls in January and early February. The site lies just SW of Rugby and is accessed from the A426, two miles south of Dunchurch. A permit is required for access and is available at £3.70 per year from Severn Trent Water Authority, Avon House, Demountford Way, Cannon Park, Coventry, CV4 7EJ - ☎ 01203 416510. Alternatively, a permit may be obtained at short notice from Kites Hardwick Filling Station, 500 yards south of the reservoir entrance on the A426.

❶Brogborough Lake/Stewartby Lake - Bedfordshire
These two brick pits in the Marston Vale regularly attract up to five Glaucous Gulls in winter but their appearances are highly erratic. They visit the sites between late December and mid March and require a telescope and a great deal of expertise to locate, amongst the 20,000 or so gulls regularly roosting. Both sites lie SW of Bedford on the A421. The roost at Brogborough is best viewed from the SW side, at various pull-ins along the Lidlington road, whilst that at Stewartby is best viewed from the western end, walking from the large lay-by on the A421. Both sites are used for recreational purposes and gulls rarely arrive for roost more than two hours before dark.

❷Dungeness Area - Kent
In most years, 1-2 immature Glaucous Gulls are recorded in the Dungeness area between mid January and early April. They commute between the power station outfall and beach and the complex of pits on the reserve and may also be seen by the fishing boats further east along the beach.

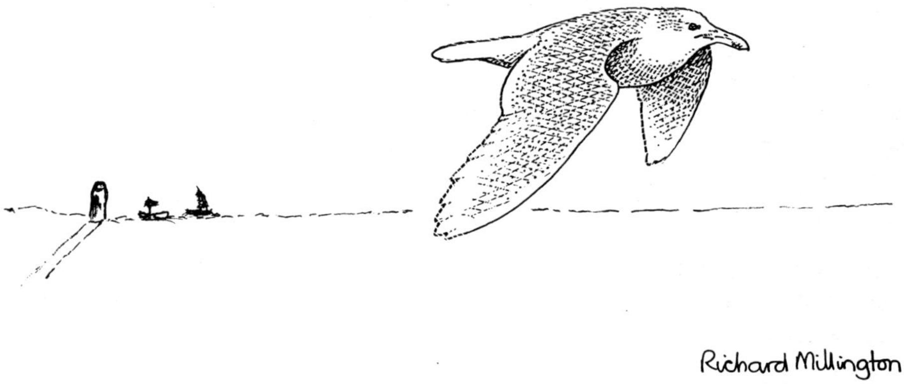

✱Selsey Bill - West Sussex
An adult Glaucous Gull has wintered annually at Selsey since 1985 and between late August and early March, can be found in the vicinity of the lifeboat station on East Beach. It regularly roosts on the railings of the lifeboat station, on the fishing boats offshore or on the sea itself, and also sits with other gulls on the roofs of nearby houses and holiday flats. Take the first left as you arrive in Selsey from the north and follow signs towards East Beach. Continue to the beach and drive south towards the Bill and park adjacent to the Lifeboat Station.

❶River Plym - Devon
Glaucous Gulls are regularly recorded in the Plymouth area between late December and late March and are best searched for at low tide on the River Plym or on the Chelson Meadows Tip on the eastern shore. The vast majority of sightings involve first or second-year birds. Park at the north end of the River Plym and walk southwards along the A374 and scan the roosting flocks. Alternatively, cross the Plym on the A379 towards Plympton and take first left at the traffic lights to the refuse tip and park in the viewpoint car park. The tip can be viewed at a distance from the top of the raised bank on the right and the estuary can be easily viewed from the riverbank.

❶Penzance Area - Cornwall
About five Glaucous Gulls are recorded in the Penzance area in an average winter but like all gulls in the area, range widely and are unpredictable in their habits. Favoured areas include the sewage outfall at Newlyn, Eastern Green Beach and the reservoir at Drift. They can also wander eastwards as far as Marazion Beach and even the Hayle Estuary, and Carbis Bay is also a regular haunt. Birds arrive in December and depart in mid March, with further passage birds noted in early spring.

Lesser Crested Tern *Sterna bengalensis*

The Lesser Crested Tern is a rare vagrant to Britain that breeds fragmentedly from Australia to the Persian Gulf and Red Sea and sporadically in the Western Mediterranean and Europe. The North African population winters mainly in Senegal and Gambia, with some birds penetrating south as far as Nigeria.

Identification is best discussed by Steve Gantlett in *British Birds* 80: 257-276 but beware the problem of hybrid birds, as several instances of successful mixed breeding has been noted.

Since 1984, a single female has returned annually to the Farne Islands (Northumberland) to breed. She normally pairs up with a male Sandwich Tern and nests on either Brownsman Island or Inner Farne. She arrives in Britain during the first week of May and after brief visits to coastal sites in Cleveland, Humberside or Tyne & Wear, generally reaches the Farne Islands by about 13th May. In recent years, heavy rain and cold easterly winds have hindered nesting and after a few weeks on the islands, the Lesser Crested Tern has often become restless and wandered south as far as Beacon Ponds, Kilnsea (Humbs) before settling down to breed. In most years, an attempt at nesting is made towards the end of the season and from mid June onwards, she can often be found sitting on eggs in the Sandwich Tern colony. If the attempt succeeds, she may remain on the island until 21st July, but more often than not it fails and she departs in early July, although may still be present in the area. The most reliable period for observation is between 18th May and 20th June and an afternoon visit to the islands is recommended. It must be stressed however, that feeding trips by the bird may take her away from the islands for over ten hours. These are mainly taken during the morning but less regularly when incubating or when the young have hatched.

The Farne Islands can only be reached by boat via Seahouses. Billy Shiel organises daily pleasure trips throughout the summer (weather permitting) and these can be booked at the pier or by telephoning ☎ 01665 720308. From late April to early September, sailings are scheduled daily at 10.00, 11.00am, 12 noon, 1.00, 2.00 and 3.30pm. Prior whereabouts of the Lesser Crested Tern is essential and when booking, information of this kind should be clearly sought and discussed.

Roseate Tern *Sterna dougalli*

The Roseate Tern is a rare breeding summer visitor to Britain with a current population of less than 85 pairs. The bird winters south to southern Africa, with most British birds wintering off Ghana. They depart this area in early May and arrive in Britain from mid May onwards. The small population is largely confined to Wales, NE England and SE Scotland, whilst a further 430+ pairs nest in Ireland.

The following are the most reliable sites for seeing Roseate Terns in Britain, including four breeding sites and three passage sites:-

❶Cemlyn Bay, Anglesey - Gwynedd
In recent years, up to five pairs of Roseate Terns have nested amongst colonies of Sandwich, Common and Arctic Terns at Cemlyn Bay, on the north-facing coast of Anglesey. They arrive in mid May and remain until late July and can be seen on the main nesting islands on the lagoon. Leave the A5 at Dyffryn northwards on the A5025. Continue for eleven miles and then turn west on a minor road at Tregele crossroads. This narrow, winding road leads past the entrance to the car park at Cemlyn Bay. Park on the west side of the lagoon at the road end at SH 330 936. and then walk back along the shingle ridge making sure to keep below the white-topped posts positioned at equal intervals along the beach. This should minimise the disturbance to the tern colony.

❷Rhosneigr, Anglesey - Gwynedd
Rhosneigr lies on the west coast of Anglesey, about eight miles south of Holyhead. It was once a regularly used site by several hundred pairs of Roseate Terns but in recent years, has failed to attract more than ten pairs of this beautiful seabird. The terns nest on the rocky islets offshore of the village and can be 'scoped from the shoreline at SH 310 741. Park just west of Rhosneigr Golf Club on the A4080 north of the town and take the public footpath north across the river and follow northwestwards to the beach. On no account attempt to walk out to the islands. They can also be seen fishing along the foreshore at high tide.

✱Farne Islands - Northumberland
Up to three pairs are known to regularly breed on the Farne Islands and at least one of these is on the tourist island of Inner Farne. Regular boat sailings visit this island daily throughout the summer and bookings may be arranged on ☎ 01665 720308. Roseates may also be seen as the boats traverse the islands. Check the flocks of roosting and bathing terns on the rocks. The birds are present from early May until mid July.

✱Coquet Island - Northumberland

A breeding population of up to 35 pairs of Roseate Terns survives on Coquet Island, about one mile offshore of Amble-by-the-Sea. These birds regularly commute to freshwater pools at Hauxley NR, where they can be seen from the hide overlooking the reserve. The birds visit throughout the day, but peak in early morning and evening. Mid May to early June and late July to early September is the peak period. Hauxley NR is situated immediately south of Low Hauxley, two miles south of Amble.

❶St Mary's Island/Tynemouth - Tyne & Wear

Post-breeding gatherings of Roseate Terns can be noted along the Tyne & Wear coast between late July and late September, and groups of up to twelve birds are regularly observed in the high tide roost at St Mary's Island, especially during late August to September. A similar number of birds can also be seen lingering at the mouth of the River Tyne and can be seen from the castle at Tynemouth roosting on the rocks or fishing between the piers. The coastguard lookout at Seaton Sluice can also be very productive.

❶Whitburn Steel - Tyne & Wear

The rocky islets and sandy beaches at South Bents, Whitburn, are a regular haunt of small numbers of Roseate Terns between early August and mid September. Park at the south end of Whitburn and follow the coastal footpath down to the Bents.

❷Dungeness - Kent

Although sporadic nesting has been known from 1-3 pairs in the past, recent sightings have only involved the odd lingering or summering birds. They favour the power station outfall at 'The Patch' or the shallow lagoon at the RSPB reserve and are generally seen in mid May to mid June or in July-August. The ternery pool at Rye Harbour NR (**East Sussex**) is also regularly visited by this species during the same period.

❷Dawlish Warren - Devon

Roseate Terns are regularly recorded at Dawlish Warren during May and June and occasionally July. They can be seen from the hide roosting on the shingle or mud or fishing offshore at the mouth of the River Exe. Check the flock of roosting Sandwich Terns.

The Warren is reached from the A379 Exeter-Dawlish road. Turn south around Cockwood Harbour bridge and turn left after a mile and a half into the caravan park complex and holiday village. Go under the railway bridge and onto the Warren and head left towards the information centre. Walk along the sand dunes to the far end of the golf course and watch from the hide.

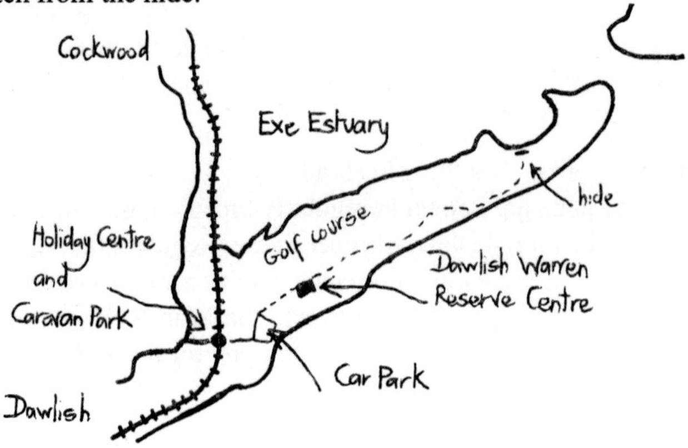

Black Guillemot *Cepphus grylle*

The Black Guillemot is a relatively common resident of Scottish waters with its highest breeding density in the Shetland and Orkney Islands. It is also commonly found in Caithness, along the west coast of Scotland and in the Western Isles, and penetrates as far south as the Calf of Man and Anglesey in the Irish Sea and to Muchalls in Grampian in the North Sea. The entire British breeding population is currently estimated to be about 37,000 individuals, about 17% of the European population.

As the species is resident, it can be seen at breeding localities throughout the year, and in NW Scotland and the Northern Isles, it can be easily encountered, especially in harbours or in sheltered coves. Away from Shetland and Orkney, where it is a relatively abundant seabird, the following is a list of sites where the bird may easily be seen.

✻Dunnet Head, Caithness - Highland
Dunnet Head is the most northerly headland in mainland Britain and lies 13 miles west of John O'Groats. The sea-cliffs to the north and east of the headland hold large numbers of nesting seabirds and amongst these are respectable breeding densities of Black Guillemots. Rock Doves can also be seen here.

Leave the A836 in Dunnet village and drive north along the B855 for five miles to the lighthouse.

Duncansby Head, Dunnet Bay and Wick Harbour are also very good localities in which to see Black Guillemots.

✻Clo Mor, Sutherland - Highland
Clo Mor is situated about four miles east of Cape Wrath, in the extreme north-west of Sutherland. The torridonian sandstone cliffs hold considerable numbers of breeding seabirds, including one of the largest colonies of nesting Puffins in Britain and good numbers of Black Guillemots. Rock Dove also breed here.

Clo Mor is reached by ferry across the Kyle of Durness (bicycles and pedestrians only) at Keoldale, 1 mile south-west of Durness and just off the A838. A connecting minibus service on the opposite side operates to Cape Wrath and one can be dropped off at the Kearvaig track, where breathtaking views over the cliffs may be obtained. The ferry operates several times daily during May to September.

✻Kyle of Lochalshe - Highland
Loch Alsh, between Skye and mainland Scotland, is an excellent site in which to see Black Guillemots throughout the year. Park by the information centre and toilet block in Kyle of Lochalshe and scan across the loch towards Kyleakin.

✻Mull of Galloway - Dumfries & Galloway
Black Guillemots can be seen in moderate numbers off the Mull of Galloway, a rocky headland some 22 miles south of Stranraer. The site is designated a Site of Special Scientific Interest and is an RSPB reserve. Leave Stranraer on the A716 to Drummore and then take the B7041 to the Mull lighthouse, where good views of the cliff can be obtained.

Nearby, small numbers of Black Guillemots can be seen in winter in Loch Ryan, north of Stranraer. The harbour area around Cairnryan, on the east shore, is best.

✱St Bees Head - Cumbria

The six pairs of Black Guillemot at St Bees Head is England's only breeding population of this seabird. They are best observed from March to June and nest on the ledges at the base of the high red sandstone cliffs.

Leave Whitehaven southwards on the B5345 and at the staggered crossroads in St Bees village, take the road straight ahead and turn right after 400 yards at the T junction. There is a car park at the end of the road at NX 961 118 from where one can climb uphill on the rather rugged cliff top path for a mile to look down on to the sea at Fleshwick Bay. It is always worth continuing for another mile towards the Lighthouse for views of more seabirds.

✱Isle of Man

On the Isle of Man, there is a good breeding population of up to 85 pairs at then or more different locations. The prime site to observe large numbers is Peel Hill (immediately west of Peel at SC 235 840) and, although a few may be present offshore here all year, the peak counts (up to 114 in recent years) are normally made in late March or early April.

✱Fedw Fawr, Anglesey - Gwynedd

A tiny population of up to six breeding pairs survive on the cliff ledges on the north coast of Anglesey. They are present from mid April until October and can be seen from the coastal footpath at Fedw Fawr. Leave Beaumaris northwards on the B5109 to Llangoed village. Turn left towards Mariandyrys and then turn right in the village on the minor road towards Plas Newydd. After about 100 yards, take a very narrow and well concealed track on the left and follow down to the end at SH 604 819 (Sheet 114). Take extreme care on this track and park sensibly at the end. The Black Guillemots can be 'scoped from the ruined house foundations near the car park or from the coastal footpath.

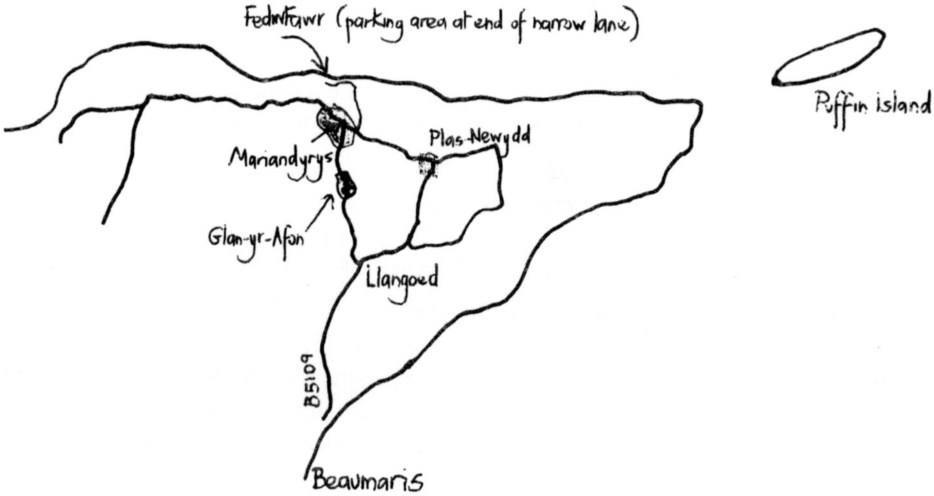

Little Auk *Alle alle*

The Little Auk is a regular winter visitor and autumn straggler to the North Sea and can be seen in vast numbers under suitable conditions passing east coast headlands between late October and early December and in smaller numbers in January. It is a species of the high Arctic seas but from time to time, strong northerly winds cause large 'wrecks' to occur.

It is regularly found off coasts of Shetland during December to February and in most years, in northern Scotland north of the Firth of Forth. Favoured sites include Fraserburgh Harbour, off Girdle Ness, Aberdeen, and off Musselburgh Lagoons.

In autumn, strong to gale force northerly winds produce large numbers off most exposed headlands south to Kent. The peak period for movements is between 29th October and 26th November, but suitable conditions in December/January will also produce sightings. The following is a summary of the best vantage points, from north to south:-

Noss Head, Wick (Highland); Tarbat Ness (Highland)); Kinnaird Head, Fraserburgh (Grampian); Peterhead (Grampian); Girdle Ness (Grampian); Scurdie Ness, Montrose (Tayside); Arbroath (Tayside); Fife Ness (Fife); Hound Point (Lothian); Barns Ness (Lothian); St Abb's Head (Borders); Newbiggin Point and Seaton Sluice (Northumberland); Whitley Bay, Tynemouth and Whitburn (Tyne & Wear); Hartlepool Headland and South Gare (Cleveland); Filey Brigg (North Yorks); Hornsea Promenade and Spurn Point (Humbs); Gibraltar Point (Lincs); Cley Coastguards, Weybourne, Sheringham, Mundesley, Winterton and Great Yarmouth river mouth (Norfolk); Southwold (Suffolk); Holland Haven and Canvey Island (Essex) and Foreness Point and Dungeness (Kent).

The best site in Britain, however, is Flamborough Head (Humbs):

❶ Flamborough Head - Humberside

Little Auks are recorded in large numbers off Flamborough Head on an annual basis and may stream past at thousands per hour following strong northerly gales. The first to be recorded are generally in the last week of October, with the largest counts occurring throughout November and into early December. Wrecks are frequently recorded and in such circumstances, many individuals can be seen close inshore off Bridlington Harbour and Promenade.

Park in the designated car park at Flamborough Head and walk south to the foghorn. Carefully negotiate the slippery footpath down the cliff to the somewhat sheltered seawatching site below the foghorn. Mornings are best, but on days of good passage, birds can be watched passing throughout the day.

Rock Dove *Columba livia*

Pure Rock Doves are now only found on coastal cliffs in the extreme north and west of Scotland, the Hebrides, Orkney, Fair Isle and Shetland. They are still relatively abundant on the Northern Isles and can be quite easily seen feeding around crofts, in stubble fields or in hayricks by the road. Elsewhere, they are highly localised and can be relatively difficult to find. The following is a list of sites where Rock Doves can reliably be seen in northern Scotland away from the Northern Isles.

✱Dunnet Head, Caithness - Highland
Rock Doves are known to breed on the sea-cliffs to the north and east of Dunnet Head and may be seen from April to August inclusive. Turn north off the A836 in Dunnet village on the B855 and follow for five miles to the lighthouse.

✱Wick Area - Highland
Reasonable numbers of pure Rock Doves may still be found between Lybster and Duncansby Head in winter and can be encountered in small groups feeding in roadside fields alongside the A9.

✱Durness Area, Sutherland
The wild and remote cliff edges and exposed headlands in the extreme north-west of Scotland still hold large numbers of pure Rock Doves. Most coastal headlands between Scourie and Tongue will produce sightings.

✱Handa Island, Sutherland - Highland
A healthy population of Rock Doves still survives on Handa Island, 18 miles south of Cape Wrath. They nest on the near-vertical cliffs of Torridonian sandstone to the north of the island as well as the lower cliffs in the south.

The island can only be reached by boat and these leave Tarbet daily (except Sunday) between 1st April and 10th September (☎ William McRae on 0971 2156). Once ashore on Handa, visitors are urged to keep to waymarked footpaths, from which there are many good vantage points to overlook the cliffs. Tarbet lies to the north of Scourie.

Elsewhere, pure Rock Doves may also be seen on Skye (fields north of Uig, particularly around Duntulm Castle, Linicro and Balgown), on North and South Uist and on the islands of Tiree and Mull.

Ring-necked Parakeet *Psittacula krameri*

The Ring-necked Parakeet is an introduced species from Afrotropical and Oriental regions which has become well established in Britain since it was first recorded in the wild in 1969. The population, restricted mainly in South-east England, is now estimated to be in the region of 2,000 birds. It is a noisy bird that can easily be detected by its screeching call and is highly gregarious, roosting in large flocks at a regularly used site. Small populations of birds have also been established elsewhere in Europe, including Belgium, Germany and the Netherlands. The best sites in Britain are in Surrey, Berkshire and Kent.

✱Elmbridge Leisure Centre, Sunbury-on-Thames - Surrey
The largest roost of Ring-necked Parakeets in Britain is to be found in the tall poplars adjacent to the leisure centre and the trees bordering the River Thames nearby. In January 1996, LGRE recorded a staggering 818 birds coming into roost between 3.15pm and dusk, the sight being truly astounding. The birds fly in in parties of up to 60 birds at a time and gather in trees by the river at first. As darkness approaches, the noisy throng then gradually flies into the trees by the road, opposite the entrance to the centre, and by dusk all are settled in. They mostly leave the roost in darkness, so that by dawn very few if any remain.

The site can be reached by driving north-east out of Walton-on-Thames on the A3050. It is signposted on the left after about 1 mile and is at approximately TQ 105 677 (Sheet 176). Park sensibly by the leisure centre and wait patiently for the birds to arrive. The last hour and a half of daylight are recommended. This is mainly a winter roost, visits between November and March being most productive.

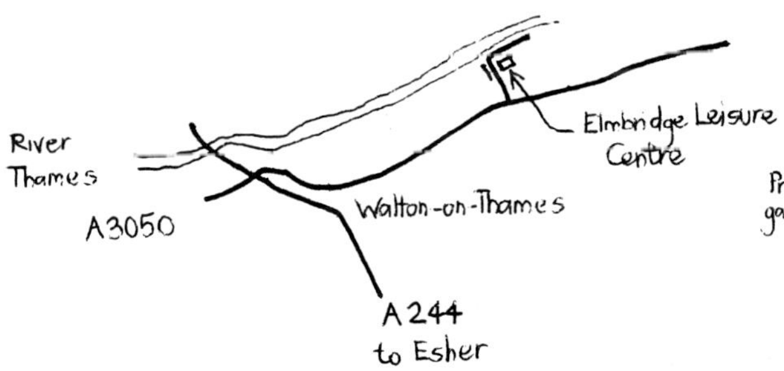

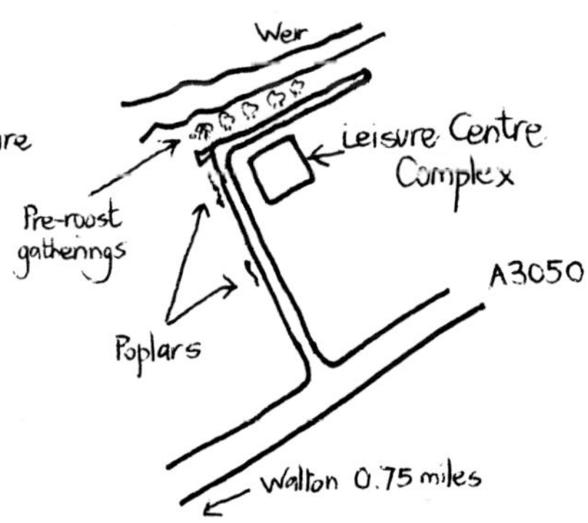

✱Wraysbury Area - Berkshire
The Wraysbury area has been a traditional site for Ring-necked Parakeets since at least 1974 with a thriving population of at least 75 birds being present. The birds congregate in small numbers in gardens throughout the village and may be located by their loud shrieking calls. They also favour the ash trees in Magna Carta Lane, west of the B376 at TQ 006 733, and can be seen in winter flighting to roost in large parties as they fly east over the lane from mid afternoon onwards.

They can also be seen at nearby Horton GP and in trees by Ditton Park, East Slough, and regularly fly over Englefield Green and Egham to the roost site at Walton.

✱Foots Cray Meadows - Kent
A population of up to 25 birds survives in Foots Cray Meadow, near Sidcup, where they favour the tall trees near the disused stables by the car park at TQ 475 716 (Sheet 177).

✱Ramsgate Cemetery - Kent
A large roost of up to 175 birds exists in Ramsgate Cemetery throughout much of the year. They favour the tall trees just south of the railway line and arrive up to 90 minutes prior to dusk. The cemetery is situated at TR 385 660 (Sheet 179). The birds disperse to outlying districts of Ramsgate and Margate during the day.

✱Northdown Park, Margate - Kent
Small numbers of Ring-necked Parakeets feed and nest in Northdown Park, Margate, and can be seen in the tall trees surrounding the large mansion or in those bordering the playing fields nearby. The park is situated at TR 378 703 (Sheet 179) and lies immediately east of the B2051 in Cliftonville. Park sensibly on the carriageway opposite the church.

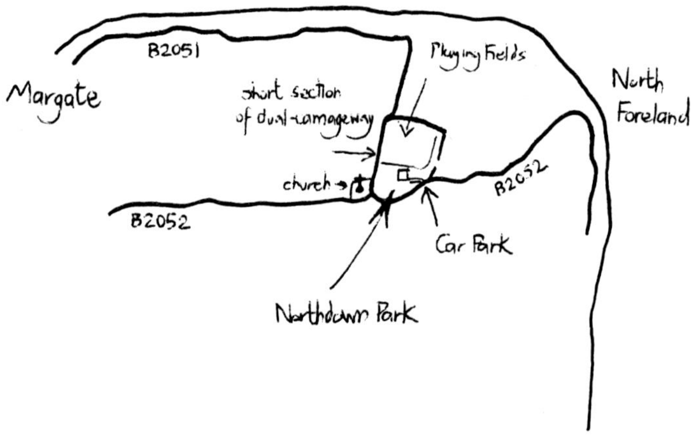

✱King George VI Memorial Park, Ramsgate - Kent
Up to 30 Ring-necked Parakeets reside in King George VI Park throughout the year and can be seen in the tall trees bordering the playing field and surrounding the park buildings. Park at the end of the minor road at TR 393 656 and walk through the gates into the park.

Away from sites in Kent, Berkshire and Surrey, Ring-necked Parakeets are very scarce in Britain and occur primarily as vagrants or very rare visitors in most bordering counties. Isolated groups of birds are present in Oxfordshire (in East Challow village and in the Ham Road and Portway allotments area of Wantage), West Sussex (a dwindling population of up to four birds in tall trees in Hollingbury Park, Brighton) and Dorset (up to eight birds around the National Trust car parks in Studland village which originated from an aviary thereabouts).

Long-eared Owl *Asio otus*

The Long-eared Owl is a very secretive species that inhabits primarily coniferous woodland where it hunts silently at dusk for Field Voles, Wood Mice, Common Rats and small birds. It is widely distributed throughout Britain with an estimated breeding population of at least 2,200 birds. These numbers are swollen by immigrants from Continental Europe in autumn and in some winters, many hundreds of birds can be found roosting in southern England. However, population levels are largely controlled by fluctuating rodent populations.

Breeding and roosting Long-eared Owls are highly susceptible to disturbance from birdwatchers and for this reason, it is not possible to detail many sites here. In fact, one of the best ways to see Long-eared Owls is to record them as migrants as they fly in across the North Sea in late autumn, especially between Filey CP (North Yorks) and Landguard Point (Suffolk). Clear days with light north-easterly or easterly winds are best.

✱Pennington Flash CP - Gtr Manchester

Pennington Flash CP is a regular winter roost for Long-eared Owls and birds can be easily viewed here without causing any disturbance to them. Use the pay and display car park at the east end of the flash at SJ 644 991 and then walk past the visitor centre and bear right on a grassy path past the famous winter feeding station that attracted a vagrant Black-faced Bunting in 1995 for about 300 yards until you reach a new hide on the right hand side of the track. Scan the willows and hawthorns on the far side of the small pond. Up to three birds are usually on view with as many again hidden in dense vegetation. Calm and sunny days are best, with wind and rain driving all the birds into dense cover. The Flash is well signposted from the A572, one mile southwest of Leigh.

✱Hobhole Drain, Boston - Lincolnshire

A regular roost of up to 16 birds may be found in thick hedgerows which border the drainage channel that leads to the River Witham. Park at Nunn's Bridge at TF 367 415 (Sheet 131) and walk 2-400 yards south down the eastern side of the Drain to view the hedgerow bordering the west bank. The birds are difficult to locate but may be found with some patient searching. They tend to favour the trees with some ivy foliage around the trunks and are often low down in the trees. The Drain is a popular place for angling and matches regularly take place here. They have conveniently hammered in white marker posts, the owls favouring to roost most recently in the vicinity of Post 69. On no account attempt to flush these birds and only view from the east side. They have been roosting there for at least twenty years and my wish is that they continue to do so. Watching from the east bank does not disturb the birds in any way.

Nightjar *Caprimulgus europaeus*

The Nightjar is an annual summer visitor to Britain that has been declining in breeding numbers over the past fifteen years, with an estimated population in summer 1995 at just under 2,000 pairs. Nearly 70% of this population is now found in just four southern counties of England: Dorset, Hampshire, West Sussex and Surrey. A further 20% are found in East Anglia, especially in Norfolk and Suffolk breckland.

Nightjars favour heathland areas on sandy or gravelly soils and require small conifer trees from which to *churr* or display. They can only be reliably seen at dusk and generally perform from about 9.45pm onwards between 27th May and 5th July. The distinctive and somewhat unique churring display is best observed and heard on calm, warm evenings, birds often failing to show on windy or wet evenings.

As Nightjar is a diminishing species, it must be stressed that all sites are regarded as sensitive and care must be taken not to disturb these birds. Do not wander off designated footpaths.

❶Nercwys Mountain - Clwyd
A thriving population of Nightjars can be found at this high level site and can be reliably seen on suitable evenings throughout June and early July. Take the minor road west (uphill) at the crossroads in Nercwys village, then left at the next crossroads and first right towards Hendrefoelas Farm. Stand at the gateway on the minor road at SJ 217 593 (Sheet 117) just prior to dusk, but remember to take plenty of insect repellent, as mosquitoes are bad at this site.

❶Gwydir Forest - Gwynedd
A healthy population of Nightjars survives in the Gwydir Forest and these can be seen or heard from late May until early July. From Llanrwst on the B5106, turn steeply uphill on a minor road into the forest at SH 794 611 (Sheet 115) and continue over Bwlch-yr-haiarn and past the Llyn Sarnau parking area to the next junction. Continue straight ahead and park in the first layby on the right hand side of the road at SH 768 587. View and listen from the roadside as darkness falls.

❶Crychan Forest - Brecon
The most reliable site for Nightjar in central Wales is Pant Mawr, where up to eight churring males can be heard from late May until early July. They favour the open clearings and moorland at the site and can be heard just prior to dusk. Leave Coelbren (15 miles north of Neath) eastwards on a minor road and follow for about five miles into the forest. Park at approximately SN 907 139 (Sheet 160) and follow footpath to left for about 1.5 miles.

❶Margam Forest - Glamorgan
Up to four churring males annually return to Margam Mountain, on the Mid/West Glamorgan border. The best area to search is the plantations and clearings around Blaen-cwmcerwyn (SS 834 902 - Sheet 170), accessed via a maze of tracks from Maesteg. Park at SS 843 904 and walk west along the public footpath to the clear-fell area.

Another good site in Glamorgan is the Mynydd-y-Glew site in Hensol Forest. The replanted clear-fell area near the car park is best. Park at ST 031 768 (Sheet 170) and walk north-east. Hensol Forest is situated four miles NE of Cowbridge.

❶Wentwood Forest - Gwent
An excellent site for Nightjar in Wales is Wentwood Forest, where up to eight churring males can be heard in June. Park at ST 423 948 (Sheet 171) and walk east to the area known as Five Paths. The clear-fell areas thereabouts attract small numbers. Leave the M4 at Junction 22 and head west along the A48. Turn right after three miles towards Llanvair Discoed and follow north through the forest to the second picnic site car park.

Elsewhere in Gwent, Nightjars can be found in the Gwyddon and Nant Carn valleys in the Ebbw Forest. Both locations lie to the east of the A467 between Crosskeys and Newbridge and can be accessed from picnic site car parks at ST 238 959 and ST 254 942.

❶Dipton Woods - Northumberland
A small population of up to ten pairs of Nightjar survives in the Slaley area of Northumberland with most occurring in the clear-fell areas of Dipton Woods. Leave Hexham southwards on the B6306 and after four miles park sensibly by the footpath at NY 960 602 (Sheet 87). Follow the footpath to the east and explore the clearings.

Slaley Forest, further south on the B6306, also holds small numbers of churring Nightjars.

✱Hatfield Moors - South Yorkshire
Hatfield Moors lie to the south of the M180 and east of the A614 and regularly attract up to 40 pairs of Nightjars. The area between Lindholme airfield and Lindholme Lake is most favourable.

Thorne Moors, to the north of the motorway, also holds large numbers of churring males. Access is from the old colliery site at Moorends. Follow the track that leads from the far side of the spoil heaps for several hundred yards.

❶Wharncliffe Crags, Sheffield - South Yorkshire
A small population of up to five churring Nightjars can be found in the Wharncliffe Woods area to the NW of Sheffield. Leave the city on the A61 and then left on the A629. Turn left at the first crossroads and continue to the T-junction. Turn left and then park sensibly by the public footpath on the right at SK 311 973 (Sheet 110). Follow the public footpath westwards to the crags.

✱Broomhead Moor - South Yorkshire
Up to three pairs of Nightjar breed in an area known as 'The Stone Circle' on Broomhead Moor, south of Stocksbridge. Follow the circular route of the Pennine Trail accessed from the minor road between Midhopestones and Agden Bridge and park at approximately SK 245 945 (Sheet 110).

❶Laughton Forest - Lincolnshire
A large population of up to 40 churring males is found in Laughton Forest between late May and early July. Leave Gainsborough northwards on the A159 and after eight miles, the forest is situated on the left. A series of tracks lead west from the main road and the one which eventually leads to East Ferry is the most productive.

Elsewhere in Lincolnshire, Nightjars can be found at Crowle Waste (8-10 pairs), Broughton Woods and in the Keb Wood area.

❶Clipstone Forest - Nottinghamshire
The most reliable site in the county is Clipstone Forest, where up to eight churring males can be heard throughout June. The forest is extensive and suitable sites change from year to year, the Nightjars favouring the clear-fell areas of the woodland. Leave Mansfield on the B6030 and continue through New Clipstone. Less than a mile after passing under the railway, park on the right by Culloden Plantation and explore the woodland to the south.

❶Cannock Chase - Staffordshire
Up to thirty pairs of Nightjar are present at this site between late May and mid August and can be relatively easily seen on warm summer evenings. The favoured localities are the clearings of Abraham Heights, Brindley Heath, Haywood Warren and the Katyn Memorial/Anson Bank area.

The Chase lies between the A34 Cannock to Stafford and the A513 Stafford to Rugeley roads, and either side of the A460 Cannock to Rugeley road. The heathlands have a well-defined series of peripheral car parks and footpaths into the area, but one recommended site is Abraham's Valley in the north-east. Park by the A513 at SK 004 207 (Sheet 128) and follow the footpath south to Haywood Warren. The open clearings either side of this ride should produce churring and displaying Nightjars at dusk.

❶Forest of Dean - Gloucestershire
A small population of less than ten pairs still survives in the clear-fell areas of the Forest of Dean. The favoured area is that around Woorgreens, although birds have been noted at East Wood, Tidenham, Oakenhill, Mallards Pike, Serridge Inclosure, Spruce Drive, Ruardean and Birchwood in recent summers. Park at SO 622 124 (Sheet 162), adjacent to the B4226, and walk north along the forest trails towards New Beechenhurst Inclosure. Nightjars are present on most clear-fell plantations in this area.

✱Salthouse Heath - Norfolk
A small population of 3-6 pairs may be found between mid May and mid July on Salthouse Heath, about three miles east of Blakeney. Leave the A149 at the Dun Cow in Salthouse village and head south up on to the heath. Drive straight over at the first crossroads and then left at the second. Park sensibly about 50 yards along and then watch from the road. The Nightjars begin performing from 9.40pm onwards and often display over the road. This site is also excellent for up to four pairs of Nightingale, as well as for Tree Pipit and roding Woodcock.

❶Roydon Common - Norfolk
Up to six churring Nightjars can still be heard and seen at Roydon Common between late May and early July. Leave the A149 at the A148 roundabout and head towards Fakenham. Almost immediately turn right towards Roydon and after several hundred yards, turn right on to a rough track traversing the common. Find a suitable vantage point and listen for churring birds.

✱Thetford Forest - Norfolk Breckland
The increased amount of clearfell and young conifer plantations in the forest has seen a tremendous increase in the number of Nightjars present in recent years with a staggering total of 177 churring males recorded in June 1992. There are numerous areas in which to see them but particularly good are clear-felled heathlands south of Mundford (immediately east of the A1065, 1-2 miles south of the Mundford roundabout), those around Grime's Graves (best accessed from the track off the minor road leading from the A1065 two miles north of Brandon) and at Santon Warren (best accessed from the Helen's Well picnic site at Santon Downham).

✱Mayday Farm - Suffolk
A large number of Nightjars are present on the clearfell areas to the west of the B1106, two miles south of Brandon. Park by the gate at Mayday Farm and walk south-westwards to view the clearings.

✱Dunwich Forest/Westleton Heath - Suffolk
Up to 60 churring males are regularly present in the coastal heathlands from Walberswick to Leiston between mid May and early August. They are best observed on Dunwich Common or Westleton Heath and start churring during the last half hour of daylight. Leave Westleton on the minor road towards Dunwich and then park on the left after about half a mile. The extensive area of heathland to the south of the road holds the largest density of birds and a number of public footpaths traverse the area. Dunwich Common is also highly productive and this entails parking in the National Trust car park at Minsmere Cliffs.

Equally high numbers of Nightjars can also be found in Tunstall and Rendlesham Forests further south. The B1078 and B1069 east out of Tunstall pass through Blaxhall Heath and Tunstall Common respectively, both sites holding moderate numbers of churring birds.

❶Mereworth Woods - Kent
Mereworth Woods is one of the best sites in Kent for churring Nightjars with up to five pairs present in some summers. Leave the M26 at Junction 2 and head south from Wrotham Heath on the B2016. Park at TQ 634 533 after about a mile and follow the main track out onto the common.

An equally good site is Goathurst Common, to the south-west of Sevenoaks, where up to three pairs are present. Leave Sevenoaks on the B2042 and turn left in the village and drive east towards Bayley's Hill. Park in the layby at the east end of the common at TQ 516 524 and follow the footpath for about 100 yards to view the area of common below the slight ridge.

❶Clowes Wood - Kent
Clowes Wood holds a population of 5-6 churring Nightjars and these can be seen and heard from the footpath that runs westwards across the heath. Leave Canterbury northwards on the minor road that leads through Tyler Hill and continue towards Chestfield. Park in the designated car park and walk west to view the heath.

✱Ashdown Forest - East Sussex
A large population of Nightjars can be found at Ashdown Forest between late May and early July. The birds are widely distributed throughout the area but favoured sites are the valley to the east of the Gill's Lap car park on the B2026 and the heathland to the north of the road between Marlpits and Camp Hill. An area known locally as the Isle of Thorns is also very good. Leave the A22 at Wych Cross and head towards Chelwood Gate. After about half a mile, park in an obvious car park on the right hand side of the road and walk south-westwards for about 300 yards. Furthermore, Nightjars can also be heard from near the obvious car park to the west of the A22 about a mile south of Wych Cross.

✱Chobham Common - Surrey
The extensive heathlands at Chobham Common attract up to ten pairs of Nightjar in summer and these are present from late May until mid July. Leave the A30 southwards at Sunningdale and after half a mile park in the monument car park on the left of the B383. A series of public footpaths traverse the common but follow the main one back towards the railway and Nightjars can be seen displaying overhead after about ten minutes walk from the car park. This site is also good for roding Woodcock, Stonechat and Dartford Warbler.

✱Thursley Common - Surrey

Several pairs of Nightjar can be found on Thursley Common and these are best searched for at dusk from the footpath that leads out across the common from the Moat car park, situated south of Elstead. The area known as Shrike Hill, to the south of the car park, is highly productive, as is the area closer to the A3.

Frensham Common, about four miles to the west, is also very good. Park near Frensham Little Pond and walk south towards King's Ridge. Nightjars can be heard churring throughout this area from 9.45pm onwards in June.

❶Bramshill Plantation/Eversley Common - Hampshire

About thirty pairs of Nightjar nest on the heathlands that lie north of Hartley Wintney and west of Yateley. Bramshill Plantation to the north of Bramshill village has about four pairs, Warren Heath to the west of the B3016 about twelve, Eversley Common to the north of the A30 about three and Yateley Heath Wood about nine.

❶Woolmer Forest - Hampshire

A large population of up to 22 churring males can be found in Woolmer Forest during summer. The footpath that runs between the B3004 and the B2131 is the most productive. Woolmer Forest lies immediately northwest of Liphook.

❶Bishop's Dyke, Beaulieu Road Station - Hampshire

Large numbers of Nightjars breed in the New Forest and these can be seen at a variety of locations, including Black Gutter Bottom, Slufter's Inclosure and Broomy Plain. Beaulieu Heath has long been a traditional site and churring birds can be heard here from mid May onwards. Leave Lyndhurst on the B3056 and immediately west of the railway station, park in the car park on the right and walk southwest towards Denny Lodge. The Nightjars favour the large area of heath to the right of the track and can be best seen from the tumulus in the middle or churring from trees around the periphery.

Further east along the B3056, the heathland on either side of the road at Pig Bush and Ferney Crofts have small numbers, as does Yewtree Heath to the north of the minor road to Ipley Manor.

❶Parkhurst Woods - Isle of Wight

There are two traditional breeding areas for Nightjars on the Isle of Wight with healthy populations present in Parkhurst Forest and the forestry area on the western chalk downland around Brighstone. The majority of birds favour clear fell areas or plantations with trees less than four metres in height within coniferous belts.

Parkhurst Forest lies immediately west of Newport and borders the notorious prison grounds to the east. It can be accessed from Mark's Corner in the north or from the A3054 in the south. Park at SZ 482 900, about 0.7 miles west of the A3020 roundabout in Newport.

❶Grovely Wood - Wiltshire

3-5 pairs of Nightjar are generally present at Grovely Wood each summer and can be found in the young conifer plantations at the site. Take the road from the east outskirts of Wylye south over the railway to Dinton (on the B3089 west of Wilton) and at the summit of the downs, where there is a sharp double bend, turn left on to a narrow lane which runs east along the edge of the wood. After less than a mile, park by the gate. Continue on foot along the central track which traverses the entire length of the wood.

Elsewhere in Wiltshire, Savernake Forest regularly attracts up to three pairs (accessed from the A4 Hungerford road east of Marlborough via the Forest Hill entrance) and at least one pair can be seen at Loosehanger Copse and Shearwater Wood.

❶The Quantock Hills - Somerset
A healthy population of Nightjars can be found on the heaths and young conifer plantations in the Quantock Hills between late May and mid July. Leave the A39 at the Windmill Inn in West Quantoxhead and head towards Bicknoller. Almost immediately in Staple, turn left up a steep and narrow no-through road leading to a public car park at Staple Plain. The young conifer plantations on the northwest facing slopes should produce Nightjars at dusk.

❶Shapwick Heath - Somerset
Up to twenty churring males have recently been recorded at Shapwick Heath and these can be heard at dusk from the footpath that traverses the area from Ashcott Corner village. Leave Glastonbury west on the B3151 and turn south on a minor road in Meare village towards Ashcott. The reserve is on the right hand side of the road after a mile.

✱Arne NR - Dorset
At least 25 pairs of Nightjars breed annually on the heaths surrounding Arne NR and can be seen on suitable evenings during May to July. The Arne Peninsula lies in the south-west corner of Poole Harbour and consists largely of lowland heath, gorse and boggy meadows. Leave Wareham southwards on the B3075 and turn left on to a minor road in Stoborough village. Continue eastwards for about four miles and then park in the designated reserve car park. A series of trails lead out across the peninsula.

✱Studland Heath - Dorset
Studland Heath is a huge area of lowland heath, interspersed with areas of gorse and sphagnum bog, which attracts large numbers of Nightjars in summer as well as supporting at least 35 pairs of Dartford Warblers. The Nightjars favour Godlingstone Heath to the west of the road, this area being easily reached via a maze of public footpaths leading off the road north of Studland village.

❶Wareham Forest - Dorset
Wareham Forest supports a large population of Nightjars and these can be seen or heard on suitable evenings from late May until mid July. The forest clearings and heathland margins are best and access may be attained via a number of routes. A number of footpaths and forest trails traverse the area but recommended walks are from Sherford Bridge east to Organford Bridge, from Sherford Bridge west to Stroud Bridge, from North Lodge north to the A35 near Morden and on the Sika forest trail from the Forest Centre at Cold Harbour.

Nearby, Bovington Heath, Affpuddle Heath and Puddletown Forest attract good numbers.

❶Canford Heath - Dorset
Canford Heath is another reliable site for churring Nightjars and this can be accessed along footpaths from the lay-by on Gravel Hill, Broadstone, at SZ 016 960 (Sheet 195).

❶Aylesbeare and Colaton Raleigh Commons - Devon
The Nightjar is a relatively abundant species on the commons of east Devon with at least 100 pairs breeding in about nine sites. Aylesbeare Common is very attractive to them and they can be seen at dusk hawking for moths above the gorse covered heath. A metalled road leading north from the A3052 near the Half Way Inn gives ample viewing of the site. Park in the car park to the south of the main road.

Venn Ottery Common, immediately to the north, also has up to five pairs, whilst Woodbury Common and Colaton Raleigh Common to the south, have much larger numbers. A number of car parks on the B3180 allow for easy access to both of these sites.

✱Haldon Forest area - Devon
The Great Haldon Forest supports a breeding population of up to 80 pairs of Nightjar and these can be found in any of the clear fell areas on the estate. One of the best sites is the raptor watchpoint, along the minor road from the A38 towards Doddiscombsleigh. The Nightjars *churr* from the valley below.

Wryneck *Jynx torquilla*

The Wryneck is a regular passage migrant to Britain appearing between mid April and late May and from mid August to early October. It also sporadically nests in Britain although this is largely confined to the Scottish Highlands. They feed principally on ants and their larvae and can often be found in large gardens or in paddocks or dune-scrub on migration.

Most Wrynecks that visit Britain are on their way to or from breeding grounds in Scandinavia and like all migrants from this region, are best discovered when winds are in an easterly direction. Large numbers of Wrynecks leave northern Europe, especially if conditions are clear, from late August onwards, and with a light easterly breeze and overcast or wet conditions in eastern Britain, large falls of birds can be expected between the Northern Isles and Kent. As the autumn progresses, Wrynecks take on a more southerly and westerly bias, and are more likely to be found in Dorset, Cornwall or Scilly.

The following is a list of the more regularly visited sites by Wrynecks but like all scarce migrants or vagrants to Britain, can be recorded virtually anywhere under suitable conditions.

❶Fair Isle
Wrynecks are regularly recorded on Fair Isle in spring and autumn passage, and peak in occurrence between 15th August and 19th September. Ideal conditions are light south-easterly winds and overcast conditions, with most being found in The Plantation, the area around North Haven and in the crop fields and gardens at the south end of the island.

❶North Ronaldsay - Orkney
Like Fair Isle, large falls of up to ten Wrynecks may occur when wind conditions are light to moderate south-easterly and it is overcast or drizzling. The best area by far is Holland House, where the heavily vegetated garden there offers shelter and food for tired migrants. The peak period for sightings is between 17th August and 20th September.

❷St Abb's Head - Borders
The gorse covered slopes surrounding the Mire Loch are an excellent place to find migrant Wrynecks at St Abb's in autumn. Light north to south-east winds from the third week of August will almost certainly result in sightings.

❷Hartlepool Headland - Cleveland
The gardens and park land at Hartlepool Headland are regularly visited by Wrynecks between late August and mid September, particularly following easterly winds.

Nearby South Gare is also very good, with birds favouring the slope above Cabin Rocks, along the southern side of the peninsular, about 600 yards west of the lighthouse.

❶Flamborough Head - East Yorkshire
Wrynecks are regularly recorded at Flamborough Head, with up to eight birds being present under suitable conditions in late August. Choice of sites is varied but most favoured is the densely vegetated valley at South Landing and the area of scrub bordering the golf course at Bay Brambles, opposite the Head car park. The sparsely vegetated slope north of the lighthouse and east to the Foghorn Signal Station can also attract them.

✱Spurn Point - East Yorkshire
Spurn Point is one of the most reliable sites in Britain for seeing Wrynecks, especially in autumn. Following easterly winds in late August, up to six Wrynecks may appear in the Observatory Trapping Area and these may stay for several days, often performing well in the dunes. One of the best areas is The Warren, adjacent to the observatory buildings, whilst the bushes by the canal and car park often attract birds and the dense buckthorn at The Narrows and Point provide sightings. Ask the observatory warden for on-site directions when visiting.

❷Gibraltar Point - Lincolnshire
The Observatory area at Gibraltar Point NR regularly attracts Wrynecks from late August until late September and these can be found at a variety of locations. The small plantation near the centre can be good as well as the complex of dune vegetation and the area around the mere.

❶Holme Paddocks and Dunes - Norfolk
Wrynecks have traditionally turned up in the Holme area from late April to May and from mid August until late September. They have a particular fondness for the relative solitude and peacefulness of the paddocks, bordering the chalets and dunes near the first car park. They can be watched at close range feeding on ants in the paddocks from the main coastal footpath leading from the golf course to The Firs. They also visit the nearby chalet gardens adjacent to the main access track. Further east, at The Firs NOA/NNT reserve, Wrynecks may be found in the dunes around the plantation. Please note that a permit is required to gain access.

Burnham Overy Dunes, further east (and north of the coastal A149) is also attractive to Wrynecks, and can be found in the dune complex anywhere west of Holkham Pines, especially around the boardwalk.

✱Blakeney Point - Norfolk
Blakeney Point is the premier site in Britain for migrant Wrynecks and birds can be virtually guaranteed in August and September when the wind is in the east. They favour the Plantation and the tea room Lupin beds at the Point, as well as the dunes and scrub at The Hood and the open area of grassland at Halfway House. In windy or wet conditions, many of the Wrynecks shelter in the *Sueda* and are difficult to see without flushing, but if fine and dry, they will feed quite happily out in the open on the dunes.

Blakeney Point is reached via Cley Coastguards or boat from Morston Quay and takes at least four hours to cover sensibly.

Hoopoe *Upupa epops*

The Hoopoe is an annual migrant to Britain which overshoots in spring from the Mediterranean and is displaced from further east in the autumn. The average annual incidence is of about 110 birds, with most occurring between mid March to mid May and from mid August to mid October. They favour large gardens, sand dunes or areas of short turf in which to feed, and are recorded at a number of south coast sites on virtually an annual basis. Obviously with such a scarce migrant, it is not possible to predict actual sites at which they will appear as this largely depends on the weather conditions and time of year. The following sites are offered as a useful guide, as they are those most regularly visited by migrant Hoopoes.

❷Portland Bill - Dorset
Portland Bill attracts 1-5 Hoopoes in an average spring and these birds can be extremely elusive and mobile. Although they fly in at the bill, they inevitably move further up the island to feed and a number of localities are particularly favoured. The clifftop fields around the stopery and the black barns, Eight King's Quarry and the East Cliffs, the Tout Quarry Sculpture Park, the grassy slopes at Verne Common and the fields and cliff paths surrounding the Pennsylvania Hotel are all regularly visited between 17th March and 15th May. Warm southerlies or light to moderate south-easterlies are the favoured conditions, especially if accompanied by rain or heavy showers.

❷Porthgwarra - Cornwall
The gardens, fields and clifftop paths at Porthgwarra are a regular haven for Hoopoes in spring. They appear following southerly or south-easterly winds and may remain for several days. They can be elusive, but normally find a quiet, sheltered garden in which to spend several hours feeding. The best period for overshooting is between 20th March and 15th May. Park in the designated pay and display car park at the end of the very narrow and winding lane.

Kenidjack Valley, to the north of St Just, can also be very good for Hoopoes in April. Check the valley slopes and fields around the pumping station.

❷Prawle Point - Devon
Barely a spring passes without a Hoopoe appearing at Prawle, and this generally occurs following a warm southerly airflow between mid March and mid May. The short turf near the coastguard lookout is one of the favourite areas but basically anywhere along the coastal footpath is good. Leave the A379 at Frogmore or

Chillington and follow the minor road to East Prawle village. Continue through the village passing the green on the left on a very winding and narrow lane to the National Trust car park at the Point. Walk south from the car park towards the sea and then turn right uphill towards the Coastguard cottage gardens. Cross the stile and check the heathland above the cliff. This area can also be good for spring migrants and breeding Cirl Bunting.

Start Point, nearby, also regularly produces Hoopoe in spring. Turn sharp left in Stokenham village and drive to car park at the top of the Point, passing Start Farm valley on the right. Walk through the gate into Start Lighthouse access area and turn immediately right down the coastal public footpath. Follow the valley down to the sea or walk down the lighthouse lane, checking around the lighthouse compound and rocky crevices.

Slapton Ley is also another attractive area for Hoopoes and they have been regularly noted feeding alongside the road north of Torcross, at Strete Gate or in the Field Study Centre area from late March to mid April. The Ley lies adjacent to the A379, between Dartmouth and Kingsbridge, Strete Gate being at the north end of the Ley, the Field Study area being across the bridge from the Monument and the area by the road straightforward.

❷Beachy Head - East Sussex

Beachy Head has long been renowned as one of the best sites for migrants along the south coast. Its diversity of habitats and its huge area means that many thousands of tired migrants may stop to rest. Hoopoes are regularly recorded at the site and favour a number of traditional locations:

1) Birling Gap, where they can be seen feeding around the hotel, the gardens on the west slope or in the fields immediately inland of the Gap;
2) Belle Tout Wood, the grassy slopes adjacent to the wood or on the cliff coast path;
3) Hodcombe Farm, the grassy slopes and criss-crossing tracks opposite the farm or the fields adjacent to it or
4) the coastal footpath and car park area near the Lighthouse.

Beachy Head is easily accessed from Eastbourne or East Dean village.

❷Dungeness - Kent

Hoopoes are regular visitors to the Dungeness area but rarely remain for longer than a few hours. They occur from the third week of March until the second week of May and may be found at a number of localities. The Observatory area between the power station perimeter fence and the lighthouse is good as well as the Trapping Area north to the Long Pits. Denge Marsh Gully, to the west, is also very good, as well as all the open ground leading to the site. Boulderwall Farm and the RSPB reserve can also be attractive and the fields and gardens inland of Greatstone-on-Sea have seen a large number of records during the last ten years.

Sandwich Bay Bird Observatory and Estate, north of Deal, is also an excellent site for Hoopoes in spring. Once again, though, this is a vast area and birds may appear virtually anywhere. The area known as St George's Bushes and the estate gardens are favourite but it is best to ask at the observatory itself for information. Leave Sandwich town eastwards on a minor road to the estate. A moderate entrance toll is charged at the gate and the observatory lies on the right after 600 yards.

❶Winterton Dunes - Norfolk

Wrynecks are annually recorded at Winterton Dunes, particularly between 20th August and 16th September. It is a vast area but birds are most often encountered in the valley to the south of the beach car park or the dunes about half a mile north of the car park.

❷Minsmere Sluice Bushes - Suffolk

In favourable conditions, Wrynecks may be found in the bushes between Sizewell Power Station perimeter fence and the sluice at the south end of Minsmere RSPB reserve.

❷Landguard NR - Suffolk

The ringing compound at Landguard NR is an excellent site in which to see Wryneck if the wind direction is in the east. Birds can be seen in May or from mid August to mid September and often feed out in the open on the earthy banks inside the compound or the mounds to the north of the observatory.

❷Portland Bill - Dorset

Up to ten Wrynecks are recorded at Portland Bill in an average autumn and these can be found at a variety of suitable locations on the promontory. In the Observatory Recording Area, the best sites are the Observatory Quarry and the bushes at Culverwell, to the west of the main road about 300 yards north of the Obs. Further north, the fields around Reap Lane and the Avalanche Road Hump in Weston regularly attract birds as do many of the quarries on central Portland. September is the peak month for occurrences.

✱Scilly

Wrynecks are annual visitors to Scilly in autumn and up to twenty birds may be recorded between late August and 17th October. They can be seen virtually anywhere and on all islands, but are generally found on St Mary's, St Agnes and St Martin's. The granite slopes of Giant's Castle and Peninnis Head (on St Mary's) have been the most regularly visited sites in recent years.

Lesser Spotted Woodpecker *Dendrocopus minor*

The Lesser Spotted Woodpecker is a fairly common resident in Britain and Wales but is a rare vagrant in Scotland. It is very localised and is often difficult to detect and favours open parkland, riverside trees, large gardens and small deciduous woods in which to feed and nest. It is usually found in the upper canopies of trees, favouring the spindly outer extremities and are best located in spring, when the distinctive call of the male can be heard over long distances. This is a loud and quickly repeated high-pitched '*quee-quee-quee-quee-quee-quee-quee*' call., uttered from late February until mid April.

The sites offered below are selected for their reliability but are not a complete representation of each county's status. Lesser Spots can be found in virtually any suitable habitat in southern England.

Northwest England

Birds are best looked for in February and March when they are most vocal and the trees are bare of leaves. Can be very hard to find at other times of year. In **Merseyside**, try the southern end of Stapleton Wood, Caldy (public footpath through wood to SJ 232 855) or Dibbinsdale LNR near Bromborough (SJ 337 813). In **Cheshire**, sites where birds regularly hold territory in early spring include Burton National Trust Wood (southern perimeter walk above gardens and church at SJ 315 743), Stanney Woods Country Park near Ellesmere Port (near car park which is alongside the A5117 at SJ 397 738), Marbury Country Park near Northwich (near the bird hide at SJ 650 7666) and Rostherne Mere NNR (permit needed for hide at SJ 743 838). **Greater Manchester** sites include Etherow Country Park (walk upriver from car park in Compstall at SJ 965 908) and Haigh Country Park near Wigan.

Northeast England

A very scarce bird in the north of the region with Yorkshire the only county likely to produce regular sightings. The willow-carr and tall trees surrounding Hornsea Mere are a regular haunt of birds from January to April, whilst Sprotborough Flash, near Doncaster, is probably the best site within the region. Up to two pairs can be located in spring in the tall trees and scrub bordering the flash. In the Sheffield area, reliable breeding sites include Renishaw Park (SK 432 783), Ecclesall Woods (SK 325 830), Wortley Top Forge (SK 309 999) and in the woodland surrounding Langsett Reservoir (SE 200 007). In the Bradford District, the broad-leaved woodland surrounding Bolton Abbey and nearby Strid Woods are very good and the relatively small plantation at Otley Gravel Pits has regularly produced sightings. Elsewhere in the county, pairs or calling males can also be seen on a regular basis in spring at Newburgh Priory, Sutton Park, Wheldrake Ings YNT, Escrick, Strensall Common, in the York area and in Fawcett Park.

In **Lincolnshire**, the most reliable sites are at Linwood Warren (east of Market Rasen and south of the minor road to Little London), Deeping St James, Burton pits (immediately north of the A57 two miles west of Lincoln), Bradley Wood (south of the A46 in Grimsby and Bradley village), Weelsby Woods (immediately south of the A46 in west Cleethorpes) and Boultham Park (two miles SW of Lincoln).

Midlands Region

Lesser Spotted Woodpeckers are relatively common throughout the Midlands region and can be found in small numbers in each county.

In **Nottinghamshire**, regular sites include Bevercotes, Wollaton Park (in grounds of Wollaton Hall just west of Nottingham), Old Moor Wood, Bestwood CP (on northern outskirts of Nottingham, off the A683), Clipstone Forest (mixed woodland accessed from A614 or B6030 near Clipstone), Attenborough NR (between Nottingham and Long Eaton), Clifton Wood (woods and open farmland in river valley, accessed from Clifton village with gate near old school), Thorndale Plantation, Sherwood Forest (large woodland in Edwinstowe area accessed by numerous public footpaths), Carburton (roadside trees viewable from B616 Carburton to Norton road), Bulcote, Colwick CP (situated east of Nottingham) and Lound NR (former gravel working site near Retford), but most reliable is Clumber Park, where birds can often be seen near the ford on the eastern side or in the extensive woodlands generally.

The best sites in **Derbyshire** are Allestree Park and Shipley Park, where up to four pairs breed at both sites. Allestree Park is situated in the northern outskirts of Derby and is accessed off the A6, whilst Shipley Park is northwest of Ilkeston and best reached north of Mapperley village. Visits during March to April are strongly recommended.

In **Leicestershire**, the species is fairly widespread, with regular sightings in the breeding season from Burley Wood (immediately north of the A606 two miles east of Oakham at SK 890 095), Oadby (wooded glades and large gardens around the village at SK 625 005), Queniborough (brookland birch scrub and trees to the east of village at SK 648 122), Swithland Reservoir (Buddon Wood to the north of the reservoir, accessed from Mountsorrel), Launde Park Wood (north of the A47 near Belton at SK 797 044) and Scraptoft (woodland around the hall at SK 648 058).

It is a difficult species to find in **Northamptonshire** but in winter and spring, at least four birds are known to inhabit Fawsley Park, south of Daventry. Leave Daventry on the A361 and two miles south of Badby, turn left along a minor road towards Everdon. After 1.5 miles, park by Fawsley Hall and search the old trees around the lakes to the south of the road. There is a solitary building ion the middle of the field and the trees around here will often hold at least one territory.

The species is fairly common but localised in **Staffordshire**, with regular sightings from woodlands surrounding Blithfield Reservoir, Belvide Reservoir (the wide avenue of woodland 1.5 miles SE of Belvide approaching Chillington Hall), Consall Wood (north of the A52 and accessed from Wetley Rocks), Copmere (two miles west of Eccleshall south of the B5026), Himley Hall (accessed off the B4176 between Himley Church and Dudley, east of the A449 Wolverhampton to Kidderminster road) and Tittesworth Reservoir, north of Leek.

In **Warwickshire**, regular pairs are known to breed at Ansley Hall (west of Nuneaton on the B4112) and Coombe Woods, whilst in neighbouring **West Midlands** county, Edgbaston Park and Sandwell Valley RSPB reserve are the most frequently visited sites.

It is well distributed in **Worcestershire** with regular sightings from Astley Burf, Kemerton, Bayton, Birlingham, Callow Hill, Chase End Hill, Dowles Brook, Drayton, Eckington, Hadley Poplars, Hanley Castle, Hornhill Wood, Old Hills Common, Shrawley Wood, Tibberton and Westwood.

Although widely reported in **Bedfordshire**, there are no regular sites at which they can be seen and birds range widely over a large area. Perhaps the most reliable is Swiss Gardens, Old Warden, where at least two birds have been seen displaying in recent springs. Leave the A1 on the Biggleswade northern roundabout and drive west towards Old Warden. Immediately past the Shuttleworth Aircraft Museum, park sensibly at the side of the road. The small plantation to the south of the road is Swiss Gardens.

In **Buckinghamshire**, the species is widespread and found in at least fifteen locations. The most reliable of these are the Little Brickhill woodlands and Ashridge Forest. Leave Little Brickhill on the minor road east to Woburn and park by the county boundary sign. Cross the stile and explore the mixed woodland to the north of the road. Furthermore, Back Wood, to the west of the Bow Brickhill road will also produce sightings.

Ashridge Forest is situated south of Dagnall on the B4506. Turn right in Ringshall village and after 200 yards, turn left into Ringshall Coppice car park. The area of woodland thereabouts generally holds at least one pair of birds and the birds can often be heard from the car park in March and April.

Oxfordshire has a large number of sites but most productive are the Wytham Woods, where up to eight pairs may be present. Wytham lies to the northwest of Oxford and can be reached from the northbound carriageway of the A34 or from Wolvercote village. The woodland is to the west of Wytham and is accessed along a minor track that eventually leads to Swinford.

In **Gloucestershire**, the species is relatively common and fairly easy to see in spring. Nagshead Reserve is the best site with upwards of six pairs present. A discreet blue RSPB sign marks the start of a track to the north off the B4431 on the western outskirts of Parkend village and this leads to a small car park at the reserve. Various trails are well marked and there is an information centre on site.

East Anglia

In **Norfolk**, the species is widely scattered, but only reliably seen at a handful of sightings. The best site by far is Holkham Park, where up to six pairs are known to breed. Park by the hall entrance in the designated car park and walk south through the gates. Turn immediately right inside the gates and walk west for about half a mile to the monument. The area of trees around the monument and the lake to the west will produce calling male Lesser Spotted Woodpeckers between 15th March and 4th May, especially prior to 10.00am.

Another good site is Lynford Arboretum, about 1 mile north-east of Mundford. Take the minor road east of the A1065 and park in the designated car park just past Lynford Hall. Check the arboretum trees to the right of the road and the woodland that borders the brook to the south.

The species is widely distributed throughout **Suffolk** and may be encountered at several deciduous woodlands in the coastal belt and around Ipswich. Probably the most reliable site is the woodlands on the RSPB reserve at Minsmere, where up to six pairs are known to occur. They favour the woodlands at the west end of the reserve, particularly those to the north of the entrance road, and are best seen and heard during early morning (in March and April). Less occasionally, birds can be seen in the narrow belt of trees leading down to the Tree and Island Mere hides on the reserve.

The species can be widely encountered in **Cambridgeshire**, with regular breeding season reports from Trumpington, Chippenham Fen, Stapleford, Great Shelford, Milton Country Park, Wandlebury, Whittlesford GP, Bluebell Wood, Castor Hanglands, Monks Wood and Waresley Wood.

Southeast England

In **Essex**, the species can be found in a large number of suitable woodlands with up to six pairs nesting in Hainault Forest and Danbury Common. Hainault Forest lies to the east of Chigwell and can be accessed via a number of public footpaths from the minor road to Lambourne End or from the A1112. The broad-leaved woodland at Danbury Common lies to the south of Danbury, the main car park allowing straightforward access. Norsey Wood, just north of Billericay, has at least two pairs present in spring, and Thorndon Park, near Brentford, and Fridaywood, south of Colchester, are also excellent sites. Early morning visits in March to May are essential.

The species is localised in **Hertfordshire**, with the best site being Bricket Wood and Common. Elsewhere, they can be seen or heard in spring at Ashridge Forest (trees around the monument and along access road), the tall elm trees surrounding Tringford Reservoir, Tring, Cassiobury Park (in west Watford), Digswell (in riverside trees north of Welwyn), Panshanger Park (east of Welwyn Garden City and south of the B1000) and Whippendell Woods (south of Chandler's Cross).

Bricket Park lies to the east of the M1 and is best reached from the minor road south of Bricket Wood railway station. Alternatively, leave the A41 400 yards north of Junction 5 of the M1 along a minor road that runs parallel with the River Colne. Follow the narrow lane across the motorway bridge and park after 400 yards. The parkland and riverside trees are excellent for Lesser Spots early morning in spring.

Up to three pairs are regularly recorded on Hampstead Heath (**Gtr London**) in spring, this site being best approached from the Vale of Health car park on the B519 at TQ 264 865 (Sheet 176).

The species is widespread in **Kent** and may be seen at a large number of wood and parkland localities. Several pairs are known to breed in Tudeley Woods (south of the B2017 immediately east of Tonbridge), Blean Woods (SW of the A290 NW of Canterbury) and Church Wood (accessed from Rough Common village, west of Canterbury) and sightings are relatively frequent at Stodmarsh and Westbere and in Hamstreet Woods. Larkeyvalley Wood, SW of Canterbury, is also very good and may be accessed from the minor road in Thanington (on the A28). The woodpeckers frequent the trees at the south edge of the wood. This latter site is also good for Nightingale and Hawfinch.

In neighbouring **Sussex**, one of the most reliable sites is Burton Mill Pond, south of Petworth. Leave Petworth on the A285 and turn left after two miles on a minor road to Sutton End. After half a mile park by the pond on the right and explore the alders and birches around the car park and western side of the pond or take the footpath southwest towards St Michael's Burton Park. Calling/drumming birds can often be found in the vicinity of the first crossroads, in the tall sweet chestnut trees.

Elsewhere, Rackham Woods and Greatham Common, north of Rackham Church (two miles west of Storrington), the 100 Acre Wood and Wych Cross area of Ashdown Forest and Fore Wood at Crowhurst are very reliable sites.

The riverside trees at Boulter's Lock, Maidenhead, and the park along Ray Mead Road are one of the most reliable sites in **Berkshire**. Early morning visits in March and April are essential. Ray Mead Road (the A4094) runs parallel with the River Thames and there is a public car park opposite the Boulter's Lock Hotel.

In **Surrey**, Richmond Park has at least two pairs (mainly in the Pen Ponds area and north sector) and the woodland surrounding Frensham Little Pond at least three. The Birch woodland surrounding the Stew Ponds on Epsom Common also regularly attract 1-2 pairs.

With large tracts of woodland throughout the county, it is not surprising that **Hampshire** has one of the highest densities of the species in Britain. There are many sites but most reliable are Fleet Pond and Denny Wood in the New Forest.

Fleet Pond lies to the south of the B3014 north of the town and is best accessed from the southern end. The mixed woodland bordering the south shore holds at least three pairs of birds and these can be easily heard (and seen) early mornings in March and April.

Denny Wood is the large tract of woodland to the south of Beaulieu Road Station, three miles east of Lyndhurst on the B3056. Park in the car park under the pines just west of the railway bridge and walk south over the heath to the wood. Veer left in the wood and follow the footpath for about 400 yards to an area of mature trees, some of which are laden with ivy around the trunks. This area is excellent for Lesser Spotted Woodpeckers in the early morning.

Wales

In North Wales, the most regular site for Lesser Spotted Woodpeckers is Wepre Wood about a mile west of Queensferry (SJ 295 680). Access is best obtained from the minor road south of the A548.

The Oak woods in the beautiful Vale of Ffestiniog hold several pairs of Lesser Spotted Woodpeckers but these are difficult to locate in the extensive area. The A487 and B4410 pass through the valley and the main block of woodland is best explored by parking at SH 659 410 (Sheet 124) and following the main track.

The Ynys-hir RSPB reserve, six miles SW of Machynlleth (**Powys**), regularly attracts up to three pairs of Lesser Spotted Woodpeckers in spring and these can be seen or heard in the small wood around the information centre and car park. Take the minor road west of the A487 at Eglwsfach and the reserve is well signposted.

Further south, the RSPB reserves at Gwenffrwd and Dinas can be very good. Both sites hold extensive areas of hanging Oak woods and alder carr which are attractive to the woodpeckers. Leave Llandovery northwards on an unclassified road to Rhandirmwyn and continue towards the Llyn Brianne dam. The Dinas RSPB information centre is at Nant-y-ffin, four miles further up the valley, and is open 10.00am to 5.00pm from Easter until the end of August. Access to the Gwenffrwd reserve can be organised from here.

Another good site in Carmarthenshire is Stradey Woods at SN 487 015 (Sheet 158). Leave Llanelli west on the B4308 and the woodlands are on both sides of the road after half a mile. A number of public footpaths traverse the area. The Castle Woods immediately west of Llandeilo (Dinefwr Castle Woods at SN 614 217) are also recommended.

1-2 pairs may be found in the oak woodlands at Cwm Clydach, particularly during early mornings in March and April. Leave the M4 at junction 45 and take the A4067 Brecon road. Turn left at the crossroads in Clydach village and follow the unclassified road through the valley and park in the designated car park at SN 682 053 (Sheet 159).

An excellent site in Radnorshire is the oak woodlands in the Wye-Elan valley. Dyffryn Wood is possibly the best site and this may be accessed from the A470 just south of Rhayader at SN 980 672 (Sheet 55).

Southwest England

In **Wiltshire**, one of the best sites is the open parkland and deciduous woods surrounding Corsham Lake, particularly at the eastern end. Park in the Church Square in Corsham village and walk north-eastwards across the parkland towards the Mynte crossroads on the A4.

The woodlands of the Quantock Hills are the most reliable areas in **Somerset**, particularly the area around Friarn. Leave the A39 immediately west of Nether Stowey towards Bincombe and continue uphill towards Crowcombe. Park at approximately ST 176 386 (Sheet 181) and follow the footpath north to Bin Combe or explore the woodland to the south of the road.

The species is fairly common in **Dorset** and can be seen at a number of reliable sites. Particularly good are the tall trees and scattered bushes in Upton Country Park, just west of Poole (immediately south of the A35), the birch woodland immediately north of the reserve car park at Arne RSPB (four miles east of Wareham), the open woodland and birch scrub around Bloxworth Heath, in Wareham Forest (immediately east of the minor road at Sugar Hill between Bere Regis and Wareham at SY 881 926 (Sheet 194), the riverside trees east of Sherford Bridge (on the B3075 between Sherford and Wareham), the western fringe of Puddletown Forest around the nature trail to Hardy's Birthplace (west of Puddletown on the A35 and accessed from Higher Bockhampton), the extensive parkland and open woodland at Canford Heath, north-east of Broadstone, north Poole (accessed from various points, including Gravel Hill, Broadstone at SZ 016 960, Wheelers Lane, Bearwood, at SZ 044 964, Francis Avenue, Alderney, at SZ 046 952 and from Gravel Hill, Merley, at SZ 012 972) and in Harewood Forest (walking north along footpath at SU 405 450 on minor road between B3400 at Andover Down and B3048 at Longparish).

The most reliable sites in **Devon** are Yarner Wood and Dunsford Woods - Yarner Wood is an RSPB reserve situated west of Bovey Tracey and accessed from the B3344 at Reddaford Valley and Dunsford Woods extend to the southwest of Dunsford, both Bridford Wood and Cod Wood straddling the B3212 being productive.

Woodlark *Lullula arborea*

The Woodlark is a scarce but increasing breeding bird in Britain with an estimated population of 700-750 pairs. It breeds in open country with areas of bare ground and very short grass and has a particular preference for lowland heaths, rabbit-grazed grass-heaths, derelict pasture, tree and shrub nurseries and clear-felled conifer plantations. Its distribution is widely scattered but largely confined to southern England, with stable populations in East Anglian breckland, the New Forest and on the Surrey heaths.

Although Woodlarks now breed as far north as Nottinghamshire and Lincolnshire and also in Bedfordshire, Buckinghamshire, Wiltshire and Devon, the strongholds still remain in Norfolk, Suffolk, Hampshire, Berkshire, Surrey and Dorset. Woodlarks sing and display from mid March until late April and this is the best time in which to see them. The following is a compilation of the most reliable sites but please be careful as to not disturb these rare breeding birds.

➊Weeting Heath - Norfolk

In recent years, a small but increasing population of Woodlarks has become established at this site and can be seen from the westernmost hides at the reserve. The males can be heard singing from late March until late April and family parties of birds can be seen feeding on the heath up until early June.

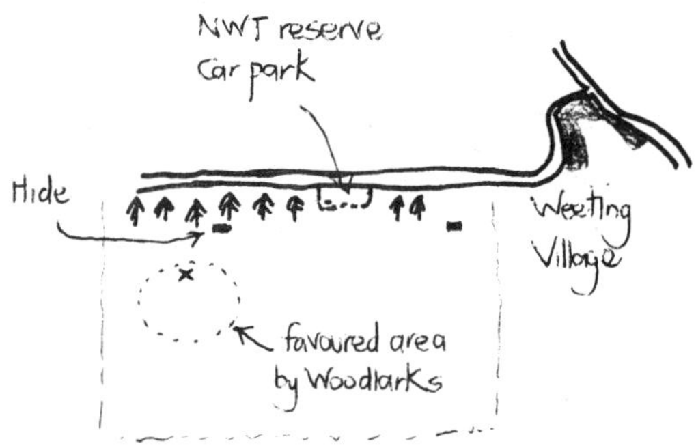

The NNT reserve lies to the northwest of Brandon on the minor road to Hockwold. After about 1.5 miles from Weeting village, look for the designated car park on the left hand side, somewhat hidden from view in the belt of conifer trees alongside the road. Permits can be obtained from the warden's hut, as well as on-site directions for the reserve's specialities, including Woodlark and Stone Curlew.

Several clear-felled areas north of Weeting and east of the B1106 also have singing Woodlarks.

➊Santon Downham - Norfolk

The large clear-felled plantations north of the river at Santon Downham hold sizeable populations of Woodlarks. Park in Helen Wells picnic site on the north side of the river (signposted from the minor road north from the village immediately south of the railway crossing) and explore the heathlands to the north of the railway. Alternatively drive half a mile north of the crossing and park by the first crossroads of Forestry Commission tracks and check the clear-felled areas for singing birds. They particularly like the rows of piled up logs and branches bordering the clearings.

Nearby, the clearings to the east of the A1065, one mile south of the Mundford crossroads are very good. Please keep to the footpaths to avoid any disturbance to the birds.

✱Mayday Farm - Suffolk
Mayday Farm is a traditional stake-out for Woodlarks and is known to hold at least twelve pairs in the vicinity. They are best looked for early mornings in April and May when their melodic song can be far-carrying. Park by the farm track to the west of the B1106 (two miles south of Brandon) and walk southwest for about half a mile. A well-marked nature trail skirts you round the main heath and Woodlarks may be encountered throughout its length.

❶Westleton Heath - Suffolk
Up to eight pairs of Woodlarks can be found on Westleton Heath, immediately east of the village. Park by the sandy ride on the left and check the heathland on both sides of the road to Dunwich. A few pairs can also be found on Dunwich Heath, immediately south of Minsmere Cliffs car park.

❶Thursley Common - Surrey
A healthy population of Woodlarks survives on Thursley Common and these may be seen or heard from the boardwalk or nature trails leading west of the Moat car park, two miles south of Elstead village. From the car park, walk south towards Shrike Hill and skirt around the edge of the bog on the main footpath and check the tall conifers and small saplings to the left. This area is excellent for singing Woodlark.

At nearby Frensham Common, Woodlarks may be encountered singing from the tops of the trees or in the sky above the fields and tree nurseries to the north and east of the car park about 0.25 miles east of the Little Pond.

❶Beaulieu Heath - Hampshire
One of the best sites in the New Forest is Beaulieu Road Station, where up to four pairs of Woodlarks may be present in spring. Leave Lyndhurst on the B3056 and park in the car park immediately west of the railway bridge. Walk south towards Denny Lodge and then turn left on to a track across the railway bridge. The bare heath immediately east of the railway is good for Woodlark as well as the pine-belt and heath to the right of the footpath. A similar area of burnt heather north of the car park is also regularly visited.

❶Wareham Forest - Dorset
A small population of Woodlarks inhabits the forest clearings and heathland margins of Wareham Forest, particularly the area east of Sherford Bridge towards Organford Bridge. Park by the B3075 Sandford to Morden road.

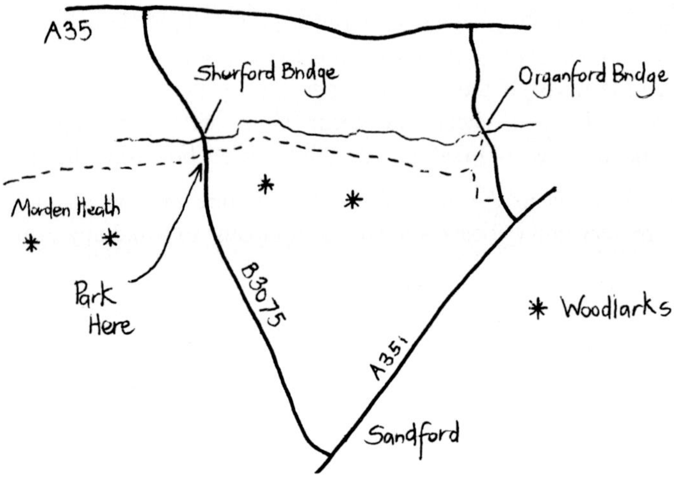

Shorelark *Eremophila alpestris*

The Shorelark is an annual autumn migrant to eastern Britain, with small numbers remaining to overwinter. In recent years, less than fifty birds have chosen to remain, and these are generally distributed from Northumberland south to north Kent. There are only a handful of regular wintering sites.

❶Saltfleetby - Lincolnshire
In recent winters, a small group of up to seven birds has remained on the dunes and saltings between Theddlethorpe Dunes and Donna Nook. They are typically wide-ranging and difficult to locate but may be seen from the coastal footpath that leads north or south from Saltfleet Haven.

❶Titchwell/Thornham Point - Norfolk
From early December until late April, Shorelarks may be encountered feeding on the saltings between the Tern Hide and Thornham Point, along Titchwell Beach. Flocks vary in size but in good years may number as many as 50 birds. However, they are typically wide-ranging and may be extremely difficult to locate.

They also visit nearby Brancaster Beach, where they range up to 1.5 miles east of the golf clubhouse, as well as the two main areas of saltings on both sides of Holkham Gap.

❶Blakeney Point/Cley Beach/Salthouse - Norfolk
Following regular sightings in late October and November, variable numbers of Shorelarks remain to winter in the Cley area and are generally present from December until early May. They are highly mobile and can be found virtually anywhere between Salthouse Beach and the outer shingle ridge on Blakeney Point, but favour the small pools and saltings to the west of the beach car park at Salthouse, the shingle ridge east of Arnold's Marsh and near North Hide at Cley and the outer ridge of Blakeney Point.

Other sites used by wintering Shorelarks in recent years, but not annually, have included the clifftop fields north of Filey North Cliff Country Park (North Yorks), the coastal fields between Port Mulgrave and Staithes (North Yorks), Spurn Point and nearby Beacon Ponds, Kilnsea (East Yorks), the dune-grass around Benacre Sluice (Suffolk), the saltings south of Aldeburgh Beach (Suffolk), the saltings east of Bradwell Bird Observatory (Essex) and the beach east of Reculver Towers (Kent).

Richard's Pipit *Anthus novaeseelandiae*

The Richard's Pipit is an annual passage migrant to Britain which breeds in western Siberia. Up to ten are regularly recorded in spring and up to 100 in autumn, particularly between 29th September and 18th November (peaking in mid October). Rough coastal grassland is favoured and birds are most often located by their distinct and rather explosive series of notes uttered in flight. Most occur on the east coast between Shetland and Kent, but good numbers are also recorded west to Scilly. Like all passage migrants, they are prone to make landfall at virtually any site on the east coast but there are certain locations that seem to attract birds on a regular basis.

➊Spurn Point - East Yorkshire
The fields to the west of the Observatory and Warren and north of the canal are regularly visited by 1-3 Richard's Pipits in late autumn, but are generally seen only in flight as they are indirectly flushed by observers. Clubley's Field, between the Crown and Anchor pub and the canal scrape car park, also regularly attracts them. View from the coastal footpath.

➊Burnham Overy Dunes - Norfolk
The extensive dune system at Burnham Overy regularly attracts Richard's Pipits in autumn, between mid October and early November. Park sensibly on the A149 and walk north along the public footpath to the dune complex. The area known as Gun Hill (to the west of the boardwalk) and the dunes up to 500 yards west of Holkham Pines are the favoured areas.

➊Weybourne/Sheringham - Norfolk
One of the most reliable sites in mainland Britain for Richard's Pipit is the coastal fields between Weybourne Beach car park and Sheringham Golf Course. The birds appear following easterly winds in mid to late October and may remain for several days. The favoured area is the cultivated fields and short grassland to the east of the coastguard cottages.

✱St Mary's Airfield/Golf Course - Scilly
The best site by far in Britain for Richard's Pipit is St Mary's, on the archipelago of Scilly. From late September until early November, Richard's Pipits are present on almost a daily basis and offer observers unrivalled views elsewhere. They favour the prominent headlands of Peninnis Head and Giant's Castle but generally end up on the short grass of the airfield or the golf course. The rough grazing fields at Trewince and Telegraph are also ideal feeding areas and flocks of up to six birds have been recorded in easterly autumns.

Water Pipit *Anthus spinoletta*

The Water Pipit is a winter visitor and passage migrant to Britain, with a wintering population of about 150 birds. It is generally found in freshwater habitats, particularly on damp meadows, watercress beds, sewage farms and gravel pit margins. They are best observed in early spring (mid March to mid April), when the whitish breast of breeding plumage is tinged salmon or rosy-pink. However, Scandinavian Rock Pipit *Anthus petrosus littoralis* is very similarly marked in spring and may cause confusion (see an identification paper by Chris Heard in the *South East Region Bird Report for 1991* and Vinicombe *et al*, *Macmillan Field Guide to Bird Identification*, pages 159-161).

Prime Sites

In Northwest England, there are several traditional sites which in recent years have had a wintering population of 25+ birds. These numbers are boosted by further arrivals of northbound migrants in late March/early April. Wintering birds reappear in mid November.

1).On the coast, birds are most easily seen along the Dee Estuary shoreline. Ten metre tides which inundate the inner salting can bring 10-15 birds into view (often standing on floating tide wrack) near the Old Baths car park at the north end of Parkgate Parade (SJ 273 790) but beware that several hundred Rock Pipits may also be present! Nearby up to six more birds favour the stream which runs into the salting at Neston Old Quay and the waterlogged grazing marsh and pools just south of here towards Marshlands Road in Little Neston. Wellingtons are essential to follow the narrow, winding stream into the marsh for up to 200 yards (SJ 285 765) and much patience is needed to obtain good views of the birds on the ground.

2). The two best inland sites within the region are in **Greater Manchester**: at the new Hope Carr NWWA reserve southeast of Leigh with up to six birds in the 1995/96 winter (birds favour small islands on pool in front of hide at SJ 665 990) and Pennington Flash Country Park southwest of Leigh where regularly 1-2 birds feed on the spit in front of Horrocks' Hide.

In northeast England, the best site is in **West Yorkshire** at New Swillington Ings (SE 380 285). Up to three birds normally winter at the site and a further six or more visit in early spring. View from the hide overlooking Astley Lake, accessed via Fleet Lane off the A642 in Oulton, southeast of Leeds. The nearby sewage farm at Knostrop also regularly attracts 1-3 wintering birds.

The best site in the Midlands is Ditchford Gravel Pits in **Northamptonshire**, where up to six birds can be found between November and late March. They frequent the shallow edges and margins of the main pool at SP 950 692 (Sheet 153) and can be watched creeping along the edge from the footpaths. In Higham Ferrars village, find Wharf Road and continue to the end and park at SP 953 687. Continue on foot over the A45 footbridge and walk northeast along the edge of the main lake.

Up to 75 birds regularly winter in East Anglia, with coastal marshes in **Norfolk** and **Suffolk** being particularly favoured. The 'Serpentine Lakes' at Cley Marshes NNT reserve regularly attract up to twenty birds between late November and early April. The flooded meadows to the east of the East Bank and the wider meandering stretch of water immediately south of Arnold's Marsh are where they are most likely to be seen, but they may be found on the small pool outside Bittern Hide or in the roadside vegetation and margins between West Bank and the Information Centre.

The main lagoon at Titchwell RSPB reserve has recently attracted 1-4 birds in winter and these are normally found on the small islands and vegetation visible from the hides overlooking the lagoon.

The beet factory lagoons at Cantley and Wissington are also highly attractive but both sites may only be visited by permit holders.

In **Suffolk**, the best wintering site is Minsmere RSPB reserve scrape, where up to six birds may be found feeding amongst island vegetation from any of the hides overlooking the area (East, North or West hides). A marked increase in birds is noted from mid February to early April.

Water Pipits winter in much larger numbers in Southeast and Southwest England and may be encountered at a large number of wetland sites. Essex, Hampshire and Dorset are the major strongholds of the species in winter in Britain. Up to thirty birds may regularly be found between late October and mid April feeding on the remnant saltings close to the old stone barges along the River Thames Foreshore at Rainham Marsh (TQ 517 800 - Sheet 177). They can be found anywhere along the foreshore between the factories and the refuse terminal jetty as well as on the nearby silt lagoons and filtering beds. Follow the minor road south from Rainham railway station and continue to the river.

The best and most reliable site in **Kent** is the water meadows and ox-bow lake at Stodmarsh, about six miles east of Canterbury. Turn left in Stodmarsh village by the Red Lion pub and park in the designated car park. Follow the well-marked trail to the Lampen Wall and continue for about 1.25 miles to the river and adjacent floodmeadows. The birds frequent this area but can be very difficult to see in the tall vegetation. There is generally 4-8 birds present between December and mid April.

A regular site in **Hertfordshire** is Rye Meads Sewage Farm, but access is restricted to Rye Meads Ringing Group and permit holders only.

An excellent site is Stoke Water Meadows, north of Guildford (**Surrey**), where up to fifteen birds are regularly present between late October and the first week of April. Leave the A3 westbound at the A3100 turning to Guildford and turn right at the second roundabout towards Jacobs Well. Immediately after crossing the A3 road bridge, turn left into a small cul-de-sac and find a sensible place to park, avoiding residents' driveways. Follow the footpath along the east side of the River Wey and view the watermeadows on the opposite side after 800 yards (TQ 007 524 - Sheet 186).

Elsewhere in **Surrey**, the sewage lagoons at Beddington Sewage Farm attract up to three wintering birds but this site may only be visited by permit holders.

Hampshire has a number of good sites for wintering Water Pipits, including the Lower Test Marshes, Titchfield Haven NR, Farlington Marshes and Camp Farm and Brockhampton sewage farms.

The wet grassland and meadows at the Lower Test Marshes regularly attract up to 30 wintering birds between late October and late March and these can be viewed from the road bridge or the footpath by the A35 at Totton.

The wet meadows at the north end of the Meon river valley are the best area in which to see Water Pipits at Titchfield Haven. Up to twelve are normally present and these can be viewed from the public footpath that runs south from Titchfield village at SU 542 055 (Sheet 196). The main lagoon at Farlington Marshes NR regularly attracts 1-3 wintering Water Pipits and these may be located in the grass and vegetated margins to the north of the lagoon.

Brockhampton Sewage Farm lies to the south of Havant and usually attracts 1-5 Water Pipits during December to March. Park in the raised car park at SU 708 050 and 'scope the sewage farm pools and complex to the north.

Camp Farm sewage works is in Aldershot at SU 884 527. View from the minor road that runs alongside. Up to five birds are generally present in winter.

Like Hampshire, **Dorset** also has a substantial wintering population of Water Pipits and these may be seen at a number of localities along the south coast. Probably the best site is Lodmoor CP, immediately east of Weymouth and north of the coast road. The birds favour the flooded meadows and shallow margins on the reserve and may be watched from the numerous hides overlooking the area. Park by the entrance to the refuse site along the A353 at SY 687 808 and explore the reserve on the right.

Equally as attractive as Lodmoor is Waddock Farm cress beds, north of the River Frome, near Moreton, where up to six birds are generally present between November and March. Take the B3390 between Affpuddle and Warmwell and park by Waddock Farm at SY 798 909 (Sheet 194). The extensive cress beds lie immediately to the west of the road.

The sewage works at Wareham also attract up to three birds in winter and Stanpit Marsh, in Christchurch Harbour, occasionally provides sightings on Priory Marsh.

In Southwest England, one of the better sites is Chew Valley Lake (**Avon**), where regularly up to twelve birds winter between November and mid April. Although singles are normally to be found along the north and east shores, the favoured area is the long grass and pool margins around the south end of the lake and are best seen from Stratford Hide. At dusk, small numbers can often be seen high over Herriott's Bridge as they come into roost in the main reedbed. In March and early April, they often feed in a large grassy field at Moreton Bank on the west side of the lake, visible from the tarmac footpath between Stratford and Moreton Hides. A flock of up to a dozen birds is often present, with many acquiring their distinctive summer plumage. In recent years, this spring flock has regularly commuted to Heron's Green Pool, to the west of the car park on the B3114.

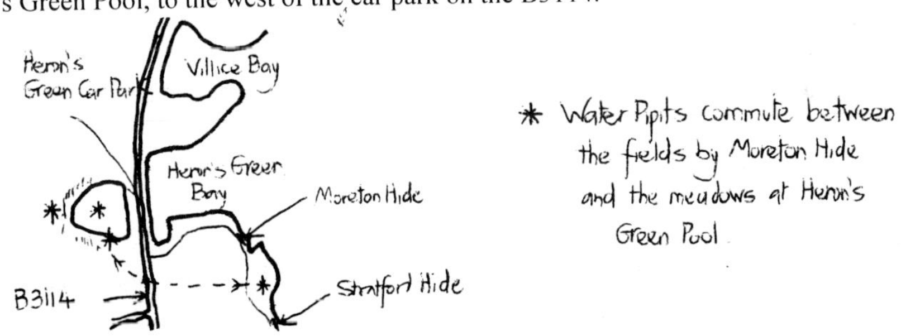

In **Devon**, there are two main wintering sites: the Otter Estuary and Warleigh Marsh.

The Otter Estuary lies to the east of Buddleigh Salterton and can be accessed from the car park at the eastern end of the seafront. Follow the raised estuary bank northwards to the view the boggy meadows and reedbed area north of the cricket pitch. Up to twelve birds are normally present.

Warleigh Marsh is at SX 465 625 (Sheet 201), to the west of the River Tavy. Leave the northern roundabout of the A386 at Roborough and follow the maze of minor roads to Horsham village.

Water Pipit is scarce in **Cornwall** but may be found in small numbers in winter at the east end of Copperhouse Creek, Hayle (on marsh behind Leo's superstore), at Marazion Marsh, at Landulph Marsh, on the west side of the River Tamar, and at Walmsley Sanctuary, on the River Camel.

Waxwing *Bombycilla garrulus*

The Waxwing is an annual winter visitor to Britain from its breeding grounds in northern Scandinavia. Birds appear in the Northern Isles and at various sites along the east coast south to Suffolk from late October onwards and by mid December, the movement has largely ceased. The numbers reaching Britain is highly variable, and is subject to occasional irruptions (about once every eight years), and may vary from less than fifty birds to an influx of over 12,000 birds. Waxwings, like many irruptive species, are highly unpredictable and it is not possible to detail any regular sites in which they may be reliably seen. When present, they are usually seen feeding on berry-bearing bushes, especially *Cotoneaster*, and regularly visit suburban gardens. Aberdeen and various villages along the north Norfolk and Suffolk coasts are the most regularly visited.

Dipper *Cinclus cinclus*

The Dipper is a fairly common resident of fast-flowing rivers and streams in the upland regions of Scotland, Northern England and Wales, with isolated populations in southwest England and the north Midlands. Although resident, they are less easy to locate during winter and are often to be found wintering at lower levels and often where rivers meet the sea. The following sites are a random selection of some of the better known and most reliable in a particular region or area, but they are only a small sample of the many haunts known to be frequented by this species.

Prime Sites

❶Findhorn Valley - Inverness-shire

The River Findhorn that runs parallel with the minor road to Coignafearn Lodge holds several pairs of breeding Dippers between March and September. Leave the A9 at Tomatin and head west at Findhorn Bridge towards the lodge. Park at suitable looking sections of river and search.

✱Carrbridge - Inverness-shire
The River Dulnain that runs through Carrbridge is a prime site for Dipper and these can be seen from the road bridge by the B9153 at the north end of the town. They are often downstream of the bridge or upstream of the waterfall to the west of the road.

✱Dulnain Bridge - Inverness-shire
Several pairs of Dipper survive on the stretch of river that runs through the village. Park by the bridge at the junction of the A95 and A938/9 and check both sides or view from the car park on the A938 about half a mile west of the junction.

Nearby, the bridge over the River Spey, on the road to Nethy Bridge from the A95, should produce sightings.

❶Aviemore area - Highland
The River Druie has several pairs of nesting Dipper and these can be seen from the various bridges crossing the river north of the B970 Cairngorm Ski-Lift road. The entrance to Inverdruie Fish Farm regularly produces sightings.

Further south on the B970, the roads either side of the river at Glen Feshie will produce Dipper, as well as the riverside walk from Tolvah village.

Nearby, several pairs nest on the river at Glen Tromie. Leave the B970 at Drumguish and follow the minor road upstream.

❶Glen Tanar - Grampian
Dippers can be commonly encountered on the river running through Glen Tanar, especially on the section near the House. Leave the A93 at Aboyne and drive south across the River Dee. Turn west on the B976 and turn left towards Braeloine in Bridge o'Ess. Continue southwards to Glen Tanar House and park. Follow the well-marked trail to the river and explore upstream.

❶Angus Glens - Tayside
The glens of Isla, Prosen, Clova and Esk all have populations of breeding Dippers in summer and these may be seen from the numerous roads that run alongside these sites. Glen Isla is probably the better of the glens, also offering a wide range of other moorland and crag nesting species. However, Glen Clova is the easiest to access and is reached via the B955 from Kirriemuir.

✱Glen Devon - Central
The A823 between Gleneagles and Burnfoot passes through Glen Devon and the fast-flowing river here holds several pairs of Dipper. Park at the various bridges and roadside stops and check the river, particularly the section up to two miles north of Burnfoot.

❶Tillicoultry Glen - Central Region
Tillicoultry Glen is an open valley with a fast-flowing stream running through it and this regularly holds at least two pairs of Dipper. Tillicoultry lies east of Stirling on the A91. Park at NS 914 975 (Sheet 58) and check the burn.

❶The Hirsel - Borders
Dippers can be found on the River Leet, between Hirsel House and the bridge over the minor road leading east from the A697. The main entrance to the Hirsel is signposted off the A698 Kelso to Coldstream road, on the outskirts of Coldstream. An obvious footpath leads from the country house alongside the river.

❶Falls of Clyde SWT Reserve - Strathclyde
Dippers may be seen by the gorge and waterfalls at this site. Park at NS 882 425 (Sheet 71) and explore.

➊ Newton Stewart area - Dumfries and Galloway.
The A714 north of Newton Stewart and the minor road that runs alongside the east side of the fast-flowing river are very good sites from which to see Dipper. The Wood of Cree RSPB reserve, immediately north of Minnigaff, is also very good.

✱ Brothers Water - Cumbria
Dippers can be easily encountered between late March and August from the car park at the north end of Brothers Water.

➊ River Dane - Cheshire
Along the River Dane, upstream of Congleton, particularly the stretch between Hugbridge on the A523 (SJ 930 637) and Danebridge (SJ 965 652), a concentration of at least twelve pairs of Dippers can be found. The birds can be viewed from the public footpaths which skirt the river at various points.

➊ Abergele - Clwyd/Llanfairfechan - Gwynedd (North Wales)
In winter, Dippers can regularly be seen at the mouth of the river Gele (just follow the stream inland from Abergele Beach car park for 200 yards) and at the mouth of the Afon Ddu (just look upstream from the bridge at the west end of Llanfairfechan Promenade).

Elsewhere in this region, Dippers can be seen with relative ease in summer from the Greenfields Heritage Park, near Holywell (**Clwyd**), the bridge near the woollen mill in Trefriw (Conwy Valley, **Gwynedd**) and the bridge near the visitor centre at Loggerheads Country Park (near Mold, **Clwyd**).

✱ Tregarth Bridge, near Bethesda - Gwynedd
A very reliable site near Bangor is the Afon Ogwyn, about two miles north of Bethesda. Turn left at the Rachub crossroads and park by the youth hostel and bridge. Check the river on both sides of the bridge.

✱ Aber Falls NR - Gwynedd
The fast-flowing river that runs from Aber Falls to Aber beach has a population of at least three pairs of Dipper. In winter, they may be seen at the mouth of the stream where it meets the shore but in summer they may be anywhere from just below the car park to 300 yards downstream of the Falls. Leave the A55 at Abergwyngregyn and follow the minor road by the garage up towards the falls. After half a mile, a car park is on the right.

➊ Vyrnwy Confluence - Montgomeryshire
Dippers are regularly encountered at the Vyrnwy Confluence on the Shropshire border. Suitable areas of river may be obtained from the minor road which leaves the B4393 at Crewgreen for Melverley and from a footpath that leads north from Pentre Farm (SJ 345 156).

➊ Wooler area - Northumberland
The Wooler Water between Haugh Head (on the A697) and Doddington Bridge (on the B6525) holds at least six Dipper territories and these can be seen from the various vantage points and footpaths skirting the river.

➊ Allendale - Northumberland
The River East Allen between Spartylea and Allendale Town normally attracts up to six pairs of Dipper in summer. They can be seen from various points along the B6295 as well as from the smaller roads to the west.

Elsewhere in **Northumberland**, up to six birds frequent the River Blyth, between Hartford Bridge (on the A1068 south of Bedlington) and Bebside (near the A189 at the Blyth bridge), three pairs nest at Cupola Bridge and several birds can be found in the Lilburn Burn and Threestoneburn area west of the A697, five miles south of Wooler.

River Wharf - West Yorkshire

The River Wharf between Barden Bridge and Ilkley holds a healthy population of Dippers which can be seen with relative ease. The B6160 passes close to the river in numerous places and Barden Bridge and Bolton Abbey both have nesting pairs in summer.

Sowerby Bridge - West Yorkshire

The River Calder between Sowerby Bridge and Copley, south of Halifax, is a well-known and reliable site for Dipper. The A6026 straddles the river for much of its path and the area around Copley is particularly productive.

Monsal Dale - South Yorkshire

Good numbers of Dippers frequent the River Wye between Miller's Dale and the A6 at Ashford in the Water. Leave Ashford on the B6465 and turn left after one mile towards Cressbrook. Dippers can often be seen in the stretch of river alongside the road or from the footpath that leads through Monsal Dale.

Hope Valley - South Yorkshire

Up to five pairs of Dippers are known to frequent the Hope river valley between Castleton and Brough. They can be seen from the bridge just south of the staggered crossroads in Hope village or from the footpath running alongside the river.

River Derwent - Derbyshire

A total of 45 pairs of Dippers are known to be present on the River Derwent and its tributaries between Ambergate (three miles north of Belper on the A6) and Leadmills at SK 234 806. This is a distance of about 32 miles but birds may be seen at numerous suitable-looking sites along the A6.

Hartington Mill/Dove Dale - Derbyshire

The River Dove between Hartington Mill and Dove Dale holds large numbers of wintering and breeding Dippers and these can be seen from the footpath running alongside this stretch of river. Milldale will produce sightings from the road as will the minor road to Dove Dale Castle and car park. The area is northwest of Ashbourne.

Coombes Valley RSPB reserve - Staffordshire

At least one pair of Dippers nests at Coombes Valley RSPB reserve and this site can be reached from the A520 Leek to Stone road. Turn left after three miles on a minor road to Cheddleton Station and continue towards Basford Green. Turn left up the narrow and very steep lane and the reserve is on the right after 1.5 miles.

They can also be seen nearby on the River Churnet, at Rakes Dale and at Tittesworth.

Dowles Brook - Worcestershire

Up to three pairs of Dipper nest at Dowles Brook, in the Wyre Forest. Leave Bewdley on the B4194 towards Kinlet and after about a mile the road crosses the brook. Park sensibly on the left and follow the footpath westwards along the old disused railway line. This takes you through the Fred Dale NR, crosses over the brook and then opens out onto Dry Mill Lane. This follows the brook for about a mile to Knowles Mill.

Bond's Mill, Stonehouse - Gloucestershire

A pair of Dippers have been regularly seen at Bond's Mill (SO 814 043 - Sheet 162) for about five years, although January to April appears to be the best period for sightings. Leave the A419 in Stonehouse (immediately west of Stroud) towards King's Stanley. The Mill should be on the left after 300 yards.

❶Nag's Head RSPB reserve - Gloucestershire
A pair of Dippers regularly nest on the stream that flows through Nag's Head reserve and can be seen with patience throughout the summer. Turn north off the A48 at Lydney on the B4234 to Parkend. Turn left onto the B4431 and on the western outskirts of the village, a discreet blue RSPB sign directs you to the start of the track to the reserve car park and information centre.

The stream at nearby Cannop Ponds (two miles north on the B4234) also has breeding Dipper.

❶Nant Irfon - Breconshire
Dippers can be reliably and fairly easily seen on the river that runs through the Irfon Valley. Leave the A483 in Llanwrtyd Wells by the unclassified road signposted Abergwesyn. There are several Forestry Commission picnic areas close to the river, and these provide good vantage points. Beyond the village, there are several further good pull-off points.

❶Talybont Reservoir - Breconshire
The minor road that leads north from Merthyr Tydfil meanders up the very beautiful and scenic Taf Fechan. After passing Pontsticill and Pentwyn Reservoirs, the road drops down a steep decline to Talybont. Dippers can be found on the river to the right of the road and the reservoir itself is an excellent site for breeding Goosander.

❶The Rheidol Valley, Aberystwyth - Cardiganshire
The Rheidol Valley that runs east from Aberystwyth is an excellent site in which to see Dipper. Take the minor road to Capel Bangor railway station and explore the river at various vantage points.

❶Dinas Reserve - Carmarthenshire
The Dinas RSPB reserve and Gwenffrwd reserve in the Upper Tywi Valley hold good populations of breeding Dippers. Leave Llandovery on the minor road to Rhandirmwyn and bear right to Dinas Reserve at SN 788 472 or turn right past the bridge to Cwrt-a-Cadno and Gwenffrwd at SN 749 460. The fast-flowing river can be checked from various vantage points. Red Kite, Wood Warbler and Pied Flycatcher can also be seen from late April to August.

❶Cwm Clydach -West Glamorgan
The Cwm Clydach holds several pairs of Dippers along its stretch and these can be seen from the RSPB reserve or from the minor roads that traverse the river in the area. Leave the M4 at Junction 45 (north of Swansea) and head north on the B4603 to Clydach. Turn left at the main crossroads in the centre of town and follow the minor road through the valley to the reserve car park at SN 682 053. The reserve is open at all times.

❶The By Brook - Wiltshire
The By Brook between Box and Ford, west of Chippenham, attracts at least three pairs of Dipper in summer and these can be seen from the small bridge on the road between Box Hill and Colerne or from the minor roads near Slaughterford village. Parts of the river may be viewed from public footpaths.

❶Warminster area - Wiltshire
Up to five pairs of Dipper breed on the River Wylye between Longbridge Deverill and Norton Bavant. There are numerous vantage points in which to view the river including the B3095 and Norton Bavant village, the B3095 at Boreham, the A350 and the minor road at Crockerton Green.

➊Pensford area - Avon
The River Chew between Chew Magna and Keynsham has a population of at least three pairs of Dipper and these may be seen in summer from Pensford Bridge (on the A37 north of the village), Publow and Woollard Bridge and in Compton Dando. Pensford lies five miles south of Bristol.

Elsewhere in Avon, Midford Brook is a reliable site. This is the stretch of river that runs upstream of Midford, towards Monkton Combe. Midford lies south of Bath on the B3110.

➊River Tone, near Wellington - Somerset
The River Tone between Bradford on Tone and Payton regularly attracts 3-4 Dippers in summer and these may be seen from the minor roads that cross the river at various points or from the public footpaths that lead along the riverbank.

➊Maiden Newton - Dorset
One of the best sites in Dorset is the River Frome, between Maiden Newton and Toller Fratrum (about six miles NW of Dorchester on the A356), where up to two pairs are regularly present. They range from the mill in Maiden Newton to the bridge in Toller Fratrum and are best seen between April and August.

✱Lynmouth - North Devon
An extremely good site for Dipper is the East Lyn river between Lynmouth and Watersmeet. Birds are resident at this site and it is not unusual to find up to eight birds on this short stretch of river in January. The A39 traverses the river along the entire section and a public footpath on one side allows easy viewing.

➊Dunsford - Devon
The River Teign between Steps Bridge and Clifford Bridge regularly attracts Dippers and these may be seen from the footpath that runs north through Cod Wood alongside the river. Park by the B3212 west of Dunsford.

➊River Dart, Buckfastleigh - Devon
At least two pairs of Dipper breed on the River Dart, between Austin's Bridge and Staverton Bridge. The river can be checked from a few vantage points to the south of the A384 (Buckfastleigh to Dartington road).

➊Bodmin area - Cornwall
The Dipper is relatively common on the River Fowey and its tributaries and may be encountered at a number of localities. Traditional sites include Ninestones, Draynes, Lamelgate, Respryn, Pantersbridge, Ashford Bridge and Mount. The river runs alongside the A38 between Dobwalls and Bodmin and this section may also provide sightings, but to explore, the minor roads to the north of the A38 have to be traversed.

To the NW of Bodmin, the tributaries of the River Camel are also productive, including Polbrock Bridge and Hoopersbridge.

➊Idless, Truro - Cornwall
A traditional site for Dippers is the River Allen at Idless. Leave Truro (the B3284) northwards on a minor road to the village and check the section between the bridge and the village, as well as the visible sections of the River Kenwyn to the south.

Nightingale *Luscinia luscinia*

The Nightingale is a fairly common summer visitor to Britain which is found primarily in the woodlands and dense scrub of East Anglia and southeast England. It is an incredibly skulking species, but is best located by its very loud, guttural song, uttered from mid April until June.

Prime Sites

✱Whisby Pits - Lincolnshire
The woodlands and dense vegetation surrounding Whisby Pits attract up to 24 singing male Nightingales in May. The pits lie to the southwest of Lincoln and are best accessed from the minor road that leads west from the A46, about half a mile north of its junction with the A1434. The Nightingales flourish in the wood north of the railway and can be seen and heard from the railway bridge.

❶Bourne Woods - Lincolnshire
A population of some 10+ singing male Nightingales is present in Bourne Woods, west of Bourne on the A15, from late April until August. Leave the A15 at Cawthorpe and drive to the far end of the road and park. A number of footpaths lead into the wood from here.

Nearby, Temple Wood is also highly productive, with up to eight males in spring. Leave the A15 at Aslackby and drive westwards to the end of the road and park. Follow the main footpath into the wood.

❶Barrowden area - Leicestershire
A healthy population of Nightingales can be found in the mature woodlands and commons in the Barrowden area, southwest of Stamford. Luffenham Heath (SK 959 020), Barrowden Fox Covert (SK 962 023) and Coppice Leys (SK 955 013) all support singing birds in May and June and can be accessed by public footpaths.

Barnsdale Wood lies at SK 909 089 and regularly attracts five or more singing Nightingales in May and June.

❶Glapthorn Cow Pastures - Northamptonshire
The small reserve at Glapthorn regularly attracts 6-8 singing Nightingales in May and June and these can be heard from the numerous footpaths running through the site. Leave Oundle on a minor road north-westwards to Glapthorn village. Continue west and after one mile, the reserve is on the right.

Nearby, Fermyn Woods and Souther Wood (east of Brigstock) regularly attract up to four singing males, and Fineshade Wood (east of the A43 two miles south of Duddington) three birds.

❶Kineton - Warwickshire
The woodlands near Kineton (10 miles south of Royal Leamington Spa at SP 360 510) are the best site in the Midlands in which to see or hear Nightingale. There are normally up to six singing males present in May and June.

❶Waterloo Thorns, Sandy - Bedfordshire
Up to five singing Nightingales can be found in Waterloo Thorns, near Tempsford, east of the A1. Leave the A1 at Tempsford on the Potton/Everton road and after a mile park on the left and walk north to the small woodland.

Elsewhere in Bedfordshire, Nightingales can be found at Home Wood, Northill (SW of Sandy) and in the old pits area of Priory Country Park, immediately east of Bedford (accessed from Barkers Lane off the A428).

✱Calvert - Buckinghamshire
A small population of Nightingales survives at Calvert, south of Steeple Claydon. The entrance to the BBONT reserve lies on the Edgcott-Gawcott road adjacent to Calvert Lake at SP 682 252 (Sheet 165), the Nightingales singing from the dense scrub and vegetation between the railway line and the lake. May is the best month in which to visit.

Elsewhere in Bucks, Nightingales can be found at Finemere Wood and in Salcey Forest.

✱Highnam NR - Gloucestershire
Up to 26 singing males are regularly recorded at Highnam NR, west of Gloucester. Leave Gloucester on the A40 and after two miles, the woods and reserve are to the north of the road. A number of footpaths traverse the area.

Good numbers can also be found in the Wormington area and at Frampton on Severn.

✱Salthouse Heath - Norfolk
Nightingales can be relatively easy to see on Salthouse Heath from late April until late May. Leave the main A149 coast road in Salthouse village at the Dun Cow public house and drive south inland. At the crossroads, go straight across and then park sensibly near the second crossroads. The birds can be found in this area, activity and singing being mostly undertaken prior to 10.00am and after 6.00pm.

Several more singing birds can be located on Muckleburgh Hill, immediately west of Weybourne. At least two birds have territories bordering the road at this site and others may be heard along the various tracks that lead from the road.

❶Barnhamcross Common - Norfolk
Barnhamcross Common lies on the Norfolk/Suffolk border, immediately south of Thetford, and in recent years has attracted at least six singing Nightingales in May and June. Access is fairly straightforward, with most birds favouring the river valley on the east side of the A134.

Elsewhere in Breckland, Thompson Common holds a healthy population of singing birds. Park by the A1075 near Stow Bedon and explore the woodland westwards towards Thompson Water on the various footpaths thereabouts.

✱Minsmere RSPB Reserve - Suffolk
The RSPB reserve at Minsmere regularly attracts up to 35 singing male Nightingales between late April and early June, with the vast proportion in the open scrub either side of the nature trail between the information centre and Island Mere hide. The reserve is well signposted from Westleton village.

Equally impressive numbers can also be found in the Walberswick, Dunwich Common and Blythburgh areas, particularly in dense vegetation and thickets.

North Warren RSPB reserve, on the B1122 north of Aldeburgh, also attracts up to fourteen singing males.

✱Wolves Wood RSPB reserve - Suffolk
Recent coppicing and suitable management of Wolves Wood has seen the population of singing Nightingales rise from four to 18 birds in just ten years. The reserve lies on the A1071 Hadleigh to Ipswich road, 1.5 miles east of Hadleigh, and is well signposted.

✱Little Paxton GP - Cambridgeshire
The thickets and dense vegetation bordering the gravel pits to the east of Little Paxton are an excellent haven for Nightingales, with 12-15 singing birds regularly present in May. Leave the A1 southbound at Little Paxton and take the first left towards the nature reserve. Continue eastwards through the village to the designated car park and explore the trails at the south end of the reserve, especially that which runs up the west side of the main pit complex.

Elsewhere in Cambridgeshire, the local nature reserve at Woodwalton Fen, east of the A1 northwest of Huntingdon on the B1090, is very good (access by permit only), as is Ash Wood, Perry, and the thickets along the SW shore of Grafham Water and Castor Hanglands, to the NW of Peterborough (on the minor road between Ailsworth and Helpston).

✱Fingringhoe Wick - Essex
One of the best sites in Britain for Nightingale is Fingringhoe Wick NR, where up to 40 singing males may be present from late April until June. Leave Colchester on the B1025 and after three miles, turn left towards Fingringhoe village. Follow South Green road and continue east towards the reserve entrance. The reserve is closed Mondays.

Nearby, Fridaywood, south of Colchester, has a population of at least 20 singing males. These tend to congregate in the riverside vegetation along the Bounstead Brook and the Roman river valley, as well as the woodland and coppices surrounding Park Farm. Park at Fridaywood car park east of Bounstead Road, between Layer-de-la-Haye and Maypole Green.

Danbury Common, south of Danbury, also has at least 20 singing males in May and June. Park along the minor road south of Horne Row and explore the common from the maze of footpaths.

●Epping Forest - Greater London/Essex
The woodland around Connaught Water and the edge of the Chingford Plain is good for singing Nightingales in spring. Leave the A104 at the A1069 junction and park by Connaught Water a short way down on the right. The Nightingales can be found along the river valley and the coppices within this area.

✱Blean Woods, Canterbury - Kent
A staggering population of at least 50 singing males can be found in Blean Woods, west of Canterbury. The best area is that around Church Wood RSPB reserve, accessed and signposted from the minor road at Rough Common.

Nearby in the Stour Valley, singing birds can be found at Stodmarsh (in the area around the car park boardwalk and the start of the Lampen Wall and in nearby Trenleypark Wood) and along the riverside walk at Fordwich.

●New Hythe/Eccles - Kent
A large population of up to 40 singing males can be found around the complex of gravel pits and riverside vegetation north of New Hythe and around Eccles. Holborough Marshes is also very good.

✱Fagg's Wood, Hamstreet - Kent
In south Kent, one of the best sites for Nightingale is Fagg's Wood, to the west of Hamstreet, in Orlestone Forest. Leave Hamstreet northwards on the A2070 and turn left at the first crossroads. Turn left after half a mile and then park in the designated car park on the right. Follow the footpath south of the car park where up to ten Nightingales should be heard during early morning and evening.

●Woods Mill, Henfield - East Sussex
The thickets and dense vegetation on Oreham Common and neighbouring Woods Mill and Hoe Wood, east of the A2037 and south of Henfield, are very good sites for Nightingale. Park at Woods Mill and explore from the road or from the public footpaths in the vicinity.

Elsewhere in Sussex, Beauport Park, northwest of Hastings (between the A21 and A2100) is a good site, as is the roadside vegetation along the minor road between Ditchling and Streat (east of Keymer) and Abbots Wood, near Hailsham (along minor road to Arlington).

❶Pingewood and Crookham Common - Berkshire
Crookham Common, south of Thatcham (SU 523 643), regularly attracts several singing Nightingales between late April and May and these can be seen from numerous footpaths running along the river and over the common. Leave the A4 at the eastern Thatcham roundabout and continue south past Thatcham station along the minor road to the common. Pingewood lies to the north of Burghfield and regularly attracts up to twelve singing Nightingales in spring.

Dinton Pastures Country Park (SU 780 725), north of Winnersh on the B3030, is also very good, particularly in the scrub bordering Black Swan Lake.

❶Bookham Common - Surrey
Bookham Common, to the west of Leatherhead, is one of the most reliable sites in which to see or hear Nightingales in Surrey. The area is extensive, but best accessed from the footpath east of Bookham railway station.

✱Flagpond Copse, Botley Wood - Hampshire
A large population of at least fourteen singing birds can regularly be found in Flagpond Copse, north of Fareham. Leave the A27 northwards on a minor road at Titchfield and head north towards Wickham. Park at SU 552 098 (Sheet 196) and follow the public footpath to as far as the pylons cross the wood.

❶Combley Great Wood - Isle of Wight
The most reliable site on the Isle of Wight is Combley Great Wood, where up to five singing Nightingales may be recorded in May. Park at SZ 536 884 and walk east to view the wood.

❶Clangar Wood - Wiltshire
Up to six singing males may be found at Clangar Wood in May, singing from deep thickets within the woodland. Leave Trowbridge on the A363 and join the A350 at North Bradley. Continue south for about a mile before turning left towards the woodland. A small car park is set back from the road and a wide track leads from here through Picket Wood to Clangar Wood to the east.

✱Inglestone Common - Avon
Inglestone Common is a superb site for Nightingales with as many as 13 pairs present during a good season. From the centre of Wickwar, take the B4060 north towards Wotton-under-Edge, and then fork almost immediately right down a narrow lane, signposted to the common. Follow this lane for about a mile to where it crosses a cattle grid on to unfenced common land with scattered houses and farms. For the next mile, the road skirts the main block of woodland and Nightingales thrive in the dense thickets hereabouts.

❶West Sedgemoor/Shapwick Heath - Somerset
Small numbers of singing Nightingales may be heard at West Sedgemoor and Shapwick Heath in May. At West Sedgemoor, the Nightingales frequent the deciduous woodland that clothes the steep escarpment to the south. Leave Langport on the A378 and about two miles west of Curry Rivel, turn right to the RSPB reserve at Swell Wood. A clearly-marked woodland trail allows full exploration. Shapwick Heath lies to the west of Glastonbury and is accessed from Ashcott Corner, on the minor road south of Meare.

✱Lydlinch and Deadmoor Commons - Dorset
Lydlinch Common is possibly the best site in Dorset for Nightingale, the site being easily accessed from the A357 between Lydlinch and the junction with the A3030 Sherborne road, and from the A3030 between the above junction and Brickles Wood. Deadmoor Common can be reached from Woodrow near Fifehead Neville by taking the footpath northwest at Elm Tree Cottage and Hambledon, and then across the field and the stile to the wood towards Holbrook Farm. Good numbers can also be found on the Lulworth Army Ranges, but this area is largely restricted from access.

Bluethroat *Luscinia svecica*

The Red-spotted Bluethroat is a regular passage migrant to Britain, with up to 125 recorded in spring (400 in an exceptional year) and 40 in autumn. Most are found in the Northern Isles or on the east coast north of Norfolk in spring and in the Northern Isles and along the south coast west to Scilly in autumn. The optimum period for occurrence is from mid to late May in spring and from mid September to mid October in autumn.

The White-spotted Bluethroat also visits Britain in small numbers, but much earlier in spring and primarily along the south and east coasts north to Yorkshire. They occur from mid March to mid April, with rarely more than four sightings per spring.

Both forms favour dense vegetation in ditches or reedbeds, but may be found in dune scrub and *Sueda* bushes on migration. They are very skulking and often very difficult to detect as they run about the ground.

Like all scarce migrants, Bluethroats can be recorded virtually anywhere along British coasts under suitable conditions, but a number of sites continue to record them on a regular basis. They are best seen in spring following easterly or north-easterly winds, and in autumn following south-easterly winds. Perhaps the best areas are the Shetland mainland and Fair Isle, where large falls of birds may be found in late May and early June, and in Orkney (mainly on North Ronaldsay and Stronsay). Further south, Fife Ness and the Isle of May attract large numbers of birds, and under the same easterly conditions, the Farne Islands, Holy Island and St Abb's Head are highly productive.

Away from these areas, prime sites include Filey North Cliff Country Park, Flamborough Head and Spurn Point (East Yorks), Donna Nook/Theddlethorpe Dunes and Gibraltar Point NR (Lincs), Holme Dunes and Blakeney Point (Norfolk) and Minsmere Sluice Bushes and Landguard NR (Suffolk).

Along the south coast, the top fields, Observatory Quarry and hut fields at Portland Bill (Dorset) are regularly visited in September and St Agnes and St Mary's (Scilly) in October.

Cetti's Warbler *Cettia cetti*

The Cetti's Warbler was first identified in Britain in 1961 and had even bred in the country (in Kent) by 1972. The population rapidly grew to a peak of over 300 pairs in 1984 but a succession of severe winters in eastern England saw numbers plummet and the population has since shifted its nucleus from south-east England to south-west England. A total of 388-409 singing males were located in 1993 suggesting a renewed population expansion.

Cetti's Warblers favour dense scrubby margins of freshwater areas, including reed swamps and reedbeds, rivers and gravel pits, and are extremely difficult to see. Their presence is often only detected by their powerful and rather explosive burst of song, uttered throughout the year and at all times of day. The British population is resident and birds may be located throughout the year.

The current breeding range of Cetti's Warbler is restricted to southern England, with birds extending north to Norfolk and Suffolk and Warwickshire, and west from Sussex and Buckinghamshire to Cornwall and Pembrokeshire. A number of the most reliable sites are detailed below.

❶Brandon Country Park - Warwickshire

A pair of Cetti's Warblers has successfully reared young at Brandon CP over the past two years and it is possible to see these birds from the main nature trail that runs through the reserve. The birds favour the thickets and scrub bordering Central Marsh and are best seen from mid April to July. Access is by permit only and this can be obtained from the information centre adjacent to the car park. The reserve lies on the south side of Brandon Lane, three miles SE of Coventry, and is best approached from the A45 Northampton to Coventry road. Entry to Brandon Lane is from the eastbound carriageway only and is almost immediately on the left. The entrance to the reserve is on the right after about a mile.

Elsewhere in the Midlands, the only regular sites are in **Northamptonshire**, where single birds have been present intermittently at Ditchford GP (in the dense area of undergrowth and scrub bordering the south end of the main pit) and Stortons GP, to the west of Northampton.

✱Strumpshaw Fen - Norfolk

Strumpshaw Fen RSPB reserve is one of the best sites in Norfolk to see Cetti's Warbler, with a population of up to six birds. They favour the reedbed and scrub adjacent to the hide by the main entrance as well as various suitable sites along the riverside trail. The reserve is situated about ten miles east of Norwich. Leave the A47 at Brundall roundabout and continue east through the village and under the railway bridge. Turn first right towards Buckenham and the RSPB reserve is on the right after 400 yards.

On the opposite side of the River Bure, Rockland Broad is also very good for Cetti's Warblers. Park by the boat moorings east of Rockland St Mary and walk along the east side of the river to the broad. There are several territories bordering this footpath.

Surlingham Broad, to the west, is also populated, and birds here can be seen from the footpath and boardwalk that border the eastern side of the broad.

❷Oxwich NR - West Glamorgan

Oxwich Marsh is a massive area of reed swamp on the Gower Peninsular which, since 1977, has held an increasing population of Cetti's Warblers. Leave Swansea on the A4118 and after seven miles, turn left (south) to Oxwich village. The road cuts through the freshwater marsh and eventually comes out at the reserve car park and information centre. A footpath leads from here along the edge of the marsh. The Cetti's Warblers can be seen and heard from the road as well as from the trail.

➊Teifi Marshes - Pembrokeshire

An increasing population of Cetti's Warblers can be found in the fen-carr bordering the west shore of the Teifi river valley, between Cardigan and Cilgerran. A minor road from behind Cardigan Castle takes one alongside the marshes and birds can be heard from here. Several footpaths also traverse the area and may provide views of this particularly skulking species.

In nearby **Carmarthenshire**, Cetti's Warblers have been recently recorded in typical fen-carr habitats at Witchett Pool (SN 285 075), Dyfatty Marsh, Burry Port (SN 457 009) and Ffrwd Fen (SN 420 025).

➊Thorney Deeps - West Sussex

Up to ten singing birds frequent the extensive reedbed bordering the sea wall on the west side of Thorney Island. Park to the south of Prinsted Yacht Basin and walk west along the public footpath to the sea wall. Continue south to the northern channel and listen. Birds will be regularly heard here but rarely seen but in April and May, breeding activity may force birds to the outer periphery of the reedbed and scrub.

1-2 singing males are also present at a number of other Sussex localities, Pulborough Brooks RSPB reserve possibly offering the best chance of a sighting.

✱Titchfield Haven NR - Hampshire

Titchfield Haven is one of the best sites in Britain for Cetti's Warbler with an estimated population of at least 26 singing males. Birds can be seen (or heard) from various parts of the reserve but access is strictly by permit only (obtainable from the reserve centre immediately east of the Meon Bridge in Hill Head). Birds can also be seen in the deep channel that runs alongside the western boundary of the reserve, walking north from the coast along the public footpath leading from the narrow bridge on the minor road between Hill Head and Titchfield village. Park by the toilet block adjacent to the caravan site along the Haven beach and walk north for about half a mile.

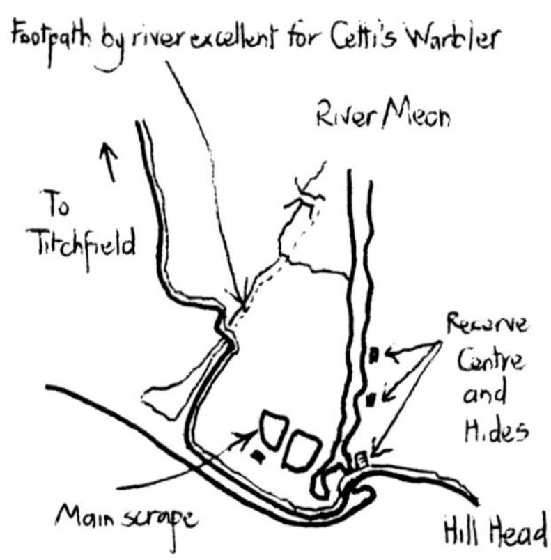

The Lower Test Marshes, north of the A35 Totton bridge, also have a thriving population of at least 14 pairs, and these may be recorded from the main footpath that traverses the area from Testwood village.

➊Alresford Pond - Hampshire

Alresford Pond is a large lake surrounded on three sides by reed swamp, and on its remaining shore by Alder and Willow carr. This latter area has recently attracted up to seven singing Cetti's Warblers and these can be heard from the B3046 (Broad Street) which borders the lake. Leave New Alresford north on the B3046 and the lake is on the right after several hundred yards.

➊Brading Marsh - Isle of Wight

The most reliable site for Cetti's Warblers on the Isle of Wight is Brading Marsh, an extensive and private area of marshland, rough pasture and drainage ditches and streams to the west of Bembridge. There are several public access points to this area including footpaths (a) from Bembridge Windmill west along the Bembridge Trail past Bembridge Airport to Centurion's Hill, and then north-west to Wall Lane in Brading; b) from the B3395 west of Bembridge Airport; c) from the end of Wall Lane in Brading and d) from the end of Vicarage Lane in Brading north-east across fields to the B3330 just west of St Helens at Carpenters Farm.

✱Westhay Reserve - Somerset

Westhay Moor, NW of Glastonbury, is an excellent site in which to see Cetti's Warblers, with several pairs nesting on the RSPB reserve to the north of Westhay village. Park in the main reserve car park off the B3151 and follow the main footpath to the hides, South Hide in particular attracting several birds in the vicinity.

The Lower Brue Valley is also very good, especially the gravel pits and fen-carr habitat (Apex Pits) to the south of Highbridge, near Burnham on Sea.

✱Radipole Lake NR - Dorset

A large population of Cetti's Warblers exists in the reed swamp and willow-carr habitat found over much of Radipole Lake RSPB reserve, north of Weymouth. A trail leads north from the information centre and RSPB shop into the reserve and several territories of Cetti's Warblers are entered within the first 200 yards. Several more are located along the Buddleigh Walk to the right and even more along the footpath to the North Hide. Early mornings are best or in early spring when birds are displaying, but birds may be encountered at this site throughout the year.

Good numbers of birds can also be found in the reedbed and alder-carr at Lodmoor CP, adjacent to the A353 immediately north-east of Melcombe Regis. The birds favour the reedbed to the north or the dense vegetation along the central causeway, accessed from the coast road. Park by the entrance to the refuse tip and explore the trails to the right of the road.

✱Slapton Ley - Devon

Slapton Ley is the premier site in Britain for Cetti's Warblers, with no less than 40 singing males present in spring. It is also one of the better sites to actually watch Cetti's Warblers, as birds often show throughout the day at this site. The best area is the Higher Ley, where a multitude of birds frequent the area under the bridge and the marsh and reedbed to the north of there. At the north end of the Ley, turn left towards Slapton and park by the bridge and field centre. A footpath leads north to the Higher Ley and another takes one around to Ireland Bay, but Cetti's Warblers will give rewarding views from the bridge. They can also be seen and heard from the footpath that skirts alongside the Lower Ley, from Stokeley Bay to the west of Torcross village.

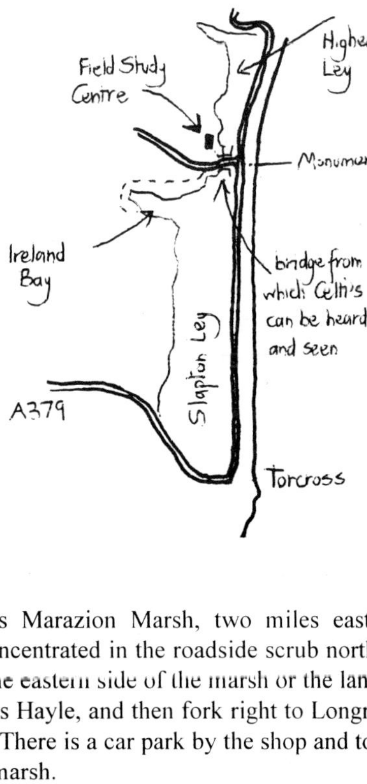

✱Marazion Marsh - Cornwall

The best site in Cornwall for Cetti's Warblers is Marazion Marsh, two miles east of Penzance. Several pairs are present and these are concentrated in the roadside scrub north of Longrock Pool or along the footpath that traverses the eastern side of the marsh or the lane to the caravan site. Leave Penzance on the A30 towards Hayle, and then fork right to Longrock and follow the coast road towards Marazion village. There is a car park by the shop and toilet blocks on the beach or one at the eastern end of the marsh.

Elsewhere in Cornwall, Par Beach Pools and Bude Marshes NR also hold several singing birds throughout the year.

Although none breed in NW England, recent wintering birds in 1995/96 at Marton Mere LNR, Blackpool and Leighton Moss RSPB (both Lancs) hint at possible future colonisation of the region.

Savi's Warbler *Locustella luscinioides*

The Savi's Warbler is an extremely rare breeding bird i ιin with maybe less than ten singing males being recorded in recent years. In 1995 produced just five singing males in May-July and none were recorded at reguıar known haunts.

Savi's Warblers favour large, wet reedbed sites with adjoining fen vegetation, and in the past, have visited Minsmere RSPB reserve and Hickling Broad on a regular basis. They are now exceptionally rare and are considered vagrants in virtually every county in Britain.

The song of Savi's Warbler is somewhat similar to Grasshopper Warbler and can often be mistaken for several species of nocturnal insects. It is much lower pitched and faster than Grasshopper Warbler, and has a very dull buzzing feel to it. It is also repeated at far less regular intervals and in much shorter phrases, and like all sounds of this nature, is very difficult to pinpoint in location and distance.

There are four main sites for Savi's Warbler, but none of these has been visited in recent years. They are best recorded from late April to late May or from mid June to July and generally sing from dawn for about two hours, briefly during mid afternoon and from 6-8.00pm until dusk.

Stour Valley - Kent
The Stour Valley between Stodmarsh and Westbere was once a traditional site for singing Savi's Warblers, but recent years have seen a dramatic decline in occurrences. From late April to late May, it may be possible to hear the distinctive deep-pitched and dull-buzzing reeling of the male from the reedbed east of the Lampen Wall at Stodmarsh or from the new footpath that leads to the Marsh Hide there, or from the footpath that leads through the rank vegetation south of the railway crossing in Westbere village. The footpath that runs alongside the north bank of the Stour east of Fordwich village may also reap results.

Hickling Broad - Norfolk
The footpath that runs from Potter Heigham church to the lagoon at Swim Coots was once a traditional site for singing Savi's Warbler, with 1-3 birds regularly being seen and heard from late April onwards. There has been no records however, in the last three years. Park by the church in Potter Heigham and walk north along the footpath to join the Weavers Way. Head east as far as the hide or check west as far as the houses at the west end of the track. This site is also very good for singing Grasshopper Warbler, as well as for Barn Owl, Marsh Harrier and Bearded Tit.

Minsmere RSPB reserve - Suffolk
The Island Mere reedbed has, in the past, attracted up to three singing Savi's Warblers at once, but this is largely a thing of the past. If present, they favour the rank vegetation that borders and infiltrates the main reedbed to the left of the Island Mere Hide. The birds rarely sing after 6.30am or before 8.00pm and are always extremely difficult to see.

1-3 singing males were also regularly recorded at Walberswick NNR until recently and these could be heard from the footpath that leads from the Westwood Lodge road out towards Westwood Marshes and the raised plateau of Dingle Hills. Park alongside the lane leading west from the B1387 in Walberswick village and follow the well-worn footpath to the wind pump and then towards the marshes. Evenings are best.

Aquatic Warbler *Acrocephalus paludicola*

The Aquatic Warbler is a very scarce passage migrant which occurs in Britain after leaving breeding grounds in Poland and Russia from mid August until late September. It inhabits reedbeds primarily on migration and is rarely encountered outside of the mist-net, and moderate numbers are trapped and ringed at a number of south coast localities each autumn. A small number of reedbed sites have adjoining *Juncus* vegetation, where Aquatic Warblers are known to feed infrequently, and these are the best sites in Britain in which to see this bird.

❶Marazion Marsh - Cornwall

Marazion Marsh is the best, if not only, regular site in Britain in which to see Aquatic Warblers. An average autumn sees about fifteen appear between 10th August and 20th September, with the peak passage from 24th August until 10th September. They favour the *juncus* pools adjoining the main reedbed and the ditch that runs alongside the eastern edge of the reedbed as far north as the railway. They show well during the first three hours of daylight, often perching up in full view on vegetation as they attempt to dry out after feeding amongst the wet sedges, and again from early evening.

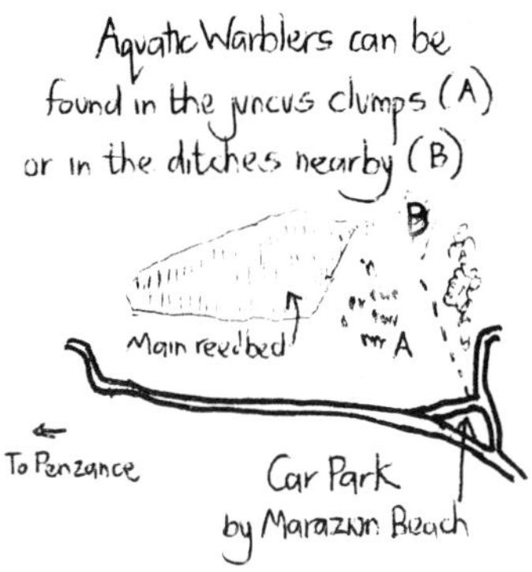

Marazion Marsh lies east of Penzance and is accessed from the coast road that runs from Longrock to Marazion. Park in the car park at the eastern end of the marsh and cross the stile by the stone bridge and walk north for about 100 yards to the main area of vegetation.

❷Elmley RSPB reserve - Kent

Elmley reserve has recently attracted Aquatic Warblers on an annual basis and the area of short reeds and sedge vegetation behind Counterwall Hide has been particularly favoured. With patience, observers may be rewarded with excellent views at this site as birds, normally responding to passing birders, flit up and perch in full view at the top of the reeds for a few seconds. Aquatic Warblers respond well to '*pishing*' and can be very inquisitive. The first two weeks of September have been the peak period for sightings, especially following south-easterly winds.

Elmley Reserve lies on the Isle of Sheppey and is accessed from the A249 one mile north of the Kingsferry Bridge. Turn right on to a rough track which after two miles leads to Kings Hill Farm. Park here and walk east to the reserve, turning left to the two hides after about a mile.

❷Lodmoor CP - Dorset

Aquatic Warblers are almost annually recorded at Lodmoor CP, where they are often disturbed from the rank vegetation and sedgebeds along the footpath leading to Tollhouse Hide, parallel with the road to the refuse site.

Aquatic Warblers are annually trapped in varying numbers at Kenfig Pool (**Mid Glamorgan**), Icklesham Reedbed (**East Sussex**), Farlington Marsh (**Hants**), Keysworth, Poole Harbour (**Dorset**) and South Milton Ley (**Devon**) but are rarely recorded in the field at these sites.

Marsh Warbler *Acrocephalus palustris*

The Marsh Warbler is a rare breeding bird in Britain with a population of up to 60 singing males in June and July. The population is restricted to the Avon Valley in Worcestershire and to parts of Kent and Sussex, but appears to be expanding its range in south-east England and retracting it in the Midlands.

It favours relatively damp areas with dense vegetation, and has a particular preference for nettle beds, Meadowsweet and Willowherb. Birds arrive from wintering grounds in Africa from about 21st May and are largely settled in by 20th June. They have a loud and mostly mimicked song and utter this from inside small bushes or hedgerows or from just below the tops of stems.

✱St Margaret's Bay - Kent

Marsh Warblers are increasing their tenuous stronghold in Kent and have started to expand west and northwards. The nucleus of population is still largely restricted to the east coast and Stour Valley, but singing males have recently been recorded as far north as Norfolk and west into Hertfordshire. St Margaret's Bay has long been an attractive site for this species and several singing males can be encountered in late May and throughout June in the steep and well vegetated valley immediately inland and north of the lighthouse. These birds can easily be 'scoped from the footpath at the seaward side of the ridge and must not be disturbed. *Do not leave the footpath under any circumstances and avoid the track that runs through the bottom of the valley.*

Park in the designated car park in St Margaret's at Cliffe village and walk south to the lighthouse at South Foreland. The valley is immediately apparent inland of the lighthouse and is traversed by a number of tracks.

✱Eckington Bridge - Worcestershire

The area of willows and nettles which abound on the north side of the River Avon near Eckington Bridge regularly attract 1-3 singing male Marsh Warblers from late May until August. The area is a reserve of the Worcestershire Nature Conservation Trust and the Marsh Warblers are wardened throughout the summer. Leave the A4104 south onto the B4080, and after one mile, park in the car park on the south-east side of Eckington Bridge. Walk west along the south shore of the river for about 500 yards and 'scope the rank vegetation on the far side of the river immediately east of the railway bridge. Marsh Warblers can often be seen in the bushes and hedgerows in this area as well as in the willows bordering the river. *On no account should you cross the river over the railway bridge.*

If this site fails to produce any singing males, another site nearby is often wardened by the *Marsh Warbler Study Group* and access instructions are duly broadcast by *Birdline Midlands* on a special phone-line ☎ *0891 664409*. A small fee may also be charged to help with protection measures.

Marsh Warblers can also be encountered on migration at numerous east coast localities in late May/June and are featured on Birdline and the relevant regional bird lines. Favoured haunts include Spurn Point and Flamborough Head (East Yorks) and North Ronaldsay (Orkney), Fair Isle and south Shetland Mainland.

Icterine Warbler *Hippolais icterina*

Icterine Warbler is a scarce passage migrant to Britain which breeds from Scandinavia, the Low Countries and north-east France across central Europe to western Siberia and winters in Africa south of the Sahara. Numbers fluctuate in spring depending on the weather conditions but may reach as high as 300 birds in an exceptional year but only 20 in an average one. The majority are recorded in autumn, between 15th August and 8th October, and peaking at the end of August through to mid September. Most are displaced on easterly winds to the east coast (from Shetland to Kent) but south-westerly/southerly winds also produce birds in Dorset and Scilly.

The identification and separation of Icterine Warbler and Melodious Warbler can be problematical and are best illustrated in Lars Jonsson, *Birds of Europe*, pages 434-435, and Harris, Tucker and Vinicombe, *The Macmillan Field Guide to Bird Identification*, pages 181-185.

There are many localities that turn up Icterine Warblers on a regular basis, but like most scarce migrants, there is no particular pattern to occurrences and sightings are largely dependent on weather conditions at the time. The following perhaps offer a 50% chance of connection if the wind is south-easterly and overcast and it is in late May/June or mid August to mid September.

❶ Blakeney Point - Norfolk

A classic spring or autumn fall will undoubtedly see 1-3 Icterine Warblers displaced on Blakeney Point. Birds generally arrive there from early afternoon onwards and may make landfall at various points along the 3.5 miles stretch of coastline. Favoured areas to search are the *Sueda* bushes around Halfway House and The Hood, the vegetation on Yankee Ridge, the small Plantation near the west end and the Lupin beds behind the tea-room. Remember however, that the best conditions to produce the birds are often the worst conditions to birdwatch in, and views of these birds may be frustratingly brief and difficult.

Blakeney Point can be reached via a three hour hike across shingle from Cley Coastguards or by boat from Morston Quay (check tide times and sailings).

In similar conditions, birds may well be found in the sparse vegetation and elder bushes at Gramborough Hill, just east of Salthouse Beach car park.

❷Wells Woods/Holkham Meals - Norfolk
A traditional site for Icterine Warblers in autumn is the stretch of coastal woodland that runs from Wells Channel to the west end of Holkham Pines. Up to eight birds are found during an average autumn and these tend to be concentrated in only two or three locations: a) in the birches and sallows at the back of 'The Dell' (an area of low-lying scrub that lies 400 yards west of the boating lake at Wells); b) the birch woodland surrounding the drinking pool (an area concealed from the main track about half a mile west of the boating lake) and c) the birch woodland to the west of Meals House at the western end of the pines. Birds, however, are always difficult to find and normally associate with large roving feeding flocks of tits and warblers.

In suitable conditions, Icterine Warblers can also be found at Holme, normally around the car park or in the Firs at the NOA reserve.

❶Spurn Point - East Yorkshire
In moderate to fresh east/north-easterly winds in both late spring and autumn, Icterine Warblers are likely to appear at Spurn Point. Although the sea-buckthorn offers secure protection and seclusion, birds may be found behind the coastguard cottages on the Point, in the area known as 'The Narrow Neck', the Warren adjacent to the Bird Observatory, the bushes by the Canal scrape, the bushes and hedgerows bordering Beacon Lane, the trees surrounding Kilnsea Churchyard and the bushes at Sammy's Point, Kilnsea.

❷Flamborough Head - East Yorkshire
Up to fifteen Icterine Warblers may be found in the Flamborough area in an average year, the majority of these being located in the dense vegetation at South Landing or in the hedgerows surrounding the Timoneer Hotel. The Bay Brambles, between the Fog Signal Station and the Golf Course, can also be productive, as well as the Old Fall Hedge and Plantation.

❷Filey North Cliff Country Park - North Yorkshire
Icterine Warblers are regularly recorded at Filey NCCP and are usually found in the top scrub inland of the camping area and car park, Arndale Ravine (opposite the cafe) and in the Glen and adjacent gardens to the south of the sea front. Late May to mid June and from mid August to early September are the peak periods of occurrence and the wind must be in the east.

❷South Gare - Cleveland
The Cabin Rocks and the bushes and scrub bordering the road to the South Gare Lighthouse regularly attract Icterine Warblers (and Wrynecks and Barred Warblers) under suitable conditions. The site is reached from Redcar via the northwest continuation of Tod Point Road.

The area north of the Tees is also very good in suitable conditions, the gardens, tennis courts, bowling green and allotments area of Hartlepool Headland regularly harbouring Icterine Warblers, as well as Seaton Carew churchyard and the few bushes at North Gare.

❷Holy Island - Northumberland
Holy Island is renowned for attracting large falls of migrants including Icterine Warblers, when wind conditions are in the east and the weather is overcast or wet. Although birds can be found in virtually any area on the island, favoured sites include the few bushes on The Snook, the hedgerows leading up to the Straight Lonnen and the gardens of the Priory and elsewhere in the village.

Holy Island can only be reached at low tide, and it is essential to consult with local tide tables before crossing.

Icterine Warblers are also regularly recorded in the Northern Isles, particularly on North Ronaldsay (Orkney), Fair Isle and in the gardens at Scatness and Sumburgh Head (Shetland). June is often the premium month for sightings.

Melodious Warbler *Hippolais polyglotta*

The Melodious Warbler is an annual scarce migrant from breeding grounds in Mediterranean France and Spain, which appears in very small numbers in spring and more commonly (up to 45 records) in autumn. The vast majority are located at sites along the South Coast, between Kent and Scilly, with smaller numbers up the west coast north to the Isle of Man and Bardsey Island (Gwynedd). Although migrants can occur at Portland Bill as early as mid July, the main passage period occurs between 9th August and 13th September, with straggling birds west to Scilly and Cornwall throughout October. Melodious Warbler is a vagrant to the east coast and extremely rare north of Margate (Kent). The following are those sites with more than a 60% chance of turning up a Melodious Warbler in warm southerly/south-westerly winds in autumn.

❶Portland Bill - Dorset

Portland Bill is the premier site in Britain for passage Melodious Warblers in autumn, and birds may be seen from 9th August until 17th September. They generally turn up in the Hut Fields, the bushes and scrub around the Observatory Quarry, the Observatory Garden, the scrub and blackberry bushes by the Pulpit public house and in the bushes and vegetation at Culverwell. Although many birds remain for less than one day, some individuals can remain for up to ten days and often afford observers some extremely good views.

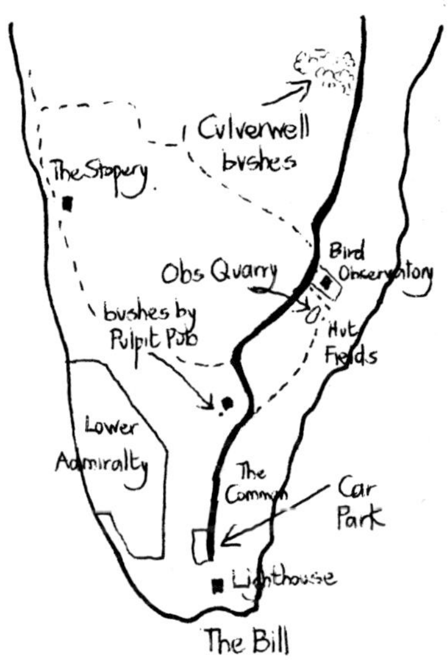

❷Nanquidno Valley - Cornwall

In recent years, Melodious Warblers have become almost annual in Nanquidno valley, and may be found in the sycamores surrounding the houses in the middle of the valley or in the ivy-clad scrub on the hillside below the house at the far end of the valley. They can also be found in the sallows on the opposite side of the stream. The period 29th August to 15th October is most productive.

Nanquidno lies immediately north of St Just Aerodrome. Take the road west from the St Just to Sennen road and continue down into the valley. Park sensibly at the edge of the lane just beyond the overhanging trees.

Melodious Warblers are also regularly found at Porthgwarra (in the thick vegetation of 60' Cover half way along the main valley) and in Kenidjack Valley, Cot Valley and at St Levan.

The only other reliable sites are the south-facing headlands of Prawle and Start Points (Devon), Durlston CP (Dorset) and Dungeness/Denge Marsh Gully, but records from these sites are less than annual. They are, however, annual on Scilly, with 1-4 birds regularly well-seen in October.

Dartford Warbler *Sylvia undata*

The Dartford Warbler is a scarce breeding resident of dry lowland heath in southern England and is largely confined to Surrey, Hampshire and Dorset, with smaller populations in Sussex, the Isle of Wight, Devon and Cornwall. The number of singing males located in summer 1993 was a staggering 1,026 birds, largely resulting from a consecutive run of mild winters.

In breeding habitat, Dartford Warblers are most easily located by their scolding '*churr*' or in spring by the male's rather scratchy song delivered from the top of a convenient gorse clump. They are resident but are best seen in the breeding season, when singing males can easily be seen in April and May and feeding family parties can be encountered from late June through to August. At all potential breeding sites, be careful not to cause any disturbance to these birds and never stray away from the main footpaths. Dartford Warblers can be seen very easily at all of the sites mentioned below from public footpaths.

Prime Sites

●Frensham Common - Surrey
A small population of Dartford Warblers survives on the neighbouring heathlands of Thursley, Hankley and Frensham Commons. They inhabit the heather and gorse clumps at each site and are best located in spring, when the characteristic scratchy song of the male or the harsh, grating scolding call can be heard.

For Thursley Common, park in The Moat car park south of Elstead and take the south-east footpath skirting Shrike Hill to the common. The birds can be seen from the numerous footpaths traversing the area as far south as the ridge of gorse bordering the A3.

For Frensham Common, park by the Great Pond and follow the bridle path across the A287 to check the extensive gorse heathland on both sides of King's Ridge. View only from the footpaths.

✱The New Forest - Hampshire
The Dartford Warbler is a relatively abundant resident of the New Forest with a current population well in excess of 600 pairs. Although they can be found in virtually any area of suitable habitat, there are a number of sites with fairly healthy populations and which offer viewing facilities from public footpaths.

1). **Ashley Walk and Hampton Ridge**: this area of high gorse-clad ridges of heathland interspersed by boggy valleys is ideal habitat for Dartford Warblers and birds may be seen from any of the footpaths that traverse this area. For optimum results, park at the east end of Frogham village (two miles SE of Fordingbridge) and walk east along the footpath to Hampton Ridge and then north through Pitts Wood Inclosure to Ashley Walk and the B3078.

2). **Slufters Inclosure**: Dartford Warblers can be encountered relatively easy at Slufter's, with birds present in gorse and heather on both sides of the road. Leave the M27 at Junction 1 and continue north-westwards on the B3078. Turn left after two miles at Long Cross and then drive south back towards the A31. Turn right after 1.5 miles on a road that runs parallel with the A31 and then turn left at the T-junction towards Bolderwood. Park on the right after 300 yards and follow the footpath around the pond to the south of the road or the footpath that skirts the gorse and heather on the north side. Do not leave the bridle paths.

3). **Beaulieu Road Station**: the extensive areas of gorse and heather to the west of Beaulieu Road Station hold good populations of Dartford Warbler. Park in the car park south of the B3056 immediately west of the railway bridge and take the footpath that runs south-west towards Denny Lodge. The area to the north of the car park at Pig Bush, a mile further south-east, is also very rewarding.

✱Avon Valley, Christchurch - Dorset

Dartford Warblers can be found in good numbers close to the northern outskirts of Christchurch at St Catherine's Hill and Town Common. Most are concentrated on the eastern slopes of the hill and the low-lying heathland on the common. Both locations can be reached from the junction of the B3073 (Fairmile Road) and Marsh Lane (signposted to Dudmoor Leisure Centre) in Christchurch, by following St Catherine's Lane north up the hill. It can also be reached from Marsh Lane, off the Fairmile Road, by following the footpath north of Coward's Marsh along the disused railway line.

✱Studland Heath and Arne Heath - Dorset

Dartford Warblers are fairly abundant in the Studland area and are best seen from the footpaths that lead along the west side of Littlesea or lead out towards Godlingstone Heath. Park by the entrance to the observation hide at the south end of Littlesea and explore the gorse scrub in the immediate vicinity. Do not leave the footpaths.

The Studland Peninsula can be approached either by ferry from Sandbanks to Shell Bay or from Studland village.

The RSPB reserve on the Arne Peninsula also offers a wide range of heathland species including Dartford Warbler. Leave Stoborough (south of Wareham on the B3075) eastwards on a minor road to Arne village and park in the designated car park. A bridle path leads from here to Shipstal Hill, where large areas of gorse heathland can be viewed.

✱Woodbury Common - Devon

The East Devon Commons to the north-east of Exmouth maintain a healthy population of Dartford Warblers and these may be seen at a number of locations, including Woodbury and Aylesbeare Commons. Leave the A3052, six miles east of Clyst St Mary, and drive south on the B3180 towards Budleigh Salterton. After two miles and just beyond Woodbury Castle, a car park appears on the left. An obvious bridle path leads eastwards across the common and Dartford Warblers may easily be seen from here.

Aylesbeare and Harpford Commons lie to the north of the A3052, immediately east of the Half-Way Inn junction. About half a mile east of the junction, turn right towards Hawkerland and park immediately in the car park on the left. Cross back over the A3052 and follow the tarmac road which runs north across the reserve. Dartford Warblers can be easily seen on both sides of the road but for better views, it is advisable to take a 'scope.

Barred Warbler *Sylvia nisoria*

The Barred Warbler is an annual scarce migrant to Britain, extremely rare in spring but regular in autumn. Following east or northerly winds in early August, the first birds are generally recorded on Fair Isle and Shetland from about 10th August onwards, and by 20th August, small numbers filter south to Norfolk, Yorkshire and Cleveland. The peak arrival is from 26th August to 15th September, primarily on the east coast, with numbers later penetrating westwards and peaking along the south coast and on Scilly throughout October. In recent years, up to 130 birds have been recorded, and several sites have become regular stop-over points. Barred Warblers are generally secretive in their habits and can be remarkably skulking. They favour dense clumps of elder scrub or hawthorn thickets.

✱The Northern Isles

Barred Warblers can be seen virtually anywhere in the Northern Isles between 10th August and 15th September, particularly on Fair Isle (around North Haven or in gardens at the south end), on North Ronaldsay (mainly the Holland House Garden) and Stronsay (both Orkney), on the Out Skerries, on Unst (in gardens around Haroldswick and Northdale) and in the Virkie/Scatness/Sumburgh Head area (all Shetland).

➊St Abb's Head - Borders

One of the most reliable sites in mainland Scotland is St Abb's Head, where Barred Warblers may be found in late August/September in gorse scrub around the edges of Mire Loch and in gardens around St Abb's village.

The dense scrub and vegetation at Fife Ness also regularly attracts Barred Warblers in autumn.

➊Hartlepool Headland/South Gare - Cleveland

The gardens, tennis court area, bowling green and allotments area at the south end of Hartlepool regularly attract Barred Warblers along with other scarce migrants in autumn, when weather conditions are conducive to falls.

The few bushes by the Cabin Rocks and breakwater at South Gare are also very productive from late August onwards. Leave Redcar on Tod Point Road and continue towards the end.

➊Filey North Cliff Country Park - North Yorkshire
Barred Warblers are regular visitors to Filey NCCP and may be found in the dense scrub and vegetation in the Top Fields or in Arndale Ravine. Numbers peak from late August through to mid September and may number up to four birds.
Further south, the promontory of Flamborough Head is a very good site. Birds make landfall in the Bay Brambles, immediately north of the Fog Signal Station, in South Landing or in the Old Fall Hedgerow. Light easterly winds are best.

➊Spurn Point - East Yorkshire
Spurn Point is renowned for attracting falls of Barred Warblers and other scarce migrants in autumn, especially following fresh south-easterly winds with rain. Barred Warblers favour the elderberry scrub and may be found on The Warren, adjacent to the Bird Observatory, in Beacon Lane, Kilnsea, or in the Canal Bushes.

➊Holme Dunes - Norfolk
Barred Warblers are regularly recorded at Holme between 18th August and 17th September. They favour the dense clumps of elderberry and may be found in the scrub by Broad Water (on the NOA reserve at The Firs - permit access only), in the dune-scrub between the golf course and The Firs reserve or in the gardens and paddocks bordering the chalets along the reserve entrance track. Park in Holme Beach car park (adjacent to Hunstanton Golf Course) and walk east along the coastal track or drive into the reserve.

Similar numbers of birds may also appear in the elderberry scrub at Burnham Overy Dunes, and this area may be accessed from the main A149 coast road about one mile east of Burnham Norton or from the western end of Holkham Pines.

➊Blakeney Point - Norfolk
In moderate to fresh north-easterly winds from mid August, Barred Warblers may be found along the ten mile stretch of coastline between Blakeney Point and Cromer. The Plantation, Tea-room Lupin Beds and *Sueda* bushes from Halfway House to The Hood are all likely areas of landfall and longer stays may be expected of birds in the few bushes at Gramborough Hill, Salthouse, and on Muckleburgh Hill.

➊Winterton Dunes - Norfolk
Barred Warblers are annually recorded at Winterton from late August throughout September and are most often recorded in the well-vegetated valley to the south of the main beach car park. They commute between the bramble scrub in the South Dunes to the elderberry scrub on the slope bordering the gardens and holiday complex, They can also be found in the spartan vegetation to the north of the car park.

To the north of Winterton, the thick scrub surrounding the Shangri-La bungalow at Waxham can be very productive, but birds are particularly elusive here.

➊Landguard NR - Suffolk
The thick scrub and vegetation that abounds in the Landguard Point ringing enclosure is a haven for migrating *Sylvia* warblers and other tired migrants and Barred Warbler is one of these species which is found on a regular basis. Although the initial discovery of one is largely restricted to the mistnets, birds often fall into a pattern of favouring the same bush or area of bramble and can eventually be seen from outside the compound on the public footpath.

Landguard NR lies at the south end of Felixstowe and is accessed from a minor road bordering the docklands. There is a designated car parking place at the site.

Pallas's Leaf Warbler *Phylloscopus proregulus*

Pallas's Leaf Warbler is a rare but regular scarce migrant to Britain with between 25 and 150 birds appearing annually in late autumn, from about 16th October to 20th November. Occurrences are influenced by weather conditions over the Continent, but periods of south-easterly winds (or winds between north and east) will produce varying and often fluctuating numbers. As with all Siberian migrants, there is a bias towards the east coast, and birds may be found at virtually any coastal site between Fraserburgh (Aberdeenshire) and Dungeness (Kent) when conditions are right. In fact, recent years have seen some exceptional influxes occur. As the autumn progresses, occurrences are more likely to occur south of Norfolk and there is a preponderance of sightings along the south coast, from Beachy Head (East Sussex) to Scilly. It is not possible to provide any guaranteed sites for this species but under suitable conditions and at the correct time of year, a 75% chance of likelihood can be offered from the following haunts.

❶Holkham Pines - Norfolk

The stretch of coastal woodland between Wells Channel and the west end of Holkham Pines is possibly the best site in Britain in which to see or find a Pallas's Warbler in autumn, with up to ten birds at a time being located under suitable conditions. They tend to join up with roving tit or warbler flocks and are most often seen in the birch scrub. The area known as 'The Dell', about 400 yards west of the Wells Boating Lake car park, is regularly visited, as is the area of birch woodland between the Kissing Gate and The Dell in the first 200 yards from the car park and the birch scrubland surrounding the 'Drinking Pool', about half a mile west of the car park. To the west, the pines and scrub around Meals House and pond are good, as are the trees bordering the track west to the far western end of the Pines.

Park in the main car park at the end of the coast road north of Wells next the Sea, immediately inland of the Lifeboat Station, or at the end of Lady Annes Drive at Holkham Gap.

The small plantation at Holme Firs NOA reserve has also repeatedly attracted Pallas's Warblers during the last twenty years, but a permit is required for access.

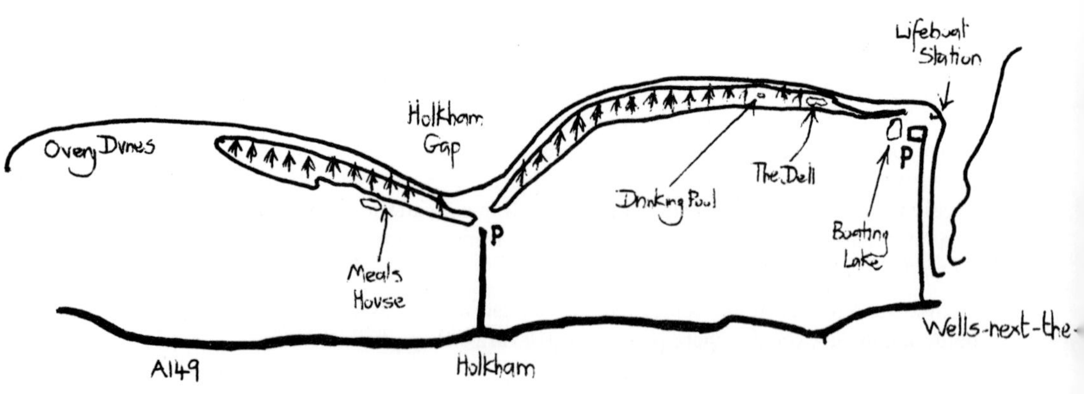

Pallas's Warblers may also be seen under suitable conditions between mid October and late November at Cruden Bay (Aberdeenshire), Fife Ness (Fife), Hartlepool Headland (Cleveland), Filey NCCP (North Yorks), Flamborough Head (East Yorks) (particularly in South Landing and around Timoneer Hotel and in gardens around the head car park), Spurn Point (East Yorks) (in the Canal Bushes, the bushes around the Bluebell Pub and around Kilnsea Churchyard), Winterton village (in the sycamore trees around the holiday complex to the south of the village), Lowestoft (Suffolk) (in the sycamores and parkland vegetation in Belle Vue Park and Sparrows Nest Theatre north of the town and along the disused railway line at Corton), Warden Point, Isle of Sheppey (Kent), North Foreland (Kent) (in the trees at Northdown Park, Margate, or alongside the golf course), Sandwich Bay (Kent) (in St George's Bushes, the main plantation or in gardens on the estate), St Margaret's at Cliffe (Kent) (in the valleys north and south of the village), Dungeness (Kent) (in the Moat or the trapping area), Beachy Head (East Sussex) (primarily in Belle Tout Wood or in gardens at Birling Gap), Selsey Bill (West Sussex) (in the small wood known as the Severalls immediately south of Church Norton beach), St Catherine's Point (Isle of Wight), Hengistbury Head (Dorset), Portland Bill (Dorset) (in sycamores by Avalanche Road Hump, Weston, and in various quarries in Weston), Prawle Point (Devon) (in sallows around the car park or in Pig's Nose Valley nearby), Start Point (Devon) and Porthgwarra (Cornwall) (in bushes and sallows adjacent to the car park and in valley gardens). The Isles of Scilly are also productive in late autumn (from about 22nd October onwards), with birds regularly being found in the Higher Moors Clump, Holy Vale and tall elms by Newford Duck Pond, St Mary's, in The Parsonage on St Agnes and in sallows along the north side of Tresco Great Pool.

Yellow-browed Warbler *Phylloscopus inornatus*

The Yellow-browed Warbler is a regular autumn migrant to Britain from breeding grounds in West Siberia and Central Asia, which, depending on weather conditions, appears in fluctuating numbers between 27th September and 30th November (and peaking throughout October). Records are mainly distributed on the east and south coasts, with late September sightings concentrated on North Ronaldsay (Orkney), Fair Isle and the Northern Isles and mid to late October in SW Britain (particularly Scilly).

Easterly winds from late September onwards sees variable numbers of Yellow-browed Warblers appear along the east coast from Unst (Shetland) south to Dungeness (**Kent**), with numbers largely concentrated in East Yorkshire and Norfolk. They can be recorded at virtually any coastal location and favour sallows and willows in gardens, parkland or coastal scrub. The following locations have a 90% chance of occurrence in autumn providing the weather conditions are suitable:-

Northdale and Baltasound, Unst; Whalsay; Kergord Plantation and Quendale Mill (all Shetland); Fair Isle; North Ronaldsay and Stronsay (Orkney); Fife Ness Muir and Denburn Wood, Crail (both Fife); Barns Ness (Lothian); St Abb's Head (Borders); Holy Island (Northumberland); Locke Park, Redcar (Cleveland); Hartlepool Headland (Cleveland); Filey NCCP (North Yorks); Flamborough Head (East Yorks); Spurn Point (East Yorks); Gibraltar Point NR (Lincs); Wells Dell and Drinking Pool and Holkham Pines (Norfolk); Stiffkey Meals (Norfolk); Walsey Hills, Cley (Norfolk); Waxham and Winterton (Norfolk); Great Yarmouth Cemetery (Norfolk); The Naze (Essex); North Foreland (Kent); Sandwich Bay (Kent); St Margaret's Bay (Kent); Dungeness Area (Kent); Belle Tout Wood, Beachy Head (East Sussex); Hengistbury Head (Dorset); Portland Bill (Dorset); Lundy Island (Devon); Kenidjack Valley and Cot Valley, St Just (Cornwall) and Porthgwarra (Cornwall).

The best site in Britain however is the archipelago of Scilly, where an average autumn will produce at least 27 birds between 30th September and 15th November. Favoured localities are the elms on the lower battery wall and the tennis court area on the Garrison, Lower Moors, the Higher Moors Clump, Porthellick Pool sallows, Holy Vale, Newford Duckpond and Watermill Lane (all St Mary's), the Parsonage on St Agnes, Lower Town on St Martin's and Borough Farm, the Abbey Woods and the sallows along the north side of the Great Pool on Tresco.

Firecrest *Regulus ignicapillus*

The Firecrest is a rare breeding bird and regular passage migrant to Britain, with a population of at least 80 nesting pairs and peak passage numbers in March-April and September-October. The breeding population is very localised and is restricted to conifer plantations (particularly Douglas Fir and Norway Spruce) and some broad-leaved woodlands in southern England. The bird is also a fairly common passage migrant along south and east coasts and also winters in small numbers in south-west England.

Firecrests in Summer

The number of Firecrests nesting in southern England and Wales fluctuates on an annual basis, from a maximum of 175 singing males at 75 different locations in 1983 to just 46 in 25 locations in 1984 and 1985. It can be very difficult to detect in the breeding season, but the high trisyllabic call or its high-pitched song should reveal its presence in ivy-clad trees or coniferous woodland. The calls and song are very similar to that of Goldcrest but differ from that species in that the song reaches its climax towards the end rather than at the start and the 'seep' call-note is higher-pitched, more piping and shorter in length.

❷Wendover Woods - Buckinghamshire

A population of up to 46 singing males (April-June 1975) survives in the extensive conifer plantations to the east of Wendover. They are present from late April until August and can be seen in small family parties in late summer in the tree canopies. Leave Wendover on a minor road signposted The Hale and after a mile, park on the left by the Forestry Commission Cottages. A footpath leads north from here into Halton Wood (Wendover Woods) and can be followed to the ridge to the north. This forms a triangular route and passes through several Firecrest (and Goldcrest) territories, particularly in the area around the tree hide and at the various crossroads in the wood.

❶Jubilee Grove, New Forest - Hampshire

A moderate population of up to ten pairs of Firecrest can be found in the tall Douglas Fir and Cedar trees that border the Jubilee Grove car park at Bolderwood. They can be seen in the tall trees that border the road to Bolderwood Farm or at lower levels in the scrub that lies opposite the car park or that borders the paddock around the farm buildings. The birds are present from late April to August.

Leave Lyndhurst on the Stoney Cross road and turn left in Emery Down village towards Bolderwood. Continue for about 2.5 miles to an obvious car park in the fork of two meeting roads. Park here and investigate the woodland to the south of the car park or the low-lying hawthorn scrub to the northwest of the car park.

Firecrests on Passage

The Firecrest is a regular passage migrant that peaks in numbers from mid March until mid April and from mid September until early November. They can be recorded at a widely distributed number of localities, primarily along the south and East Anglian coasts. Prime sites include Dungeness and Sandwich Bay (Kent), Beachy Head (East Sussex) (mainly in gardens west of Birling Gap or in Belle Tout Wood), Portland Bill (Dorset) (mainly in the Observatory Garden, the Hut Fields and in Culverwell), Prawle Point (Devon), Landguard NR (Suffolk), Great Yarmouth Cemetery (Norfolk), Waxham (Norfolk) and Wells Woods (Norfolk).

●Dungeness Trapping Area - Kent
Firecrests can be seen at Dungeness in small numbers (up to fifteen) in winter (late November to early March), early spring (15th March to 18th April) and in autumn (25th September to 15th November). They favour the dense vegetation surrounding the Long Pits and Trapping Area (250-500 yards north of the Observatory), the Moat encircling the Observatory and the clumps of gorse bordering the Observatory access road and around the cafe and lighthouse.

●Porthgwarra - Cornwall
At least ten birds are recorded at Porthgwarra in an average autumn, between 9th September and 7th November. Favoured sites include the willows and sallows bordering the river immediately in and north of the car park at the bottom of the valley, the gardens on the eastern side of the valley and the thick vegetation at the top of the valley known as 60' Cover.

Similar numbers of birds can be recorded during the same period in several other west-facing valleys, including Nanquidno, Nanjizal, Cot and Kenidjack, as well as Church Cove on the Lizard and at Treveal, west of St Ives.

✱The Isles of Scilly
Firecrests winter in small numbers on Scilly and are regular visitors in March and from mid September until November. The largest numbers are to be found on St Mary's and St Agnes, where localities such as Holy Vale, Lower Moors, Newford Duckpond, Porthellick, the Garrison Battery Walls and The Parsonage will almost guarantee sightings.

Firecrests in Winter

A wintering population of up to 200 Firecrests is present in Britain between late November and March, and is mainly concentrated in south and southwest England, with smaller concentrations in North-west and South Wales. Up to 15 winter at Dungeness (Kent), but elsewhere, significant numbers are present in the Weymouth area (Dorset) (in bushes and cliff side scrub below Sandsfoot Castle, in gardens around The Nothe and on Portland Bill), at Prawle Point and Wembury (Devon) and at St Andrews Pool, Par (Cornwall).

Red-breasted Flycatcher *Ficedula parva*

The Red-breasted Flycatcher is an annual passage migrant to Britain, with up to fifteen recorded in spring (mid May to late June) and up to 150 in autumn (late August until early November). As expected with a Scandinavian/West Siberian migrant, it is most numerous at east coast localities, particularly on the Northern Isles and between Fife and Norfolk, with smaller numbers west along the south coast from Kent to Cornwall, but a disproportionate peak on Scilly. Arrivals take place following easterly/south-easterly winds, with a marked peak between 18th September and 11th October.

Birds can be seen at widely scattered localities under suitable conditions but the following most regularly attract the species: Sumburgh Head/Virkie/Scatness area (Shetland South Mainland), Fair Isle, North Ronaldsay (Orkney), the Isle of May (Fife), Fife Ness (Fife), St Abb's Head (Borders), Holy Island (Northumberland) (the Straight Lonnen and Vicarage Garden), Hartlepool Headland and South Gare (Cleveland), Filey (North Yorks), Flamborough Head (East Yorks), Spurn Point (East Yorks), Bardsey Island (Gwynedd), Gibraltar Point NR (Lincs), Holme Firs NOA Reserve (Norfolk), Holkham Pines (Norfolk), Blakeney Point (Norfolk), Landguard NR (Suffolk), North Foreland (Kent), Dungeness Trapping Area (Kent), Portland Bill (Dorset), Prawle Point (Devon) and Porthgwarra (Cornwall).

✱The best site, however, is the **Isles of Scilly**, where up to 30 birds may be recorded in an average autumn, between 27th September and 25th October. Favoured localities include Watermill Cove (particularly the trees in the orchard alongside the Lane), the tall elms lining the avenue beyond Newford Duckpond, Old Town Churchyard and the Garrison Pines (all St Mary's), the Parsonage on St Agnes, Borough Farm and the sallows and fields on the north side of the Great Pool on Tresco and Lower Town on St Martin's.

Bearded Tit *Panurus biarmicus*

The Bearded Tit is an uncommon resident in Britain with a breeding population of just over 600 pairs. These are largely restricted to the reedbeds of East Anglia, but extend north up the east coast to East Yorkshire and Lincolnshire and west along the south coast to Dorset. There is also an outlying population in Lancashire and sporadic breeding occasionally takes place elsewhere.

The birds often keep low in reedbeds, especially in windy conditions, but are most easily located by their distinctive *'pinging'* calls. They are most conspicuous in spring and late summer, whilst large gatherings of birds may be encountered in autumn in dispersive irruptions.

Prime Sites

✱Leighton Moss RSPB reserve - Lancashire
Up to 40 pairs of Bearded Tits are resident in the reedbeds at Leighton Moss RSPB reserve and are most easily observed on fine sunny mornings in mid October when small parties can be seen exhibiting 'eruptive' behaviour - flying around and pinging loudly high over the reeds alongside the public causeway which crosses the reserve.

Leighton Moss is accessed off the minor road between Silverdale and Yealand Redmayne, the RSPB car park and visitor centre located at Myers Farm, just east of Silverdale railway station.

✱Blacktoft Sands RSPB reserve - East Yorkshire
Blacktoft's reedbeds are an internationally important breeding site for Bearded Tits with a population of up to 100 pairs in some summers. A raised embankment and at least four hides give ample views over the reedbed and the birds can be easily seen between late March and September.

Leave Goole on the A161 towards Swinefleet and Crowle and continue on the minor road through Reedness, Whitgift and Ousefleet. About half a mile further east, turn left on the sharp right hand bend into the reserve car park. The reserve is open daily (except Tuesdays) from 9.00am until dusk.

❶Barton-upon-Humber Pits - Lincolnshire
Although five or more pairs breed in the area, large numbers of irrupting birds are recorded in October and November. The birds can be seen between the pits immediately east of the town and those at New Holland, but most regularly in the reedbeds at Barrow Haven. Leave Barton on the A1077 and then turn left on a minor road to just beyond the railway crossing at Barrow Haven. A footpath leads along the Humber bank affording views of the pits and reedbeds.

✱Titchwell RSPB reserve - Norfolk
A population of 15-30 pairs of Bearded Tits survives in the extensive reedbed bordering the south side of the freshwater lagoon. They are best seen between April and September, when birds may be seen displaying or feeding young from the main bank that leads to the hides and the beach. By October, dispersing or irrupting parties of birds may be seen high over the reedbed heading west.

Titchwell reserve lies north of the A149 between Brancaster and Thornham and is well signposted.

Much smaller numbers are also present on Broad Water, Holme, and these may be seen (or heard) in the reeds visible from the hides on the NOA/NWT reserves at The Firs - access by permit only.

✱Cley NWT Reserve - Norfolk
The reedbed viewable from East Bank and that which leads to Daukes and Teal Hides are excellent sites in which to see Bearded Tits between mid May and October. They are most easily located by their distinctive calls and can be seen flying low over the tops of the reedbeds.

❍Hickling NWT Reserve - Norfolk
Up to 45 pairs of Bearded Tits are suspected of nesting on the reserve but only a small percentage of these can be seen from reserve footpaths. Birds can be reliably seen from the Weavers Way footpath, feeding close to the track and flying low over the reeds. Park by Potter Heigham church and walk north to join the Weavers Way by the Hickling reedbed. Turn right and continue for about half a mile to the hide that overlooks Rush Hills scrape.

❍Walberswick NNR - Suffolk
The extensive reedbeds of Corporation and Westwood Marshes south of Walberswick regularly attract up to forty pairs of Bearded Tits and these can be seen from the public footpaths that lead to Dingle Hill from the minor road that leads to Westwood Lodge.

Further north, several pairs can be found in the reedbed between Potter's Bridge and Reydon Smear (TM 509 791 - Sheet 156). View from the road or from the footpath to Smear Farm.

✱Minsmere RSPB reserve - Suffolk
Good numbers of Bearded Tits may be found in the large reedbed at Minsmere reserve and are best observed from or near the Island Mere Hide or from the footpath that leads from the sluice bushes to the woodland via West Hide. Birds are present throughout the year but are more active and conspicuous from late April onwards. Minsmere reserve is well signposted from Westleton village and is open daily (except Tuesdays) from 9.00am until dusk.

❍Old Hall Marshes RSPB reserve - Essex
A population of 10-20 pairs of Bearded Tits survives in the isolated reedbeds that lie to the north of the public footpath that leads east towards Pennyhole Fleet. Access to the reserve however, involves a long and often arduous walk from Tollesbury Marina.

Elsewhere in **Essex**, Bearded Tits can be seen at Wat Tyler Country Park and Fobbing Marshes, south of Basildon. Leave the A13 at Pitsea and drive south past Pitsea railway station to the country park at Pitsea Marsh. A number of hides overlook suitable areas of reedbed.

✱Stodmarsh - Kent
Stodmarsh is probably the best site in Britain to see Bearded Tit, with birds affording superb views as they flit to and fro across the channels and reedbeds adjacent to the Lampen Wall. Birds are present throughout the year but are most active from April to June.

Stodmarsh lies in the Stour valley, five miles NE of Canterbury. Leave the A257 Sandwich road about a mile east of Canterbury and continue for three miles into Stodmarsh village. Turn left at the Red Lion public house and continue on a rough track to the reserve car park. The Lampen Wall leads out across the reedbed to the river.

➊Titchfield Haven - Hampshire

A small but increasing population of Bearded Tits can be found in the reedbed at Titchfield Haven LNR. Access to the reserve is by permit only and these can be obtained from the reserve centre by the Meon Bridge at Hill Head. A number of hides provide views over the reedbed. Breeding numbers are regularly swelled by immigrant birds in October.

Post-breeding dispersal is also clearly evident at nearby Farlington Marshes each year, when up to 40 birds may be present in the reedbed surrounding the main lagoon visible from the sea wall. These may remain at the site from early October until early January.

✱Radipole Lake NR - Dorset

Up to twelve pairs of Bearded Tits are regularly present in the reedbed at Radipole Lake throughout the summer. They are best seen from the footpath that leads from the RSPB information centre and shop to the concrete bridge by the west side, as well as from the Buddleigh footpath. Radipole Lake NR lies immediately north of Weymouth and is accessed from the roundabout west of the railway line and at the start of the A354. Park in the pay-and-display car park adjacent to the centre.

Nearby, the reedbed at the north end of Lodmoor CP holds at least three pairs, and this area is accessed from the car park at the entrance to the refuse site off the A353 or from the west end of Overcome village.

Willow Tit *Parus montanus*

The Willow Tit is a widespread and fairly common species that breeds throughout most of England and Wales, but is very localised in its abundance and restrictive in its habitat preference (mainly damp birch or willow scrub, mature woodland and hedgerows). It is very similar to the Marsh Tit but may be separated by its loud, deep, buzzing and scolding *'si-zur-zur-zur* call-note and its somewhat Wood Warbler-like clear and melodic *tsui tsui tsui tsui tsui* song phrase. Furthermore, Willow Tit has a dull black crown which sweeps back onto the mantle, wholly white cheeks, a slightly more extensive black 'bib' patch and a distinctive pale wing panel. Willow also has richer buff flanks, and shows more obvious white in the outer tail feathers. Marsh Tit has a nasal and high-pitched *pitchou-chu-chu-chu-chu* call, a rapid and typically ringing song phrase, a slightly glossy black crown, dusky ear-coverts and less contrasting pale cheeks, largely uniform upperwings and a small black bib. Willow Tit also has a rather bull-necked appearance.

Scotland

Willow Tits are rare in Scotland with less than ten pairs nesting and none north of Strathclyde. Two pairs breed in the mature woodland at Strathclyde Park, Motherwell (**Strathclyde**), with another pair at Harthope (**Strathclyde**) and regular sightings from Baron's Haugh RSPB reserve and Merryton Haugh (both Strathclyde). In **Ayrshire**, pairs are resident at Glenbuck, Dalrymple and Craigengillan, whilst in **Dumfries & Galloway**, up to three pairs breed at the Wood of Cree.

Northwest England

The Willow Tit is widely but thinly distributed across the region with main concentrations around birch coverts on relict moss land and heathland habitats. Sites which are particularly productive include Fingland Rigg Wood (**Cumbria**), both sides of the B5307, two miles east of Kirkbride; Pennington Flash Country Park near Leigh (**Greater Manchester**) - park at the pay and display car park at the east end of the flash, walk past the visitor centre and bear right to search the scrubby areas towards the canal; Marford Quarry NWWT reserve near Wrexham (**Clwyd**) - park on the verge near the railway bridge at SJ 357 563, walk through the gate at the entrance to the reserve and continue south on main track for a third of a mile to the area of birch scrub.

In **Cheshire**, up to six are regularly recorded in Mosses Wood, Chelford (five miles SE of Knutsford at SJ 824 726) and at Woolston Eyes NR, Warrington.

Northeast England

Willow Tits are common breeding residents in **Northumberland**, with Wallsend Swallow Pond and Big Waters NR, Wide Open, offering the best chance of success. It is similarly numerous in **County Durham**, where Washington WWT has several breeding pairs within its boundaries. Wintering sites in **Cleveland** include Crimdon Dene, Billingham Beck, the Castle Eden walkway, Hutton, Kirklevington, Scaling Dam Reservoir and Wynyard, whilst breeding sites include Eston Bank, Lazenby Bank, Margrove Ponds, Ormesby, Saltburn, Skelton Castle grounds and Wilton.

Yorkshire offers a large population of Willow Tits, with reliable sites including Potteric Carr NR, just north of junction 1 of the M18 south of Doncaster, particularly in the vicinity of the Loversall Hide; the woodland surrounding Kirkham Priory, east of the A64 five miles south of Malton; Wintersett Reservoir and its environs; Broomhill Ings, near Wombwell; many areas in the Dearne Valley, Barnsley; Wilsic woods; Firsby Country Park, east of the B6093 two miles east of Rotherham and at Tophill Low Reservoir.

Wales

It is a widespread resident in **Carmarthenshire**, with up to three pairs resident on the Gwenfffrwd/Dinas RSPB reserve.

In **Mid & South Glamorgan**, Willow Tits can occasionally be seen along the Glamorgan Canal, Llanwonno, Mynydd Rudry, Hensol Forest, Kenfig NR, Coed-y-Bedw and St Mellons but they are extremely hit or miss, whilst in **West Glamorgan**, regular sites include Pengwern Common, Olchfa Woods, Crymlyn Bog, Trafle Gowerton, Grovesend, Velindre NR, Rheola Forest, Afan Argoed and Oxwich Marsh. In **Gwent**, Wentwood Forest is the most reliable site.

Midlands

The Willow Tit is a fairly common resident in **Nottinghamshire** and **Derbyshire**, with sites in the latter county including Carr Vale NR, SW of Bolsover; the willow-carr surrounding Mapperley Reservoir, NW of Ilkeston and at Birdholme Wildfowl Reserve, east of the A6661 south of Chesterfield.

In **Leicestershire**, birds can be relatively easily seen at Dimminsdale (SK 375 217), Stoughton Airfield (SK 654 014) and at Strathern Wood (SK 783 316), whilst in **Northamptonshire**, the woodland in the Walgrave Arm of Pitsford Reservoir is productive.

It is fairly common throughout the West Midlands region, with several territories at Ashwood, Belvide Reservoir (**Staffs**) and others in the county at Consall Woods, Coombes Valley, Croxall Pits, Doxey Marsh, Elford, Little Haywood, Rudyard Reservoir, Stafford Castle, Tillington Cemetery, Tittesworth Reservoir and Westport Lake. In **West Midlands** county, reliable sites include the Fens Pools reserve, Valley Park and along the Brownhills nature trail, whilst in **Warwickshire**, Willow Tits can be regularly seen at Brandon Marsh NR and Ladywalk NR. In **Worcestershire**, one of the best sites is Upton Warren NR, whilst birds can also be seen at Bredon Hill and along Dowles Brook in the Wyre Forest.

In **Buckinghamshire**, several pairs of Willow Tits can be found in the woodland fringes to Hyde Lane GP, NE of Buckingham on the A422, as well as at Calvert NR, south of Steeple Claydon, whilst in neighbouring **Bedfordshire**, the best site is Tiddenfoot CP, along Mentmore Road in SW Leighton Buzzard (park beyond Cedars School and follow footpath to right for 50 yards to concrete pipe and 6-8 birds may be seen in overhanging vegetation).

The best site in **Gloucestershire** is Nags Head RSPB reserve in the Forest of Dean, where up to ten pairs regularly breed in summer, whilst in **Oxfordshire**, at least four pairs regularly nest at Otmoor and can be seen from the Roman Road or Oddington Ride. Park in Oddington village and proceed eastwards along the lane towards the river.

East Anglia

Willow Tits are widely and locally distributed in **Norfolk** with 1-2 birds present at Abbey Heath, Alderford Common, Barnham Cross Common, Blickling, Boylands Wood, East Harling Fen, Fleggburgh Common, How Hill NR, Sculthorpe Fen, Santon Downham, Strumpshaw Fen, Thompson Common, Upton Fen, Wayland Wood, West Acre and Wroxham Broad.

Although widespread in **Suffolk** there are no reliable sites but birds may be encountered occasionally at various coastal sites such as Minsmere or Benacre and are relatively frequent in woodlands in the SW of the county.

In **Cambridgeshire**, the most reliable sites are Wicken Fen NR and Fowlmere NR (situated on the minor road between Melbourn and Fowlmere), whilst in **Essex**, Bourne Brook, Markshall and Good Easter are recommended sites.

Southeast England

Willow Tits can be seen in **Hertfordshire** in the woodland that fringes Tringford Reservoir, Tring (the southernmost of the three reservoirs that border the minor road between Tring and Wilstone village) often in the trees behind the hide or near the pumping station; at Panshanger Park, south of the B1000 west of Hertford and at Whitwell, in the scrub that overhangs the river and borders the cress beds (north of Kimpton on the B651).

The species is relatively widespread and relatively common in **Kent** with good numbers present in the woodland that borders the Winkhurst Green causeway of Bough Beech Reservoir (accessed from the stile and footpath that leads towards Bore Place) and in woodlands on the north side of Bewl Water. Small numbers are also regularly seen in Hamstreet Woods, in the section accessed from Burnt Oak Lane immediately north of Hamstreet, and in Snarkhurst Wood, immediately north of the M20 at Junction 8.

It is relatively common in **Sussex** being regularly reported from up to 24 localities during the year. Favoured sites include Wepham Woods, to the north of the A27 north of Angmering, and Westdean Woods, to the northeast of Chilgrove (park at SU 844 153 and walk north along the footpath to the clearing or check the roadside trees to the east).

Willow Tits have become very scarce in **Surrey**, with Frensham Common and the Manor Fields, Coulsdon Common, Windlesham and Aldhurst being the only regular sites.

Willow Tits are relatively abundant in **Hampshire** with regular sightings at Botley Wood, Beacon Hill - Warnford, Bramdean Common, Spearywell Wood - Mottisfont, The Warren - Oakhanger, Butter Wood - Greywell, Dognersfield Lake, between Golden Pot and Well, near Tundry Pond, Wickham Common and Yew Hill - King's Somborne.

Southwest England

In **Wiltshire**, birds can be seen in the breeding season at Blackmore Copse, Coate Water, Fonthill Lake, Franchises Wood, Hens Wood, Ramsbury and Savernake Forest, whilst in **Somerset**, the only regular site is the birch woodland surrounding Wimbleball Lake.

It is a scarce bird in **Dorset** with regular sightings only coming from Powerstock Common, Kingcombe, West Bexington, Holt Heath, Garston Wood NR, Ham Common, Loscombe and Stubhampton Bottom, whilst in **Devon**, the species is more common, with regular sightings at Aylesbeare Common, Lower Tamar Lake, Dunsdon Farm NNR, Whiddon Down at Gooseford, Kennick Reservoir, Torrington Common and Welsford Moor.

In **Cornwall**, the Willow Tit has a patchy distribution, with most occurring in the damp, scrubby woodlands to the east and north of Indian Queens, with notable densities on Goss Moor, in the Redmoor/Helman Tor/Breney Common area and in the Ninestones Bridge area of the Upper Fowey Valley.

Crested Tit *Parus cristatus*

The Crested Tit is a very localised and sedentary species which is restricted to Caledonian Pine forests and mature Scots Pine forests in the Moray Basin of the central Scottish Highlands. The population is estimated to be in the region of 900 pairs, with birds breeding from Strathspey north into Morayshire, Banffshire, Nairn. Easter Ross and southern Sutherland, and west as far as Strathglass (West Inverness).

Although resident, birds can be very difficult to locate in winter, as many leave the bleakness of the woodlands to visit rural gardens for food. However, from late February onwards, they may be found at potential breeding sites and can be located by their distinctive and trilling '*prrrulululI*' call note.

❶Forest Lodge - Highland Region

The wardens at Forest Lodge regularly hang up suet and nut bags around the lodge buildings and these attract up to ten Crested Tits in winter. The track to the right of the car park that leads down to Rynettin cottages attracts up to five nesting pairs in summer and these can be seen from the main footpath. Park at NH 019 161, accessed from the minor road between Tulloch and Loch Garten.

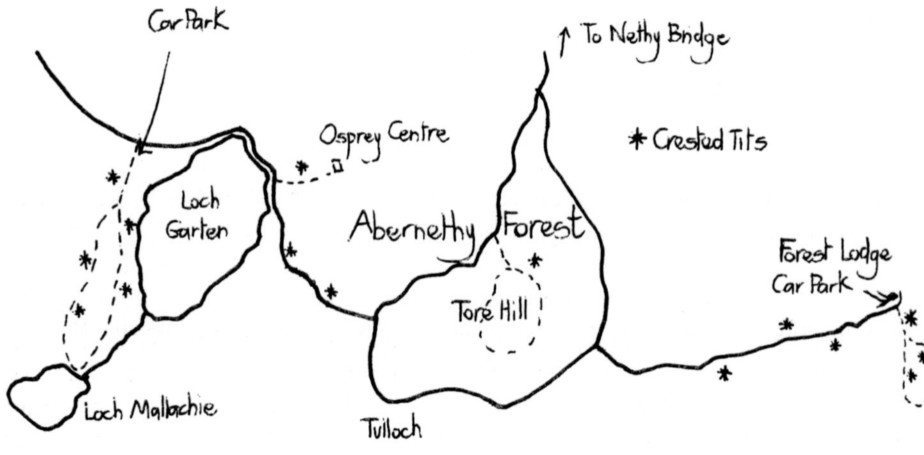

➊Loch Garten Area - Highland Region
Crested Tits can be readily located in the woodland surrounding Loch Garten from late March until October. They favour the trees that border the footpath to the Osprey Centre and Hide, the roadside trees around the loch and the trees that line the footpath that leads down the west side of the loch to Loch Mallachie.

➊Aviemore Area - Highland Region
The best site in the Aviemore area is the Mhor-mor picnic site that lies south of the road that runs from Aviemore to the Cairngorm Chairlift Centre. The layby is on the right after about two miles and Crested Tits may be found in the trees above and around the car park or in the riverside trees below. This area can also attract large numbers of Scottish Crossbills in irruption years as well as numerous Siskins.

Further back down the road towards Aviemore and immediately west of Coylumbridge Hotel, Crested Tits are relatively abundant in the woodland bordering the road. The woodland at the north end of Loch an Eilean is also highly productive for this species.

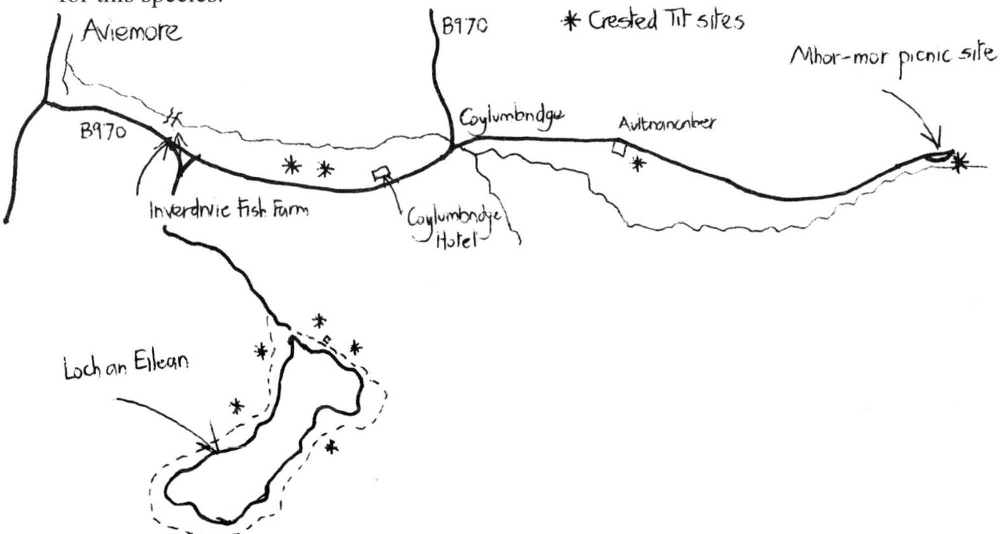

➋Carrbridge Landmark Centre - Highland Region
Crested Tits may be found in the woodland surrounding the car park at the Carrbridge Landmark Visitor Centre as well as in the trees along the trail that runs from the north end of the car park. The centre lies adjacent to the B9153.

➊Grantown-on-Spey - Highland Region
Good numbers of Crested Tits can be encountered along the various footpaths that traverse the woodlands east of Grantown. Leave the town on the B9102 and park near the gate immediately east of the golf course. Follow the main footpath south through the wood.

➊Roseisle Forest, Burghead - Morayshire
Crested Tits are commonly encountered around the picnic site in Roseisle Forest, two miles south of Burghead. Leave Burghead on the B9089 and turn right in Lower Hempriggs at the crossroads at NJ 113 646 (Sheet 28) towards the beach. The picnic site lies at NJ 104 656 and the birds may be found in winter around the car park or further afield in the wood between April and October.

Golden Oriole *Oriolus oriolus*

The Golden Oriole is a rare summer visitor to Britain as well as a scarce passage migrant. A population of up to 46 pairs is known to breed and the main passage period peaks from the second week of May through to mid June.

A small but significant population of birds has survived in East Anglia since at least 1965, where the preference is for Black Poplar plantations, a trend almost exclusive to British breeders. Although startlingly obvious, Golden Orioles are incredibly difficult to see on the nesting grounds, and their bright and very vivid colours match well with their chosen surroundings. It is normally the distinctive fluty whistling call that reveals their presence and this is best heard in early morning or in the evening. It is often best to visit sites when the birds first arrive, as the foliage on the trees is not so extensive.

There are a number of sites in East Anglia but the best known is Fordham Bridge, and to avoid disturbance to other birds and for security reasons, this is the only one to be published here. Please report any suspicious activities or persons at this site to the local Police Station or the RSPB.

❶Fordham Bridge - Norfolk

A population of up to twelve pairs of Golden Orioles frequent the Black Poplar trees that border the cut-off channel of the River Wissey, south of Downham Market. They first arrive on or about 16th May, with the majority appearing in the last week of May and the birds in full attendance by 20th June. They are highly territorial and in the first three weeks of arrival, males can be seen pursuing one another and often perching out in the open in nearby trees away from the river. Their distinctive calls can be heard from dawn until about 9.00am and again from about 6.00pm, but activity during the day is much curtailed, especially after mid June, and singing ceases after about 15th July.

Leave Downham Market southwards on the A10 and after less than one mile, turn right towards Fordham village. Cross the railway and then turn right along a minor road to the river, parking sensibly in the designated area north of the bridge. Proceed to the bridge and watch for calling birds flying to and from the poplar trees that line both sides of the river. The plantation to the south of the river is strictly private and *must not be entered.* In the first weeks of arrival, birds may be seen in the outlying plantations, especially the trees that border the lane by the farm and factory to the north of the river.

Migrant Golden Orioles may also be seen in Britain in spring, but are very difficult to predict in their arrival. Warm southerly winds from mid May onwards will see birds appear at almost any south coast locality, but favoured sites include Dungeness Trapping Area, Portland Bill and the Isles of Scilly (particularly Lower Moors).

Red-backed Shrike *Lanius collurio*

The Red-backed Shrike, until very recently, was a rare breeding summer visitor to Britain but since 1993 has failed to nest in this country. It is primarily a Scandinavian and Central/South-east European species, which migrates through Britain in varying numbers in spring and autumn. The majority are found on the Northern Isles and the east and south coasts, and peak in occurrence between mid May and mid June and from mid August until mid October. They have a preference for coastal dunes with scattered bushes, large commons and areas of open waste ground and coastal hedgerows and can normally be easily found as they choose prominent perches in which to look from. They feed mainly on large flying insects, especially bees, grasshoppers, beetles and moths.

Red-backed Shrikes appear in Britain generally following easterly winds from the Continent and may be found at a number of suitable coastal localities during the main migration periods. In spring, the bias in sightings is towards the Northern Isles, with up to 50 birds appearing on Fair Isle and in Shetland in late May/June, with smaller numbers appearing at Fife Ness (Fife), on the Isle of May (Fife), at St Abb's Head, the Farne Islands and Hauxley NR (Northumberland), Whitburn (Tyne & Wear), Hartlepool Headland and South Gare (Cleveland), Flamborough Head (East Yorks), Gibraltar Point and Donna Nook (Lincs), Burnham Overy Dunes, Sheringham Weybourne and Waxham (Norfolk) and Minsmere Sluice Bushes (Suffolk).

In the autumn, birds are more widely scattered, with occurrences peaking from late August to early October and numbers being concentrated from Cleveland south to Kent and many being found between East Sussex and Scilly. Prime sites include most of those mentioned above but also Landguard NR (Suffolk), Bradwell Bird Observatory and The Naze (Essex), Reculver, North Foreland, Sandwich Bay and Dungeness (Kent), Beachy Head (East Sussex), Selsey Bill (West Sussex), Portland Bill (Dorset), Prawle Point (Devon), Porthgwarra, Nanquidno and Cot Valley (Cornwall) and on the Isles of Scilly.

Prime Sites

❶Spurn Point - East Yorkshire

Red-backed Shrikes are annually recorded in spring and autumn at Spurn Point, with as many as five birds occasionally being present in May and June. They favour the Warren, immediately north of the Observatory, as well as the Canal Bushes, the hedgerow bordering Beacon Lane, Kilnsea, the Narrow Neck, the blackthorn scrub at The Point and Sammy's Point, Kilnsea. Occurrences generally follow light to moderate easterly winds, particularly if accompanied by rain or overcast conditions.

❷Holme Dunes - Norfolk

The paddocks behind the chalets that border the access road to Holme Firs NOA Reserve and the roadside fence posts that lead up to the reserve are excellent places in which to see Red-backed Shrikes in late May/June or August/September. The dense dune-scrub to the west of the reserve can also be very good, and it is not unusual to find 3-5 birds at this site on the same day if weather conditions are conducive to Continental arrivals.

Access to the reserve area is by permit only, obtainable from the reserve information centre on arrival. The paddocks and dunes are best accessed from the beach car park adjacent to Hunstanton Golf Course, at the north end of the village, walking east along the coastal footpath.

The same conditions that bring Red-backed Shrikes to Holme will almost certainly see birds arrive at other sites along the Norfolk coast, including Burnham Overy Dunes (in the dune scrub and bushes west of Holkham Pines), Blakeney Point (in *Sueda* at Halfway House, the Hood or Yankee Ridge, or in the Plantation, Point bushes or in the Tea-room Lupin Beds) or in Winterton Dunes (in the valley to the south of the beach car park or in the dunes up to two miles north of the car park).

Woodchat Shrike *Lanius senator*

The Woodchat Shrike is a scarce passage migrant that breeds throughout much of the Mediterranean region. Up to 30 birds regularly overshoot in spring (between the last week of March and mid June) and up to fifteen (mainly juveniles) are displaced from further east in autumn (between mid August and early October) and appear at mainly south and east coast localities in Britain, The majority remain for at least four days and may be seen perched on hedgerows or scattered bushes. Although a number of locations are regular visited, two appear to be particularly attractive.

❷Portland Bill - Dorset

Between mid March and the second week of June and from 18th August to late September, warm SW/southerly winds from France may produce a Woodchat Shrike in their wake. The areas to check include the stopery, the Top Fields around the black barn and Admiralty Hedge, Culverwell and the surrounding hedgerows and the Hut Fields and Observatory Fields area. Birds rarely remain for longer than one day though.

❷Dungeness - Kent

The vast desert of gorse scrub-covered shingle to the north of the Bird Observatory and Britannia public house has regularly attracted vagrant Woodchat Shrikes between mid April and the second week of June. They can be wide-ranging though and may wander as far north as the road to the north of the Long Pits and Trapping Area. Stays of more than one day are unusual but can last as long as a week in northerly winds.

Great Grey Shrike *Lanius excubitor*

The Great Grey Shrike is a scarce passage migrant and an uncommon winter visitor to Britain which breeds in Scandinavia. Up to twenty birds are normally located on territories in an average British winter and these remain from late November until early to mid April. A typical wintering habitat must include good hunting areas, with plenty of hunting posts and some cover in case of danger, and in most cases consists of natural heathland or extensive commons. It is well known that some birds return to the same wintering area year after year but there are few reliable sites at the time of writing. Birds can also be regularly seen on migration, with small numbers recorded on the east coast from 29th September to 12th November and returning birds in Norfolk and Yorkshire from late March through to 22nd April. Most regularly visited sites are Spurn Point (East Yorks), Holme NOA Reserve (Norfolk) and various sites in Norfolk Breckland, including Grime's Graves and Mayday Farm.

❷Westleton Heath - Suffolk

Westleton Heath has long been a traditional site for wintering Great Grey Shrikes, with one bird returning for at least six years up to winter 1994/95. The area of heath generally favoured is that to the south of the road, but like all shrikes, the winter range is extensive. Leave Westleton on the minor road towards Dunwich and after one mile, park on the obvious sandy ride on the left. Explore the surrounding heathland, particularly that on the south side of the road.

❷Mayday Farm - Suffolk

A traditional wintering site and an area regularly visited by spring migrants moving out from elsewhere is the extensive clearfell plantations at Mayday Farm. Leave Brandon south-eastwards on the B1106 and after two miles park by the farm at the side of the main entrance track. Continue on foot past the transmitter station to the open heathland and check the lines of piled-up wood litter bordering each plantation. Further similar areas are also found further west down the track.

❶Frensham Common - Surrey

Frensham Common is another traditional site for wintering Great Grey Shrikes and singles may be found in the vicinity of King's Ridge between mid December and April. Park at the north end of Frensham Great Pond and walk north along the A287 to the ridge which lies immediately east of the road. A number of bridle paths cross the area and a main footpath runs actually along the top of the ridge. Use this vantage point to locate the bird and then move closer on relevant tracks.

❶Thursley Common - Surrey

In recent years, Thursley Common has been one of the few sites to attract a Great Grey Shrike in winter, and from early December until mid April has been present in the main common area east of the Moat car park. It is typically wide ranging and may be seen on Shrike Hill at one minute and in trees under the pylons about a mile away the next.

Leave Elstead southwards towards Thursley and park after 1.5 miles in the Moat car park on the left. Follow the well marked boardwalk from the car park eastwards towards Ockley Common and where it joins a row of pine trees and veers north, scan the main bog and heath immediately east of here. The bird is usually perched aloft the dead trees or in willow clumps and may be distant. It also visits Shrike Hill, to the southeast of the car park, accessed from a separate boardwalk and trail south of the car park.

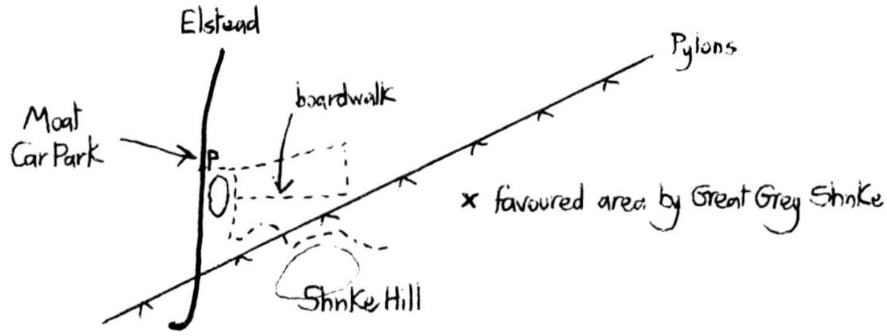

❶Ashdown Forest - East Sussex

Great Grey Shrikes are regular winter visitors to Ashdown Forest and may be present from late October until late March. A number of sites may be used in the forest but the most favoured is the well-vegetated valley to the east of the Gill's Lap car park, by the B2026. Leave the A22 at Wych Cross and drive east on a minor road towards Coleman's Hatch. Turn right at the crossroads in this village towards Newbridge and continue to the junction with the B2026. A car park is then immediately on the left and the valley is across the road.

❶Black Gutter Bottom, New Forest - Hampshire

Black Gutter Bottom and Deadman Hill, at the northern end of the New Forest, are traditional wintering sites for Great Grey Shrike, but because of the vastness of the area and its terrain birds can be extremely difficult to locate. They favour the sheltered valleys to the north and south of the road to hunt and can often be found in the vicinity of the Millersford Bottom car park at SU 193 165 (Sheet 184). Leave the M27 at Junction 1 and continue north-westwards on the B3078. Black Gutter Bottom is on the left after about five miles. Park at SU 206 165 and explore the valleys on both sides of the B3078.

Elsewhere in the New Forest, the heathlands around Beaulieu Road Station and Pig Bush car park are also regularly visited by Great Grey Shrikes in winter.

❷Woodbury Common - Devon

Woodbury Common is a traditional site for Great Grey Shrike with one bird returning for at least ten consecutive winters. The bird favours the commonland and heathland to the south of the Castle and may be located from several vantage points along the B3180.

Chough *Pyrrhocorax pyrrhocorax*

The Chough is an uncommon resident in Britain with a population of about 310 breeding pairs confined to Wales, the Isle of Man and Islay, Colonsay, Jura and Mull in western Scotland. It occurs chiefly on coastal cliffs but a number also breed inland, especially in Wales, in old slate quarries and mine-shafts.

Prime Sites

✱Islay - Strathclyde
Up to 62 pairs of Choughs are known to frequent Islay, primarily along the west coast. Favoured localities include the sea cliffs at Mull of Oa and the fields and cliff tops north of Saligo Bay. The Mull of Oa can be accessed from either Lower Killeyan or Inerval and is situated SW of Port Ellen. Saligo Bay is in the NW of Islay at NR 210 663 and is accessed from Saligo Bridge. Large spring flocks are often seen at Machrie Hotel (up to 48 in early April) and at Machir Bay (up to 30 in early June) and in autumn at Kilchoman (up to 42 in late July) and at Ardnave (up to 40 in October).

✱Isle of Man
A population of at least 75 breeding pairs and 97 non-breeding birds were estimated to be present on the island in 1993 and the following sites are the most reliable places in which to see them:

✱Manx mainland:- several pairs nest along the rocky coastline between Peel and Port Erin as well as from Santon north to Maughold. A few pairs also breed inland in deserted mine workings. A walk along the scenic clifftop path towards Spanish Head from the Manx NT car park at the SW tip of the island, or towards the sea from the picnic area on the A36 at SC 216 735 will provide good views of Chough in the summer months. Winter flocks tend to congregate in the south of the island with shoreline feeding on the tidewrack with Rooks at Langness in groups of 20-40 birds (at SC 283 673).

✱Calf of Man:- 5-10 breeding pairs - birds can be easily seen on the cliff-faces around the Calf as well as in the disused lighthouse and in artificial nest box sites in silos near the Bird Observatory. Spring flocks of 20+ birds and autumn in excess of 50 are regularly encountered, with access to the island by daily service from Port Erin between April and August.

✱Craig-yr-Aderyn ('Bird Rock') - Gwynedd
Bird Rock, at SH 644 068 (Sheet 35), holds a wintering roost of up to 70 Choughs and this can be viewed from the minor road below the cliff, two miles north-east up the Dysynni Valley from the A493 at Bryncrug, two miles from Tywyn.

✱South Stack RSPB reserve - Anglesey
Up to fifteen pairs of Chough are resident at South Stack and can be easily observed from near the car park and information centre at Ellin's Tower (SH 205 823 - Sheet 114) or from the scenic footpath that leads down some steep steps to the lighthouse. In winter, small parties of birds congregate in grassy fields bordering the approach road to the reserve and up to twenty roost daily in the Holyhead Breakwater Quarry Country Park at SH 225 833.

✱Uwchmynydd Headland, Aberdaron - Gwynedd
This site is possibly one of the best in Wales for Chough, with up to fifty birds present in the vicinity throughout the year. The birds are very approachable at the site and offer unique photographic opportunities. View from the National Trust car park on the headland at SH 140 259 (Sheet 123).

➊Lochtyn Peninsular - Cardiganshire
The cliff coast between Cardigan Island and New Quay supports about ten pairs of Choughs, two-thirds of the total Cardiganshire population. The Lochtyn Peninsular, north of Llangranog, is one of the better sites and regularly attracts up to 25 birds in late summer. The area is easily reached by the coastal path which heads north from the village.

➊Strumble Head - Pembrokeshire
Several pairs of Chough may be encountered throughout the Strumble Head peninsula and congregate in small family parties in late summer. An unclassified road is well signposted from Fishguard Harbour to Strumble Head, a car park at SM 895 414 (Sheet 157) offering ample parking space. The coastal footpath east to Carreg Gybi will almost certainly result in sightings of Chough.

On the east side of Fishguard, the Dinas Head promontory also has breeding Chough and these can be seen from the coastal footpath that leads north from Bryn-henllan village.

Further south, the St David's Head peninsular holds a sizeable population of Chough, with birds being present along the coastal cliffs north of Whitesands Bay car park and north and south of the car park at St Justinian (SM 724 252). Several pairs are also present on Ramsey Island offshore, as well as on Skomer and Skokholm islands further south.

➊Dale Airfield - Pembrokeshire
Post-breeding gatherings of up to 40 Choughs may be seen in the vicinity of Dale Airfield in September and October and these may be seen from the coastal footpath that leads south from Little Marloes Farm at SM 792 074 (Sheet 157).

Tree Sparrow *Passer montanus*

The Tree Sparrow is a rapidly declining species that was once widespread throughout much of England, Wales and lowland Scotland. It is now retracting its range at an alarming rate, possibly because of climatic conditions but more likely associated with changing farming practices and the extensive use of chemicals to eradicate farmland pests. Habitat destruction has also played a major part, with many hedgerows and ash trees now destroyed and lost. The severity of the decline can be no more striking than that noticed in the autumn passage figures recorded at Landguard NR (**Suffolk**) in the last ten years, reducing from 4,675 birds flying south in 1983 to 17 in autumn 1993.

The following is a summary of sites where Tree Sparrows can still be reliably recorded in reasonable numbers, but it will be interesting to see if these same mentioned sites still hold Tree Sparrows in ten years time. They generally nest in small groups in nest boxes specially provided in orchards or small plantations, holes in ash trees or in old farm outbuildings or barns. Occasionally, telegraph pylons and tunnels will be used. In winter, Tree Sparrows are very gregarious and form large flocks to feed on stubble or at artificial feeding stations.

Tree Sparrows are still fairly widespread in **Scotland** and maintain healthy populations in certain regions. In **Moray & Nairn**, nest boxes are normally occupied in summer at Clochan, Clackmarras, Invererne and Longmorn and two pairs normally nest at Duffus Castle. Wintering flocks include up to 60 at Miltonhill, 30 at Kintessack and 20 at Windyridge.

In **Aberdeenshire**, Tree Sparrows are most numerous in the Pitmedden and Udny Green area, east of Oldmeldrum, and around Laurencekirk, whilst in **Tayside**, a nest box study area in Blairgowrie has boosted local populations. Pairs also nest at Chapelhill Trinity Gask, Loch of Clunie and Airntully Stanley, as well as at Hatton, Fithie, Roundyhill, Forfar Loch, Windyhills, East Seaton, Montrose Basin, Craigmill Den, Seaton of Usan, Frockheim and Glen Isla in neighbouring **Fife**. Wintering flocks of up to 60 birds may also be encountered in stubble fields around Thirdpart in Crail, along the Coble shore, on the Eden Estuary, at Damhead of Torr, around Clatto Reservoir, in Muir Park, St Andrews and around Freuchie.

In **Ayrshire**, up to three pairs normally nest at Culroy (south of Ayr on the B7024), with 1-2 pairs at Mainholm smallholdings, Little Spierston, Minishant, Auchendrane, Knockdon, Grange Mains, Knockdunder Skares, Milton Farm Galston and at Blairbowie. Up to 30 winter at Mossbog.

The species is a common but localised resident in **Lothian**, with territories normally occupied in Cramond, Dalmeny, Lothianbridge, Wheatlands, Bolderston, Cairnpapple Hill, Mayfield, Prestongrange and Straiton. Winter counts of up to 80 birds are regular from the West Funton-Luffness area, Chapel, Newtongrange, Tyninghame, Drem, Whitburgh Estate and Straiton Pond, whilst in neighbouring **Borders**, flocks of up to 170 birds have been seen at Newmains in winter and of 40 at Netherbyres.

In **Dumfries & Galloway**, the only reliable site is in the Kirkton area, north-west of Locharbriggs on the A701.

The species is still relatively abundant in **Cumbria**, with nest box schemes at Aspatria and Campfield Marsh increasing numbers and birds nesting regularly at Caldbeck and Millom. In winter, flocks of up to 80 have been seen at Penrith and Low Foulshaw and of 60 in fields near Lowca and near Humphrey's Head.

Although declining quite dramatically in many areas, a few strongholds remain in the Northwest. One such area is the Fylde Peninsula (**Lancs**), where a nest box scheme has helped preserve over fifty pairs of Tree Sparrows in recent years. Winter feeding flocks of

100-130 birds can regularly be seen in North Fylde near Hambleton, near Bone Hill Farm (SD 436 465) and at Eagland Hill (SD 430 451).

In **Cheshire**, small numbers are present each summer near the car park at Inner Marsh Farm RSPB reserve (accessed from Station Road, Burton, to Burton Point Farm) and winter flocks have reached 100 on nearby arable land (between reserve and Shotwick Boating Lake). In the Northwich area, six or more pairs breed in the Forge Bed/Marbury No.1 Tank area adjacent to Neumann's Flash, and up to 50 are regular in the Forge Bed area in winter. A further regular wintering site is Birkinheath Covert (at SJ 762 836) near Tatton Park, where up to 65 birds are usually present.

In **Greater Manchester**, recent winter flocks of 30-50 birds have been recorded at Astley Moss SSI (SJ 697 977) and Worthington Lakes (SD 582 107).

In **Gwynedd**, the only regular site is of a single pair that nest in an ash tree at 'Pili Palas' Butterfly Centre near Menai Bridge on Anglesey - the tree is just north of Reception Centre.

In **Northumberland**, one of the better sites is the winter feeding station at Big Waters NR, near Wide Open, where up to 40 birds are regularly present between September and March. Nestboxes are used at West and pairs also breed at Whittonstall, Corbridge, Fenham Mill and Prestwick Carr. Larger numbers are present in **County Durham**, with Hurworth Burn Reservoir, Chilton Moor, Pittington, Wellands Farm Whitburn, Seaton Hall, Joe's Pond, High Sharperley and Kibblesworth all attracting breeding pairs. Up to 80 regularly winter at Cowpen Marsh (**Cleveland**) and 50 near Crookfoot Reservoir.

The species is widespread in **Greater Yorkshire**, with flocks of 150-230 birds being regularly counted in winter wheat or plough at Broomhill Ings (many also present at the winter feeding station outside the hide when in operation), Wintersett village (breeds in the trees around the few houses in the village), Wilsic, *HMP* Lindholme and at Misson Springs. Thorpe Marsh, Doncaster, regularly attracts a post-breeding flock of about 40 birds during August-December and nesting colonies are regularly attended at Broomhill Ings (5+ pairs), Thrybergh CP (7 pairs) and Finningley (5 pairs). In **East Yorkshire**, up to 60 birds are present in the Flamborough area (nesting in most farms and disused buildings), at Spurn Point (mainly around the Canal Scrape car park and bushes, at The Warren and in the hedgerows bordering the road between Kilnsea Churchyard and the Canal Bushes) and at Tophill Low Reservoir (along the Easingwold Strip). In **West Yorkshire**, the largest winter flocks are noted at Knotford Nook and Caldene Fields and in **South Yorkshire**, localised breeding pairs are found at Firsby, Brampton, Wentworth and Ulley. Good numbers are also present at Filey Dams NR throughout the year, with a few pairs utilising nestboxes and up to 60 feeding in winter, whilst 122 were counted at nearby Donkey Bridge in October 1993.

Large wintering flocks can be found in **Lincolnshire**, where many set-aside and rough grassland areas as well as Pheasant grain-feeding stations offer ideal feeding opportunities. Regular sites include Goxhill Marsh (up to 200 birds), Branston Booths, New Holland Pits, Market Rasen, North Cotes, Moulton Marsh, Aubourn, Deeping St Nicholas, Kirkby Moor, near Thurlby, Bourne Fen, Langtoft Pits and Billinghay.

In **Nottinghamshire**, up to 50 pairs nest on farm buildings in the Idle Valley and counts of up to 200 birds are regularly made between December and February, whilst in neighbouring **Derbyshire**, wintering flocks are regularly present at Carr Vale NR SW Bolsover (also breeds), Willington Power Station, Repton Shrubs (immediately west of Foremark Reservoir), Long Eaton Gravel Workings, Stanley Common and Bakestone Moor.

In **Leicestershire**, the species is a fairly common resident breeder, with small colonies nesting at Rutland Water Egleton Reserve, Wanlip South GP (SK 607 108), Barrow-upon-Soar (SK 577 176), Wellsborough (SK 362 024) and at Woodthorpe (SK 544 175). Large winter flocks may be encountered at Rutland Water (along the Hambleton Peninsula or at the

Egleton NR feeding station), in the Leicester Forest East/Lubbesthorpe area (SK 535 034/534 013) and at Scraptoft (SK 648 058).

It is a local resident in **Northamptonshire**, with wintering flocks of up to 90 birds being located at Pitsford Reservoir (at the new feeding station at the west end of the reservoir), Earls Barton GP (Summerleaze NR, on the minor road between Wollaston and Great Doddington), Kingsthorpe, Middleton Cheney and Merry Tom Lane, Chapel Brampton.

In the West Midlands region, Tree Sparrows are in serious decline, with **Staffordshire** perhaps holding the largest numbers. Wintering flocks of up to 100 birds may still be found at Trescott, with up to 40 at Whitemoor Hay, near Millmeece, around Blithfield and Belvide Reservoirs and at Wilnecote. Several pairs are known to nest at Sandwell Valley RSPB reserve (**West Mids**) and in **Warwickshire**, pairs regularly nest at Bascote (three), Brook End Farm, Bulkington, Chesterton, Coombe (10+), Coton, Dosthill, Jubilee Pools Coventry, Walsgrave Hill (7+), Walton, Water Orton and Whatcote. Wintering flocks of over 50 birds can be located at Chesterton, Compton Verney, Cropston Magna, Exhall, Grandborough, Priors Hardwick, Walsgrave Hill and Whitestone Nuneaton. Numbers remain scarce in **Worcestershire**, with breeding only regularly reported from Besford, Newnham Bridge, Peopleton, Pershore, Broome Pool,, Church Lench and Stoke Bliss.

In **Buckinghamshire**, Tree Sparrows are relatively common in the north of the county, with colonies in the Caldecott Lake area, around Drayton Parslow and in the Stoke Mandeville area, but much depleted in the south, with Dorney Common being the most regularly visited site in winter. Wintering groups may also be found in stubble at Little Linford, Finmere, Willen Lake and at Whaddon. The population is far more precarious in neighbouring **Bedfordshire**, where up to 20 birds may be found around farm buildings and in winter stubble at Boughton End (west of Lidlington and accessed from the A507) and small groups can be seen in Willington Woods (accessed from footpath north of Willington Dovecote from the A603 east of Bedford) and in the Tilsworth and Eaton Bray areas (nesting in holes in tall ash trees).

In **Gloucestershire**, the species has become very scarce, with most reports coming from the Severn Vale. One of the last remaining reliable sites is the New Grounds at Slimbridge, where feeding stations attract small numbers throughout the winter.

In **East Anglia**, Tree Sparrows are in serious decline, with many former sites now completely abandoned.

The best site in **Norfolk** in recent years has been the pig farm fields alongside the A1065 at Rowley Corner, four miles south of Swaffham, where up to 160 birds were present from January to April 1996. These birds also visit set-aside and stubble in fields around South Pickenham. Another regular site is the stubble fields around the drying barns north of Choseley Farm, about a mile south of Titchwell village, where 20-40 birds may be encountered, particularly in March and April. Birds are still present in the Lyng/Elsing area and around Hempstead, SE of Holt, but are difficult to locate.

The species is an uncommon resident in **Suffolk** where breeding birds are best seen around Worlingworth (south of Stradbroke) and at Minsmere and winter flocks of up to 50 birds in winter stubble or plough at Ellough, Fressingfield, Worlingworth, Sudbourne, Witnesham, Rattlesden, Haughley, Timworth and West Row Fen, Mildenhall.

In **Cambridgeshire**, up to twelve pairs nest regularly on the Ouse Washes RSPB reserve at Welches Dam (and remain to overwinter) and others can be found at Littleport and Sutton Gault, as well as Eldernell on the Nene Washes. It also breeds at Cow Lane GP and Ramsey Heights in Huntingdonshire, with wintering flocks regularly reported from Little Gidding, Woodwalton Fen, Buckden and Grafham village.

Breeding sites in **Essex** include Corbets Tey (7 pairs), Writtle (5 pairs), Walthamstowe Reservoirs (4 pairs), Hainault Forest (3 pairs), Esses Filter Beds (3 pairs) and Markshall (2 pairs), whilst wintering sites include Braxted Park (up to 30) and Margaret Roding (10-20).

The species has dramatically declined in **Hertfordshire** over the last ten years, with only one site regularly attracting birds throughout the year. This is at Beech Farm GP, Coopers Green, east of St Albans (at TL 190 090 - Sheet 166), where birds frequent the hedgerows either side of the footpath that leads past the gravel workings. Small flocks of up to 25 birds can also occasionally be seen at Tyttenhanger GP and Colney Heath.

The best site within the London area is Beddington Sewage Farm (**Surrey**) where a nestbox scheme has helped well over 60 pairs to successfully breed. These birds are resident and a flock of over 200 birds can now be regularly seen in winter. Access to the site is unfortunately restricted to permit holders only, but birds may be seen from the bridge to the north of Hackbridge Station. Park sensibly on the A237 near the station. Elsewhere in **Surrey**, the farmland and farm buildings surrounding Hookwood hold a resident population of up to 40 birds.

Tree Sparrows are still relatively widespread in **Kent** with large flocks still being encountered at some sites. The Dungeness RSPB reserve has several nesting pairs and flocks of up to 60 birds may be found around the farm buildings and barns along the Dengemarsh road, west of Lydd. Good numbers are also present at Sandwich Bay, Pegwell Bay, Cliffe village, Halstow Marshes, St Mary's Hoo and St Margaret's in winter, with breeding colonies present in Boughton Park, on the Hoo Peninsula and in orchards at Yalding.

It is a scarce species in **Sussex** being most reliably seen in the fields near Sidlesham village (immediately north of the Highleigh turning by the B2145), at Rye Harbour Wood, at Telham near Crowhurst and at Waterlot Level, Pevensey, whilst in **Berkshire**, the most reliable site is that at Slough sewage farm, where up to 50 birds may be found in winter.

Tree Sparrows are scarce and very localised in south Wales, with small numbers present in the mid Tywi valley in **Carmarthenshire** (around Ffairfach and Rhosmaen villages) and in farmland areas around Three Lords, Pendine, up to 40 in fields around Ffordd-Fawr and smaller numbers near Pennorth and at Garth, Bronllys, Garthbrengy, Llyswen and Pentregwyn Farm at Llanfihangel Nant Bran (all **Breconshire**), in crop fields adjacent to Kenfig NNR and at Aberthaw (**Mid Glamorgan**) and at Dingestow Court, Llandewi Skirrid and Porton in **Gwent**.

In **Oxfordshire**, small numbers are regularly noted at Stanton St John, Stonesfield, Radley, Hailey, Cassington, Tadpole Bridge, Rushey Lock, Northmoor, Sutton Courtenay, Wytham and Faringdon, with winter flocks appearing at Thame, Wootton Down Farm and at Beckley.

In **Wiltshire**, it is a fairly scarce resident, with the largest numbers occurring at one garden site, Wick Down, on the Marlborough Downs north of Ludgershall, where over 75 birds utilise the many nestboxes at the site. Elsewhere, birds are regularly seen at Pinkflower Gorse, Pertwood Down, Swindon sewage works,, Stanton St Bernard (north of the Kennet and Avon Canal four miles east of Devizes) and at Roundway Hill (north of Devizes).

Numbers have declined drastically in **Hampshire**, with the farm buildings at Well, two miles NW of Bentley and the A31, being one of the last remaining nesting sites of the species.

Tree Sparrows can be seen in **Avon** at the Avonmouth Sewage Disposal Works (up to 30), near Horton (up to 40) and at Marshfield and Keynsham, whilst in neighbouring **Somerset**, the best site is Ston Easton, eight miles north of Shepton Mallet on the A37.

The most reliable site in **Dorset** is Holcombe Bottom, Piddlehinton, where up to eight pairs breed and 40 remain to winter.

It is a rare vagrant in Devon and Cornwall.

Serin *Serinus serinus*

The Serin is a scarce passage migrant to Britain that has sporadically nested in southern England (Devon, Dorset and Sussex). About sixty birds are recorded in an average year, with most appearing in spring (between 13th March and 5th July) and a smaller number in autumn (between 6th September and 10th November). The majority of birds are recorded at south coast localities with a scattering of records on the east coast and in Midlands counties. Most involve brief fly-over sightings, with birds identified by their diagnostic jingling flight-call, but occasionally, singing males take up territory in a suitable area (normally an orchard or small group of coniferous trees) and remain for several days.

It is not possible to offer any guaranteed sites for Serin but a visit to the following locations in April and May may produce a sighting.

❷Portland Bill - Dorset
Serins are regularly recorded in spring at Portland Bill, with up to twenty sightings in an average year. Favoured localities include the telegraph wires around the Bird Observatory buildings, the bushes at Culverwell and the gardens and open fields to the north around Weston.

During the same period, smaller numbers are also recorded at Durlston CP, Swanage.

❷Dungeness - Kent
Between 20th March and 18th May, Serins can be regularly seen or heard as they head north or east over the Dungeness peninsular. The wind direction generally needs to be from the south, and it needs to be warm. Although most fly straight over, the odd bird settles for a few hours and may be seen singing from wires near the lighthouse and cafe, around the Observatory buildings or in the Trapping Area.

A similar passage of birds can be encountered at Sandwich Bay, St Margaret's Bay and in the North Foreland area, but rarely do any birds linger.

❷Selsey Bill - West Sussex
The small clump of trees at the old Pontin's holiday camp site immediately north of Selsey Bill is an excellent site in which to see incoming Serins throughout April and early May, with some birds remaining in the area for several days. They also visit the adjoining gardens and are best located by their distinctive flight-notes and jangling song. Turn left 600 yards before the bill seafront and drive through the residential area to the playing fields and old holiday camp site at the east end of the peninsula.

Twite *Carduelis flavirostris*

The Twite breeds in north and west Scotland and winters primarily on the east coast, between Lincolnshire and Kent. It breeds on heather moorland and in short turf of coastal cliff tops and winters in extensive saltmarshes or on arable farmland.

The following information details those sites with regular wintering flocks of 50 or more birds. Flocks tend to be very mobile in winter and easily disperse over acres of tidal saltings, and may be difficult to locate. They are generally present from mid October until late March.

In Scotland, the Twite is relatively abundant and easy to see, with flocks of 500-2,000 birds being regularly encountered at a number of coastal localities in winter. Recommended sites include Noss Head, Wick (Caithness), Strathallan (Perth & Kinross), Loch Ryan (Dumfries & Galloway) and John Muir Country Park, Tyninghame (Lothian).

Prime Sites

✸Walney Island - Cumbria
A large wintering flock of up to 300 birds may be found at Walney Island, between the Refuse Tip and a point just south of the Bird Observatory. They are often to be found in areas where cattle are feeding at approximately SD 201 634 (Sheet 96).

❍Flint Castle - Clwyd
A flock of up to 80 birds regularly winters at the edge of the narrow strip of saltings immediately upriver of Flint Castle at SJ 247 735.

✸Druridge Bay CP - Northumberland
A flock of 90-150 birds can regularly be found from late October until late April at Druridge Pools, north of Cresswell. They mainly frequent the coastal fields to the north of Blakemoor Farm and can be seen on wires along the coast road between Cresswell and Druridge.

❍Teesmouth area - Cleveland
A small flock of up to 20 birds may be found on Greenabella Marsh in winter (mainly December-February) and these may also be seen on Cowpen Marsh or along the Long Drag. Park along the A178 immediately south of Greatham Creek and walk west to view Cowpen Marsh and east to view Greenabella Marsh (to the north of the creek) and the Long Drag which runs south from the perimeter fence of Seal Sands reclamation area.

❍Scammonden Reservoir - West Yorkshire
Scammonden Reservoir is one of the best sites in Yorkshire, with flocks of up to 80 birds regularly frequenting the grassy meadows that border the reservoir. Leave Huddersfield on the A640, turning right shortly after crossing the M62 on minor roads which skirt the south side of the reservoir. Check roadside fences and wires.

Nearby, the A672 that runs from Junction 22 of the M62 to Ripponden is very good for Twite in summer. The roadside burns and grassy meadows of Rishworth Moor support several nesting pairs and these can easily be seen perched on roadside fence posts or wires.

✸Saltfleetby area - Lincolnshire
Large flocks of up to 300 Twite may be found feeding on saltings in the Saltfleetby area in winter and these range from Sea Bank Farm in the south (TF 477 901) to Donna Nook in the north. For a central vantage point, it is best to park at Rimac (at TF 467 918) and walk north or south along the coastal footpath. Similar numbers can also be seen along the coastal footpath between Northcoates Point (TA 370 035) and Stonebridge car park (TF 422 998), especially around Horseshoe Point (TA 382 018).

✱Butterwick Marsh - Lincolnshire
Flocks of up to 150 Twite are regularly seen on the saltings at Butterwick Marsh, these birds remaining from November until April. Leave the A52 three miles east of Boston on minor roads through Butterwick village parking sensibly at TF 405 440 (Sheet 131).

Elsewhere in the Lincolnshire Wash, similar flocks are also present in winter at Frampton Marsh (TF 367 385), Witham Mouth (TF 380 390) and Holbeach Marsh (TF 408 338).

✱North Norfolk coast
A total of 300-500 Twite winter in north Norfolk, with most occurring in coastal saltmarsh west of Wells next-the-Sea. Prime sites include Titchwell RSPB reserve (check the saltings to the west of the bank and the area of *Salicornia* just inland of the beach, as well as the saltmarsh lagoon north of the centre hide and the saltings east of the Tern hide), Brancaster (the car park and golf course immediately east of the clubhouse), Burnham Norton (the grazing fields at Norton Marsh), Holkham Gap (on the saltings either side of the Gap) and Wells Harbour (in the fields and west of the beach road and the saltings in the harbour and on Lodge Marsh).

Further east, flocks of Twite can be regularly seen on the saltings and in the car park at Morston Quay, along the coastal footpath at Blakeney Bank and in the fields between Salthouse Beach Road and Kelling Water Meadows. Much smaller numbers can also be found along the south sea wall at Breydon Water.

❶Dunwich/Walberswick - Suffolk
Wintering flocks of up to 60 Twite may be found in the saltmarsh between Dunwich Beach and Corporation Marshes, Walberswick. These birds are typically wide-ranging and may be difficult to locate. The coastal footpath runs between the two sites but is a walking distance of some five miles. Twites can often be seen within 200 yards of Dunwich beach car park and 500 yards of Walberswick.

❶The Naze - Essex
A wintering flock of up to 250 Twite can regularly be found in the Hamford Water and Horsey Island complex of Walton on the Naze, and can be seen from various points along the seawall or from Stone Point (only accessible at low tide). Park in the designated Naze car park (by the cafe and tower) and walk north and west along the seawall.

❶Old Hall Marshes - Essex
Large flocks of Twite may be encountered in the coastal Salicornia along the seawall north of Tollesbury Marina. The footpath leads north to Old Hall Marshes, an RSPB reserve that has no vehicular access. The seawall path that leads east of here to Pennyhole Fleet will produce small flocks of Twite between November and February.

On nearby Mersea Island, Twites can be particularly abundant on Reeveshall Marsh, accessed along the seawall from East Mersea (TM 055 155).

❶Dengie Flats, Bradwell - Essex
Large wintering flocks of Twite are often present on the Dengie Peninsular, particularly on the Bradwell saltings. In Bradwell-on-Sea village, turn right towards St Peter's Chapel and park at the end of the lane at Eastlands. Continue on foot past the chapel to the seawall and Bird Observatory. The saltings to the south of here are most productive.

❶Shellness/Swale NNR, Sheppey - Kent
The most reliable site in Kent for Twite is the saltings at Shell Ness Point and the fields around the car park at the entrance to the Swale NNR. The flock normally consists of up to 40 birds. Both sites are situated at the far eastern end of the Isle of Sheppey, following the minor coast road from Leysdown on Sea. Continue past the beach huts to the gravel car park at the end and continue on foot to the point or along the raised bank to the reserve.

Common Crossbill *Loxia curvirostra*

The Common Crossbill is an uncommon resident that breeds in highly variable numbers in a widely scattered number of localities throughout Britain. The species is highly irruptive and large numbers from Scandinavia may arrive from June onwards in certain years. Crossbills breed early (from mid February onwards) and are largely restricted to spruce or pine trees and can be very elusive, often only being detected by their far-carrying '*jipping*' calls.

Reliable sites for seeing Common Crossbill can be highly unpredictable with numbers varying from almost nil in some years to an estimated 10,000+ in others, and there are few, if any sites which can guarantee a sighting. Locations which appear to sustain populations regardless of irruptions include Kielder Forest (**Northumberland**) (try from various roadside stops in the forest or from the Raptor Viewpoint at Bakethin), Dalby Forest NE of Pickering (**North Yorkshire**) (try from various tracks that traverse the forest), Derwent Dale (**South Yorks**) (in the fir plantations that border Ladybower Reservoir and in the valley at the north end of Derwent Reservoir), Thetford Warren (**Norfolk**) (in the clearfell plantations to the west of Thetford Warren Lodge, off the B1107 NW of Thetford) and in the few trees and water tower that surround Field Barn Cottages on the road between Santon Downham and the A134.

In Northwest England/North Wales, sites which may still yield sightings in lean years include Clocaenog Forest (**Clwyd**) (the conifer trees surrounding the lake at the Bod Petrual Picnic Site at SJ 036 512, off the B5105 ten miles SW of Ruthin) and in the Gwydyr Forest in the Snowdonia National Park (at Llyn Elsi at SH 783 554 or Llyn Sarnau at SH 777 592). Also recommended are the car park on the A4080 in Newborough Forest, Anglesey (SH 412 672), Black Lake in the Delamare Forest (**Cheshire**) (SJ 538 710) and Grizedale Forest (**Cumbria**) (SD 335 945) and nearby Tarn Hows viewpoint, Coniston (SD 325 997). Once again, it must be stressed that this species can be near impossible to find in some years and incredibly abundant in others.

In the Midlands, Cannock Chase (**Staffs**) and the Wyre Forest (**Worcs**) are the most regularly visited sites, whilst in the south, sightings can often be had in the New Forest, especially in the pine trees that border the railway and surround the car park at Beaulieu Road Station, east of Lyndhurst on the B3056.

Scottish Crossbill *Loxia (pytyopsittacus) scotica*

The Scottish Crossbill is considered by Evans & Turner 1996 to be a smaller-billed race of Parrot Crossbill but by Knox (see *British Birds* 83: 89-94) to be a species in its own right. The bird is largely resident in the central and eastern Highlands of Scotland and is somewhat restricted to the Caledonian Pine forest, particularly in Speyside. It differs from Common Crossbill in having a larger bill with less overlap at the tip and a much deeper vocabulary and from Parrot Crossbill in its smaller bill and habitat preference. It is a scarce bird and fluctuates in numbers depending on the availability of its main food source, the seeds of the Scots Pine.

It is entirely confined to the central and eastern Scottish Highlands, where it is resident from parts of Tayside, Grampian, Moray & Nairn and Inverness-shire to eastern Ross & Cromarty and south-eastern Sutherland. The population is estimated to be at least 250 pairs but may rise to 1,000 pairs in irruption years created by the abundance of the pine crop.

Scottish Crossbills may be found in a large number of suitable habitat in Scotland, but care must be taken to separate from the very similar Common Crossbill. The following sites are offered as the most reliable:

❶Loch Garten RSPB Reserve - Highland Region

Scottish Crossbills may be found in moderate numbers in Abernethy Forest and are seen at a number of localities within the reserve boundary throughout the year.

a) the pines bordering the footpath to the Osprey Centre and hide and in the area by the car park and road alongside Loch Garten (NH 977 183 - Sheet 36); b) the Loch Mallachie circular trail leading from the north end of Loch Garten and extending down the west side (NH 972 185); c) the plantation at Tore Hill, accessed from the layby at NH 992 179 (Sheet 36) and d) the Caledonian forest surrounding Forest Lodge. Park at NH 019 161 and explore the circular footpath that leads past Rynettin cottages and north parallel with the river as well as the footpath that leads north from the track at NH 013 162. Capercaillie and Crested Tit are both likely to be seen from these same footpaths.

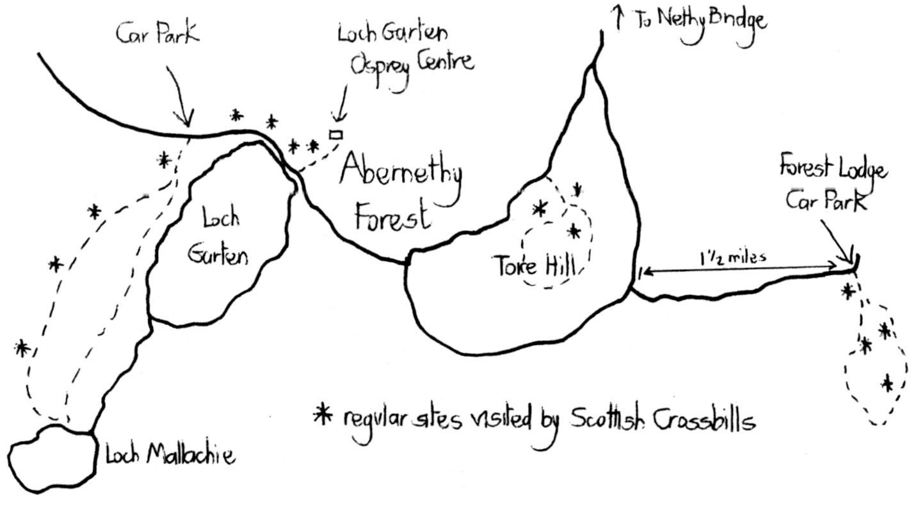

❶Grantown-on-Spey - Highland Region

The extensive forestry to the east of Grantown-on-Spey has a reasonably thriving population of Scottish Crossbills and birds may be encountered from the many footpaths that traverse the area. Leave Grantown on the B9102 and immediately after the golf course, park by the gate or roadside at NH 045 284. Follow the well-marked forest footpath for up to one mile and listen for calling birds.

❷Carrbridge - Highland Region

The woodland at the back of the Carrbridge Landmark Centre can often be very good for Scottish Crossbill. Park in the visitor centre car park on the B9153 and follow the trail that runs west from the north end of the car park.

❷Loch an Eilean - Highland

The nature trail that runs from the car park at NH 896 085 and continues clockwise around the entire shore of Loch an Eilean should produce sightings of Scottish Crossbill (and Crested Tit). Leave the Aviemore-Cairngorm Plateau road at Inverdruie and drive south along the B970. After two miles, turn left at Doune and continue to the car park at the north end of the loch.

❶Glen Tanar, Deeside - Grampian

One of the best sites for Scottish Crossbill is Glen Tanar, where several pairs frequent the Caledonian Forest that borders the river. Leave the A93 at Aboyne and follow the B976 westwards for just over a mile. A road leads left to Glen Tanar House and the reserve car park at NO 474 955 (Sheet 37). Follow the main trail to the west of the river for at least a mile until you come to the stone bridge crossing the river (at NO 462 941). The hillock beyond here with sparsely scattered pines is excellent for Crossbills.

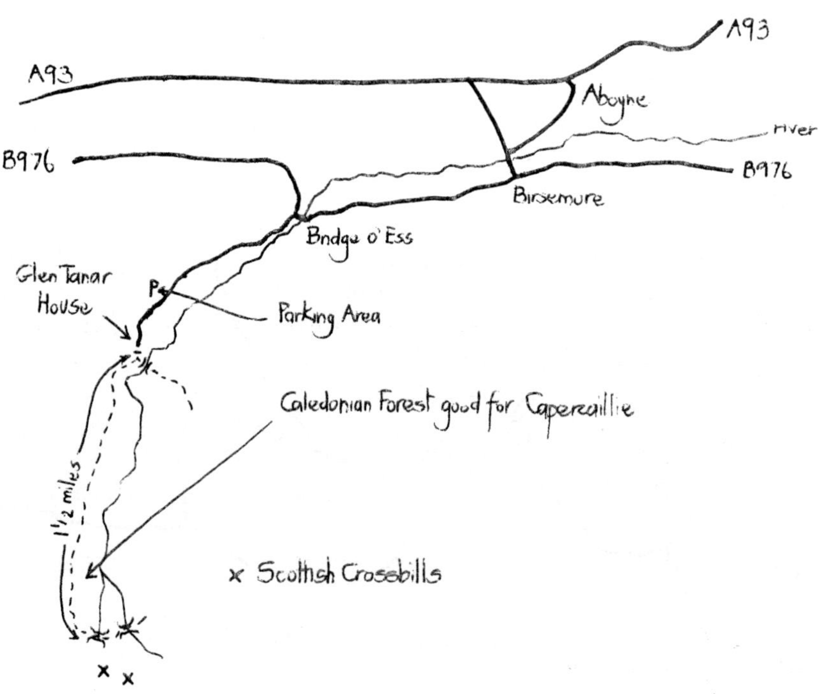

Common Rosefinch *Carpodacus erythrinus*

The Common Rosefinch is an annual scarce migrant to Britain which appears regularly in spring (late May to July) and autumn (from mid August to mid October). It is also known to make sporadic breeding attempts in Britain and in recent years has nested in Suffolk, Norfolk, East Yorkshire and at several sites in the Scottish Highlands. The species can be surprisingly skulking and on passage, singing males are often concealed deep within buckthorn or thick vegetation. They utter a clear and rather sprightly whistle, best transcribed as a three-part '*weege-weegde-vyu*', as well as a soft but sweet Greenfinch-like '*youeet*' flight-call. The majority of birds that reach Britain are sub-adult, and males rarely have the bright red on the head and upper breast that they gain in full adulthood. Autumn individuals are much streakier and are possibly one of the most nondescript and uninspiringly plumaged migrants that visit our shores. In an average year, about 40 are recorded in spring and about 75 in autumn, but the vast proportion of these are found on the Northern Isles and Scilly.

Rosefinches in Spring

Common Rosefinches are noted in spring between the 10th May and 10th July, with the majority appearing in the Northern Isles in June. Favoured localities include Baltasound on Unst, Geosetter Burn, Virkie, Scatness and Sumburgh Head on mainland, Fetlar and The Skerries (all Shetland), Fair Isle and North Ronaldsay (mainly Holland House and the croft gardens) and Stronsay (both Orkney). Further south, regular sightings in June are known from Holy Island and the Farne Islands (Northumberland), South Gare and Hartlepool Headland (Cleveland), Flamborough Head and Spurn Point (East Yorks), Bardsey Island (Gwynedd), Gibraltar Point NR (Lincs), Winterton North Dunes (Norfolk) and at Southwold and Minsmere Beach (Suffolk). Sightings generally follow a period of south-easterly winds.

Rosefinches in Autumn

Common Rosefinches are noted in autumn between the 24th August and 30th October, with up to 75 being recorded in an average autumn. As in spring, the majority of sightings are confined to the Northern Isles, where the Sumburgh area, Fair Isle and North Ronaldsay monopolise most of the occurrences. Elsewhere, small numbers are restricted to various east and south coast localities, with the best chance of connection on the Isles of Scilly in October. Here, birds congregate with flocks of Linnet and Goldfinch in fields at Borough Farm, Maypole, the Silver Trail, Penninis Head and Pelistry on St Mary's and in the Borough Farm fields on Tresco.

Hawfinch *Coccothraustes coccothraustes*

The Hawfinch is a rather secretive uncommon resident in Britain, which is widely but very locally distributed throughout much of England, Wales and southern Scotland. They favour deciduous woodland, large gardens, parks and orchards, and have a particular preference for Hornbeam stands, from which they feed on seeds. They are very shy birds and can be extremely difficult to observe, and are best seen in winter when they form small flocks in the canopy and often congregate at a regular roosting site. More often or not, however, sightings may be confined to glimpses of birds as they disappear through the canopies, often uttering their rather soft 'ticking' flight-call.

Prime Sites

❷Scone Palace - Perthshire
A small population of 2-10 Hawfinches can be found in the grounds of Scone Palace, north of Perth. They favour the Hornbeams and other berry-bearing trees at the north end and are best seen early mornings between March and May. Leave Perth on the A93 and after two miles park sensibly in the vicinity of the entrance at NO 119 269 (Sheet 58). Follow the main ride down towards the Palace and check the stretch of woodland between the two main tracks, listening carefully for the birds' distinctive 'ticking' call.

❶The Hirsel - Borders
Up to six Hawfinches are regularly seen in the woodland that borders the south end of Hirsel Lake, part of a private estate that lies one mile north-east of Coldstream. Dundock Wood, as it is known, comprises of Oak, Ash and Pine trees, with a dense understorey of Azalea and Rhododendron, and regularly attracts Hawfinches in summer, but one of their favourite haunts is the fruit trees that grow between the old stable block/visitor centre and the main house. March and April are the best months in which to see this secretive species.

The main entrance to the Hirsel is signposted off the A698 Kelso to Coldstream road at the south-east Lodge (at NT 837 934 - Sheet 74), on the outskirts of Coldstream. There is a car park near to the visitor centre at NT 827 401, about a mile from the road. A further car park is situated at NT 820 395, giving immediate access into Dundock Wood.

❷Sizergh Castle - Cumbria
Hawfinches can occasionally be seen in the mature beech and cherry trees at Sizergh Castle, alongside the A591 north of Junction 36 of the M6. Park in the National Trust car park at SD 498 879 - Sheet 97.

Small numbers can also be occasionally seen in spring at Rusland Valley, northwest of Newby Bridge. Park on the road and scan the tall beeches at SD 345 873.

❶Woodwell, near Silverdale - Lancashire
Woodwell is easily the best site in Northwest England, where up to twelve birds can easily be seen particularly in early spring. The birds can be located at SD 463 744 (Sheet 97) especially in the cherry trees behind the Alpine Garden Centre, in gardens near the Wolf House Gallery and on the ground feeding on beech mast beside the wall running along the back of the field between Woodwell track and the gallery.

❷Overleigh Bridge - Cheshire
Hawfinches can be seen in small numbers at Overleigh Bridge, south of Chester. Park near the old Dee Bridge in the Handbridge district of Chester and take the footpath south through parkland for up to a mile. A small group of 3-5 birds is noted in January-April in most, but not every, year.

❷Caer-Hun - Gwynedd
One of the most reliable sites in North Wales is at Caer-Hun in the Conwy Valley and alongside the B5106 1 mile south of Tyn-y-groes at SH 773 705. Park near the churchyard and scan the tops of the adjacent tall trees. The period January-March is best with 2-5 birds present.

❷Hulne Park - Northumberland
The most reliable site for Hawfinch in Northumberland is Hulne Park, where up to twelve birds may be found from November to April. They favour the trees near the entrance gate on the B6346, northwest of Alnwick. Early morning visits are recommended.

Birds are also regularly seen in April around Allenbanks and in Hexham and Jesmond Dene.

❷Hutton/Guisborough area - Cleveland
The disused railway line that runs from Guisborough Woods through to the west end of Guisborough and Hutton village regularly attracts 1-5 Hawfinches (previously as many as 35 were recorded) in winter and these may be seen in surrounding gardens and in the shrubbery now profuse along the raised banks. Park in Guisborough at NZ 604 148 (Sheet 94) and explore the dismantled railway line in both directions. Nearby, up to twelve Hawfinches are regularly recorded in the trees surrounding Skelton Castle, particularly in spring, and 1-2 pairs generally nest in the area. Leave Skelton northwards on the B1267 and search the tree-tops to the north of the castle or along the riverside.

❷Studley Royal Park, Ripon - North Yorkshire
Hawfinches are resident in Studley Royal Park, with up to 20 present in winter and small numbers breeding in summer. They are most reliably seen in the mature mixed woodland adjacent to Fountains Abbey, part of the National Trust grounds. Early morning visits are essential as the site is very popular with tourists and visitors. Leave Ripon westwards on the B6265 and after two miles, the Abbey is well signposted on the left.

❷Bretton Sculpture Park, Wakefield - West Yorkshire
Hawfinches can be reliably seen in Bretton Park in winter and early spring in the woodland that borders the south side of Lower Lake. Leave Kexborough northwards on the A637 and take the entrance to the park just north of Haigh village.

❷Rockley/Stainborough - South Yorkshire
A population of up to twenty birds can be found in the Rockley/Stainborough area, SW of Barnsley. They winter in the woodland behind Rockley Abbey farmhouse, accessed from the Birdwell-Dodworth minor road, immediately north of the M1 bridge. A trail from the caravan club car park leads into the wood at SE 327 019.

These same birds summer at Stainborough and can be found in the beech and lime trees behind the educational castle between April and September. Early morning visits are essential. Park in the designated car park and walk behind the castle to the wood.

❷Broughton Woods - Lincolnshire
Up to nine birds are regularly noted in Broughton Woods between December and early March. Broughton Woods are situated east of Scunthorpe, either side of the B1207 north of Broughton. Access is restricted to a limited number of public footpaths which traverse the area. Park at SE 955 093 (Sheet 112) and walk north-westwards through West Wood or at SE 955 086 and explore Gadbury Wood.

Another good site but not so in recent years is Brumby Woods and Crematorium, where up to twelve birds regularly used to winter. Park by the A18 in Scunthorpe at SE 881 104 (Sheet 112) or on the minor road at SE 878 106 and explore the trees in the crematorium. Early morning visits are essential.

➊ Clumber Park - Nottinghamshire
Winter groups of up to twenty birds may be seen between October and February at the traditional site by the chapel in Clumber Park (SK 627 746 - Sheet 120). The Hawfinches feed in the tops of the trees between the chapel and Clumber Bridge (at SK 621 739) or in the wood immediately north of the chapel. Early morning visits are best.

Clumber Park is situated SE of Worksop and accessed from the minor road off the B6034 at Carburton or from the A614 just south of its junction with the A1 and A57.

➋ Cromford Canal - Derbyshire
The relict woodland at Rose End Meadows and adjacent Cromford Canal and Wharf is the most reliable site in Derbyshire for seeing Hawfinch, with 2-3 birds present there from November to early April. Follow the footpath from the Wharf at SK 300 570 (Sheet 119) south-eastwards along the canal.

The Cromford Canal lies south of Matlock and straddles the A6 throughout its length. Park at Cromford Wharf.

➋ Fineshade Wood - Northamptonshire
Small numbers of Hawfinches are to be found in Fineshade Wood throughout the year, favouring the woodland and gardens to the east of Top Lodge (SP 979 984 - Sheet 141). Leave Duddington south on the A43 and turn left to Top Lodge after 2.5 miles.

Elsewhere in Northamptonshire, 1-4 birds are regularly seen in the trees around Barnwell village, particularly in the row of Hornbeams bordering the cricket pitch. Barnwell lies adjacent to the A605, two miles south of Oundle.

➋ Whitcliffe Common - Shropshire
Whitcliffe Common is a traditional site for wintering Hawfinches with up to twelve birds regularly recorded between December and early April. The favoured area is the line of Hornbeam trees in the gully which aligns Dinham Bridge, where flocks of birds can be seen or heard in the tree canopy. Walk from the castle in Ludlow to cross the River Teem at Dinham Bridge.

➋ Bewdley - Worcestershire
Hawfinches were present in Jubilee Gardens, in Bewdley Town Centre, for three consecutive winters from 1993-5, feeding in ornamental trees within the park. Up to eight birds were seen, primarily in early morning, between January and mid April. The site is situated just south of the town centre.

➊ Flackwell Heath - Buckinghamshire
Up to six Hawfinches can regularly be seen in the trees that border Blind Lane at the west end of Flackwell Heath, two miles SE of High Wycombe. At the junction of Blind Way and Wood Way, a footpath leads down towards an area of woodland and opens out into fields past four houses. The birds can be seen in the scattered trees in this area during early morning.

➊ Nagshead Reserve - Gloucestershire
Nagshead RSPB reserve is situated in the Forest of Dean and is an excellent site for Hawfinch, with up to ten pairs present between April and August. They favour the larger and older trees and have a particular preference for the Hornbeam trees on the reserve. Leave Parkend westwards on the B4431 and turn right to the reserve after less than one mile. Follow this entrance road to the information centre and car park and access various trails from there.

Hawfinches may also be seen in winter in the tracts of woodland and arboretum south and east of Speech House Hotel, by the B4426 halfway between Cinderford and Coleford.

❶Holkham Park - Norfolk

Hawfinches may be seen feeding on beechmast or in the tree canopy just inside the main gates at Holkham Park. They favour the trees to the SE of the entrance and may be seen from the footpaths between December and early April. They are best seen prior to 9.00am and parties of up to six birds can be seen feeding with Chaffinches under the beech trees or the single evergreen or flying overhead calling.

Leave the main coast road (A149) at Holkham crossroads and park in the designated car park on the right. Continue on foot to the gated entrance and observe from the road that runs towards the hall and lake or from the footpath that runs alongside the wall.

❶Lynford Arboretum - Norfolk

Lynford Arboretum is a very reliable site for Hawfinch between December and early April, with small parties of up to twelve birds being visible in the trees at the Lynford Hall end of the arboretum or in the large beech trees in the field beyond the lake (see sketch). They are best seen early mornings, when they can be seen on the ground feeding under the trees. It is essential to be conversant with their distinct 'ticking' calls, somewhat similar to Song Thrush or Robin, but more direct and far-carrying.

Leave Mundford northwards on the A1065 and turn right after half a mile towards Lynford Hall. Continue past the hall and park in the designated car park on the left. The arboretum lies adjacent to the hall and is accessed via a number of well-marked footpaths. The Hawfinches favour the tops of the trees and range widely over the area.

Elsewhere in the Norfolk Breckland, Barnhamcross Common, immediately south of Thetford on the A134, holds varying numbers of wintering Hawfinches and birds may also be seen in the avenue of trees that line the Brandon to Santon Downham minor road and in the remaining poplar plantations west of Lakenheath Station.

❶Staverton Forest, near Butley - Suffolk

Up to six Hawfinches may be seen in Staverton Forest, around the layby at TM 363 504 or along the public footpath that runs north through The Thicks. They are present from late March to early May and best seen early morning. Leave Woodbridge eastwards on the B1084 and continue towards Butley. The wood is situated to the north of the road, about half a mile west of the village.

❷Danbury Lakes CP - Essex

In some recent winters, flocks of up to 35 Hawfinches have been seen in the vicinity of the car park and the lakes at Danbury Country Park in Woodhill. They favour the beech and Hornbeam trees at the north end of the wood as well as the trees in the neighbouring Danbury Palace grounds. Take the Sandon road out of Danbury and park on the right at TL 770 048 (Sheet 167). Walk west from the car park towards the lakes, scanning the tree canopies and listening for the distinctive and rather unobtrusive calls. Early morning visits are essential, with January to early April the peak period for sightings. Lesser Spotted Woodpecker is also regularly recorded here.

❶Hatfield Forest - Essex

Up to 15 Hawfinches may be seen in winter in the vicinity of the car park at Hatfield Forest. They are best seen in the first two hours of daylight as severe disruption at this site from dog walkers and general public is commonplace. Leave Bishop's Stortford eastwards on the A120 and turn right in Takeley Street on the minor road to Hatfield Broad Oak. The entrance to the forest is on the right after a mile. Park by the gate and traverse the stile to the spinneys opposite, the birds favouring the stands of beech and Hornbeams to the right hand side of the track and the spinney which has the cattle feeder within it. The birds also quite often visit the road to feed, viewed from the track to the right of the car park where the sleeping policemen on the tarmac road form small puddles of water.

❷Broxbourne Woods - Hertfordshire
Small numbers of Hawfinches may be seen during January to April in the clearings and wood-edges in Broxbourne Wood. A regular area for sightings is the clearing at TL 334 073 (Sheet 166) accessed from the footpath north of the minor road between Brickendon and Hoddesdon. Alternatively, park at TL 329 069 and follow the well-marked trail north to Cowheath Wood.

Elsewhere in Herts, regular sightings are made in Post Wood, Ware, at Amwell GP, Ware and at Walkern Churchyard.

✱Bedgebury Pinetum - Kent
Bedgebury Pinetum holds one of the largest winter roosts of Hawfinch in Britain with up to 50 birds roosting between late November and early March. The birds flight in to roost in large cedars and other conifers from 3.00pm onwards and may be seen in pre-roost gatherings in nearby trees, perched high in the canopy. The Pinetum lies at the west end of Bedgebury Forest and is reached via Goudhurst in the north or Flimwell on the A21 to the south. The entrance is situated east of the B2079 at TQ 714 338 (Sheet 188). Follow the main footpath down into the steep valley keeping the ponds on your left. Climb the slope on the other side and the Hawfinches roost in the firs to the right of the main footpath and gather in the fenced off plantation to the south (at TQ 720 338).

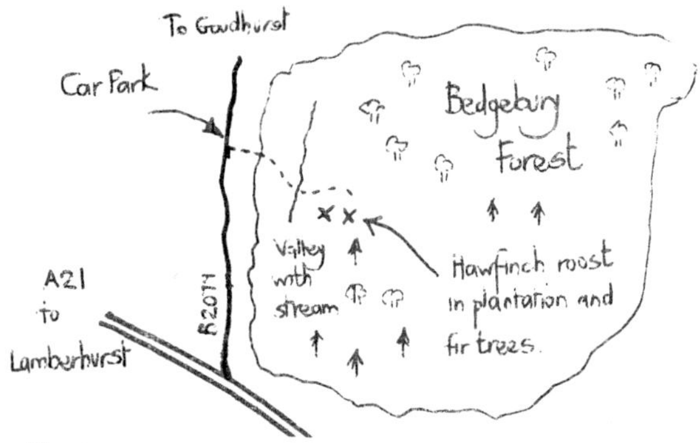

❷Walderslade - Kent
Tunbury and Round Woods, in the southern suburbs of Walderslade, have varying numbers of wintering Hawfinches and can be accessed from the road that runs from the A229 roundabout and Lords Wood or from the minor road that borders the northern edge at TQ 765 623 (Sheet 178).

Hawfinches can also be seen with some regularity in Larkeyvalley Wood, SW of Canterbury (near Thanington).

❷Windsor Great Park - Berkshire
Windsor Great Park was once a traditional site for wintering Hawfinch which, in recent years, has failed to attract any sizeable flocks. The favoured area was the berry-bearing trees adjacent to the pink house at the Blacknest Gate, on the A329 at Virginia Water. The woodland surrounding the southwest corner of the lake was much visited between December and March, and this could be accessed via the footpaths from the car park at Blacknest Gate. What are presumably birds from the same population have also been seen in neighbouring Sunningdale, visiting large gardens, orchards and stands of trees alongside the B383. No flocks have been reported since 1994.

❷Blenheim Park - Oxfordshire
Flocks of up to 15 Hawfinches are regularly recorded in Blenheim Park between December and early April, most often in the beech trees and shrubs around the garden centre. The main access to the park is from Woodstock, six miles north of Oxford on the A44. The garden centre is well signposted. Early morning visits are essential.

✱New Forest - Hampshire
The New Forest is an attractive area for Hawfinch with a wintering population of about 130 birds. There are two main sites:-

1). **Rhinefield Inclosure**: a winter roost of up to 30 birds can be seen utilising ornamental cedars and tall evergreen trees between mid November and early April. They arrive from feeding sites elsewhere in the area from mid afternoon onwards and are all generally in roost at least an hour prior to dark. They afford excellent views as they gather pre-roost at the tops of nearby trees and announce their arrival by their distinctive, but relatively difficult-to-hear, call-notes. Leave Lyndhurst south-westwards on the A35 and after two miles turn left towards Brockenhurst. After just over a mile, park by Black Water brook in the Rhinefield car park and walk back across the road to the arboretum. Stand in the vicinity of the bench just inside the gate and wait for roosting birds to fly in overhead.

2). **Pitts Wood Inclosure**: up to 30 birds can be seen wintering at this site, feeding in sheltered trees in the valley between Cockley Plain and Hampton Ridge. Park along the B3078, about three miles east of Fordingbridge, in the Ashley Walk car park and follow the main footpath across the valley towards Pitts Wood. After about a mile, the footpath leads down through an area of scattered bushes and trees and this is the favoured haunt of the Hawfinch flock.

❷Penybanc, near Llandeilo - Carmarthenshire
One of the few reliable sites in South Wales is Penybanc, near Llandeilo, at SN 61 24. 1-3 birds are seen here most years, primarily in spring. They visit various garden bird tables in the village.

❷Cattistock Churchyard - Dorset
The cherry-bearing trees in Cattistock churchyard regularly attract 1-4 Hawfinches in winter and may be seen early morning. Leave Dorchester north-westwards on the A37, turning left on the A356 and north in Maiden Newton.

Lapland Bunting *Calcarius lapponicus*

The Lapland Bunting is a regular autumn migrant to Britain, with up to 300 birds appearing between mid September and mid November, primarily in the Northern Isles, along the east coast and on Scilly. Small numbers also winter in Britain, but these are largely confined to East Anglia and the north-east coast. Wintering flocks favour stubble fields and saltmarshes, whilst passage birds regularly appear on short turf, airfields, golf courses and heather. Migrants are particularly confiding but may often only be detected by their distinctive flight calls.

Lapland Buntings in autumn may be seen at a number of regular sites including Fair Isle, Shetland South Mainland, North Ronaldsay and Stronsay, Holy Island, Flamborough Head, Spurn Point, Saltfleetby/Rimac, Cley Beach, Porthgwarra and on heather-clad promontories on Scilly, such as Wingletang Down (St Agnes) and Porthellick Down, Peninnis Head and Giant's Castle (St Mary's).

It is a scarce winter visitor, and has declined to a population of less than 200 birds in recent years, the following being the most regularly visited sites:

❷Tyninghame CP - Lothian
Up to ten Lapland Buntings can regularly be found in winter in the saltmarsh at John Muir Country Park, Tyninghame. Park in the designated car park, north of the A1087 west of Dunbar, and explore the area of saltmarsh that lies adjacent.

❶Holy Island - Northumberland
Up to 25 birds can regularly be found in the saltings at Goswick Sands, on Holy Island, from December to February. Smaller numbers can also be seen at Ross Links (NU 137 373), walking north across the saltmarsh towards the dunes.

❶Teesmouth - Cleveland
The Teesmouth area is one of the most reliable areas in which to see Lapland Bunting in winter with small flocks of birds being regularly present at three different locations:

a). Cowpen Marsh and the surrounding area, where up to 40 remain from December to March; b). Saltholme Pools, where ten are regularly seen throughout January and c). Seaton Common and adjacent stubble fields, where up to 30 remain from January to March.

All three sites are easily accessed from the A178 coast road, between Billingham and Seaton Carew. Cowpen Marsh can be reached via the footpath running along the south side of Greatham Creek towards the tip.

❶Filey North Cliff Country Park - North Yorkshire

Flocks of up to 200 Lapland Buntings have been recorded at Filey in recent years, with birds feeding in clifftop stubble fields to the north of the Brigg from December to late March. At the roundabout on the A1039 in Filey, follow the signs to the country park and caravan park to the north. Park at the north end of the car park and follow the Cleveland Way coastal footpath to view the fields.

Further south, much smaller numbers of birds can be found in some years in the coastal stubble fields at Flamborough Head or in the fields to the north of the beach car park at Barmston, five miles south of Bridlington.

❶North Norfolk coast

In recent years, a marked reduction has become apparent in wintering numbers of Lapland Buntings in Norfolk. In fact, at one of the most reliable sites, the grazing fields to the north of Burnham Norton, sightings have been almost non-existent. In the winter of 1995/96, Lapland Buntings could only be found in the fields to the north of the pig farm at West Runton (TG 187 430 - Sheet 133) and in the large stubble fields at Stiffkey Barns (TF 984 437 - Sheet 132). Sites that did regularly attract birds and may well do so again in the future include the clifftop fields between Weybourne Coastguard Cottages and Sheringham Golf Course, stubble fields inland of Bacton village and in the saltings and adjacent fields along the north and south sea walls at Breydon Water.

❶Elmley Marshes - Kent

Up to 30 birds can regularly be found wintering in stubble fields and in the grassy meadows in the vicinity of King's Hill Farm, on the Elmley Marshes RSPB reserve on the Isle of Sheppey. Leave the A249, about a mile north of the Kingsferry Bridge, on a rough track eastwards for two miles. The reserve entrance is at the farm and Lapland Buntings may be seen along the access track up to a mile before the farm, along the main track to the reserve within the first 400 yards or along the track to the right that leads down to the Brickfields Pit.

Elsewhere in Kent, small numbers can also be found in winter at Cliffe Pools, Allhallows Marshes, Capel Fleet, Sandwich Bay and in the Reculver area.

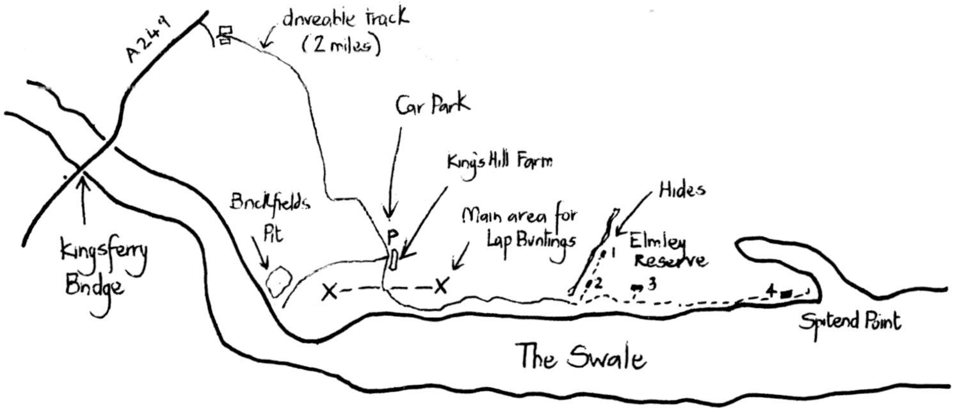

Snow Bunting *Plectrophenax nivalis*

The Snow Bunting is a regular winter visitor to Britain, which fluctuates in numbers and is largely confined to the east coast, between East Yorkshire and North Kent. It is also a scarce breeding resident of the Scottish Highlands, with up to twenty pairs nesting on the Cairngorm Plateau.

Summer

✱Cairngorm Plateau - Highland Region
Snow Buntings may be seen in the breeding season around the cairns and weather station at the peak of Cairn Gorm or along the footpath that leads south towards the peak of Ben Macdui. Leave Aviemore eastwards on the B970 and continue as far as the upper ski-lift car park. Take the chair-lift to the Ptarmigan Restaurant and walk to the cairn. Do not attempt in poor weather conditions.

In winter, these birds and others from surrounding mountain top populations scavenge on scraps from skiers and tourists in the upper car park. They remain, despite deep snow, and can be extremely confiding, and many are colour-ringed.

Winter Sites

➊Nairn/Culbin Bars - Moray & Nairn
Flocks of up to 200 birds regularly commute between the stretch of Moray coastline from Nairn Bar (NH 922 603) to Culbin Bar (NH 965 627) and are present from early December until March. Both sites are within easy access of the A96 coast road, either side of Nairn.

➊Ythan Estuary - Grampian
A flock of up to 50 birds is normally present in the dunes and golf course area to the south of the Ythan Estuary between December and February. Turn eastwards by the hotel and pub to the south of Newburgh and continue to the dunes car park at the end.

Northwest England
There are now no reliable guaranteed coastal sites in the region, but in most years, small numbers (1-5 birds) can be found on beaches

at Rossall Point near Fleetwood (**Lancs**) (SD 315 480)
at Pensarn near Abergele (**Clwyd**) (SH 935 784)
at Red Rocks, near Hoylake (**Cheshire**) (SJ 202 885).

The one regular site is the summit of Pendle Hill, above Clitheroe (**Lancs**), where 10-20 birds are regularly present in winter at SD 805 414. Much larger flocks (50-100 birds) are occasionally found by hill walkers in the Lake District National Park (**Cumbria**) at elevations up to 3,000 feet.

➊Goswick Sands, Holy Island - Northumberland
Flocks of up to 50 birds are regularly present on Goswick Sands, along the north shore of Holy Island, between late November and February. Holy Island can be reached via the A1 at West Mains through Beal and *can only be visited at low tide*. The causeway is impassable for at least two hours either side of high tide. Goswick Sands can be reached from the first car park on the left as you arrive on the island.

Elsewhere in Northumberland, the only other regular flock is at Druridge Bay CP, east of Widdrington. These birds are highly mobile and range over a wide area of coastline.

✱Teesmouth area - Cleveland
Flocks of up to 160 birds can be found in winter in the Teesmouth area, with regular flocks of 80-120 birds frequenting the coastline between North Gare and Seaton Common and up to 160 commuting between South Gare and Coatham Marsh (often in the dunes just north of the

golf course). They arrive in mid November and depart at the end of February, but may be difficult to locate due to the nature of the sites.

➊ Filey North Cliff Country Park - North Yorkshire
A flock of up to 50 birds is generally present in the clifftop fields north of the country park and these can be viewed from the Cleveland Way coastal footpath. The birds are present from mid November until early March.

➊ Spurn Point - East Yorkshire
Following the arrival of autumn migrants from early October onwards, a flock of up to 60 Snow Buntings can often be seen in the canal scrape area or by Easington Lagoon until mid December. Birds rarely winter at this site however.

➊ Saltfleetby area - Lincolnshire
A wintering flock of up to 80 birds is usually present in the Theddlethorpe Dunes-Saltfleetby area between late November and February, feeding north as far as Saltfleet Haven and south to Theddlethorpe St Helen. From ringing recoveries, it is known that birds from this site regularly winter in north Norfolk, and it is likely that some interchange regularly takes place, depending on weather conditions. The area is best accessed from the car park at Rimac (TF 467 918 - Sheet 113), with access to the dunes straightforward.

✱ North Norfolk coast
The north Norfolk coastline regularly attracts a wintering population of up to 200 Snow Buntings but these birds are highly mobile in their movements and often difficult to locate. In some winters, a flock of over 80 birds frequent the clifftop car park and golf course at the east end of Hunstanton, and these may also visit the saltings and dunes at Holme, the sandy beach and golf course to the east of Brancaster Clubhouse and the stretch of beach between the tern hide at Titchwell reserve and Thornham Point. Scolt Head Island may also be an attraction in some winters, as well as Heacham North Beach and Burnham Overy Dunes and the saltings in Holkham Bay, north of the Gap car park. Further east, a flock of up to 75 birds regularly commutes between Cley Coastguards and Salthouse Beach, but are difficult to find and are often in the section of shingle bank between Gramborough Hill and Salthouse Beach car park and Arnolds Marsh, east of Cley East Bank. Further flocks may also be encountered in the coastal fields between Weybourne Coastguard Cottages and the west end of Sheringham Golf Course.

Much smaller numbers are recorded on the Norfolk east coast, with birds occasionally wintering in Great Yarmouth along the beach between North Denes and the Lifeboat Station at the mouth of the river and at Happisburgh.

➊ Aldeburgh/Thorpeness - Suffolk
A regular wintering flock of up to 50 birds may be found along the shingle beach between Thorpeness and Orfordness. They are typically mobile and often difficult to locate and are present from November-February inclusive.

Smaller flocks are sometimes seen at coastal localities further north in Suffolk, including Sizewell Beach, Walberswick beach and the beach area between Benacre Pits and sluice and Kessingland.

➊ Sandwich Bay - Kent
The only regular wintering flock of Snow Buntings in Kent is that of up to 30 birds on the golf links and beach at Sandwich Bay. These birds arrive in December and leave in mid February and range as far north as Shell Ness. Leave Sandwich on the minor road to the estate and park by the Bird Observatory, half a mile east of the tollgate.

In recent winters, a flock of up to 25 birds has also frequented the coastal fields and shingle beach west of Swalecliffe and the coastal seawall west of Seasalter.

Cirl Bunting *Emberiza cirlus*

The Cirl Bunting is a rare breeding bird in Britain which is now confined to Devon and Cornwall. The stronghold is in south Devon, where up to 400 singing males occupy suitable territories in spring. These territories are generally found within coastal scrub adjacent to arable land or on agricultural land with hedgerows and are often on hillsides with scrub and bushes or in river valleys. Between late October and early March, Cirl Buntings form flocks of up to 40 birds to feed on cereal stubble. The distinctive rattling and quite low-toned song can be heard from mid February to June.

✱Prawle Point - Devon

Cirl Buntings are relatively abundant in the Prawle Point area and can be seen throughout the year at the site. In winter, they gather into sizeable flocks and may be found feeding in the cereal stubble adjacent to the access lane but in spring and summer, are to be found in the vicinity of the scrubby hillside that borders the car park or in the hedgerows that border the lane or skirt the coastal footpath to East Prawle. They are also present in good numbers in Pig's Nose Valley, which is reached via the footpath at the western extension of the road west of East Prawle village.

Leave Kingsbridge eastwards on the A379, turning right at Frogmore or Chillington onto minor roads south to East Prawle village. Continue through the village to the National Trust car park at the end of the road.

Cirl Buntings can also be seen from the coastal footpaths that lead from East Soar Farm, south of Salcombe, following a circuitous route taking in Soar Mill Cove and finishing up in Soar village. The birds are often in the vicinity of the hotel at the top end of the valley or in the bushes along the coastal footpath.

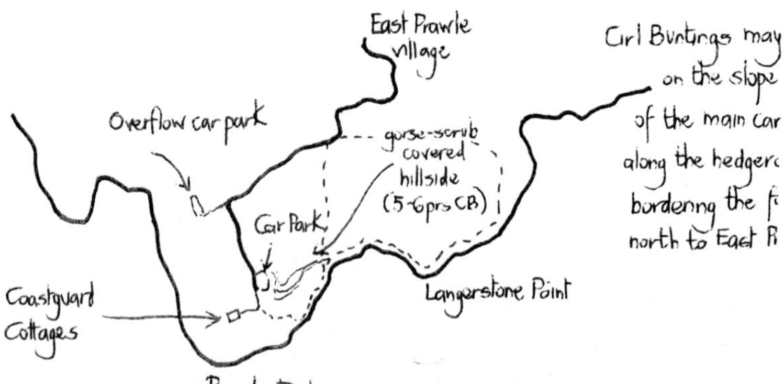

❷Slapton Ley - Devon

The car parking area at the north end of Slapton Ley, known as Strete Gate, is a traditional site for Cirl Buntings, and singing males or pairs may be observed on the scrub covered slope that borders the car park. They can also be seen in the valley that leads inland from the Higher Ley.

Slapton is about six miles south of Dartmouth on the coastal A379. Strete Gate car park is at the north end of the Ley following the sharp bends on the incline.

✱Wembury Beach - Devon

The coastal footpath between Wembury village and *HMS* Cambridge is a well-known haunt of Cirl Buntings, with up to five pairs present in summer and up to 20 birds present in winter. They favour the scrubby slope to the north of the footpath and are often present in the gorse-scrub and bushes that border the paddocks to the west of the car park. Pairs are regularly seen feeding in these paddocks, especially during early morning or in the evening. Continue on the maze of roads to Wembury village and through to the car park and toilet block at the seaward side. Walk west along the coastal footpath and either bear right to check the paddock area or continue for one mile to the hedgerow and scrubby hillside at *HMS* Cambridge.

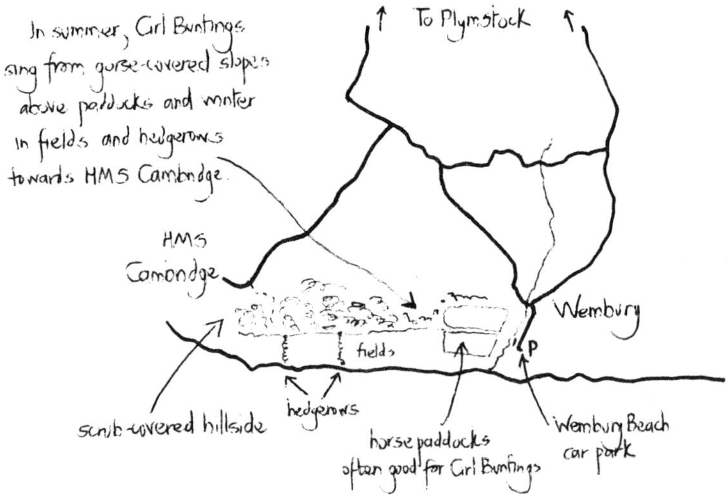

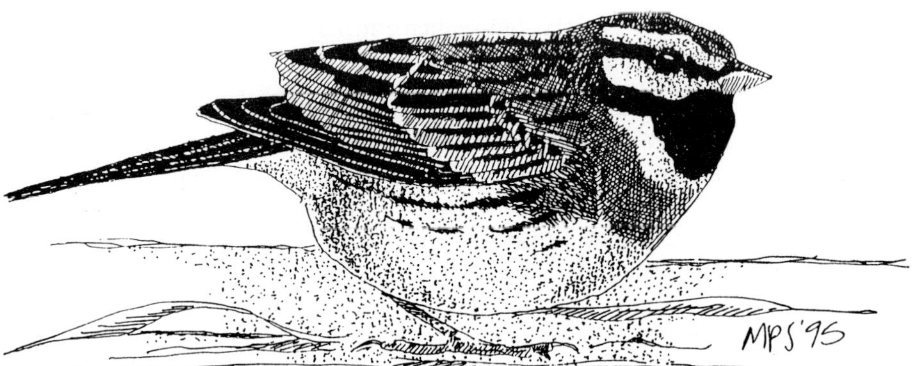

Migration and Rarity Hotspots

Britain is fortunate in that it is on a migrant flight-line and attracts birds from all over the World, including regular summer visitors from Africa and Iberia and wintering birds from Scandinavia. It also attracts vagrants from western Siberia, North America, Central Asia and the Mediterranean Region, particularly in spring and autumn.

These rare vagrants and scarce migrants can appear at virtually any coastal location around the British Isles, but statistical evidence has shown that the vast number are concentrated in a limited number of well-positioned and well-covered sites. Topping the list for vagrants is Fair Isle, followed closely by Scilly, then by Shetland Mainland and Norfolk.

The following is a detailed analysis of those top sites that regularly produce the unusual throughout the year.

Fair Isle - Shetland

Fair Isle is an isolated island situated 25 miles south-west of Sumburgh Head, on Shetland South Mainland. The island is owned by the National Trust for Scotland and covers 765 hectares and measures 3 miles by 1.5 miles. It is an island of steep sandstone cliffs, which has sheep-grazed heather slopes and grassy tussocks in the north and croftland in the south. There is a bird observatory at North Haven, at the north end of the island, where birders are welcome to stay from April to late October (booking enquiries to Fair Isle Lodge and Bird Observatory, Fair Isle, Shetland, ZE2 9JU). Apart from nesting seabirds in summer, outstanding rarities and impressive numbers of scarce migrants are the main attraction for Fair Isle to birders, and under suitable conditions, rarely fails to perform. As with all of the Northern Isles, classic conditions involve south-easterly winds with rain, and in such weather, large falls of birds may be found on Fair Isle. Most rarities are found in the turnip and oat crops around the crofts in the south half of the island, but others may make landfall in the many steep and sheltered geos around the cliff-edges.

In spring, the peak period for migration is from 12th May until 27th June, when species to expect as well as outstanding rarities are Common Rosefinch, Icterine Warbler, Marsh Warbler, Wryneck, Red-spotted Bluethroat, Red-backed Shrike and Ortolan Bunting.

In autumn, movement begins around 10th August, when the first Barred and Icterine Warblers appear, and continues through to late August, when 3-5 Yellow-breasted Buntings are often present (27th August to 20th September) and Paddyfield Warbler and Great Snipe often appear. Peak migration occurs in September and October, with varying numbers of Common Rosefinch, Bluethroat, Short-toed Lark, Barred Warbler, Ortolan, Wryneck, Red-backed Shrike, Richard's Pipit, Yellow-browed Warbler, Icterine Warbler, Red-breasted Flycatcher, Red-throated Pipit and Little and Rustic Buntings occurring. Almost annual Fair Isle specialities include Lanceolated Warbler, Citrine Wagtail, Arctic Warbler, Greenish Warbler, Olive-backed Pipit and Pechora Pipit.

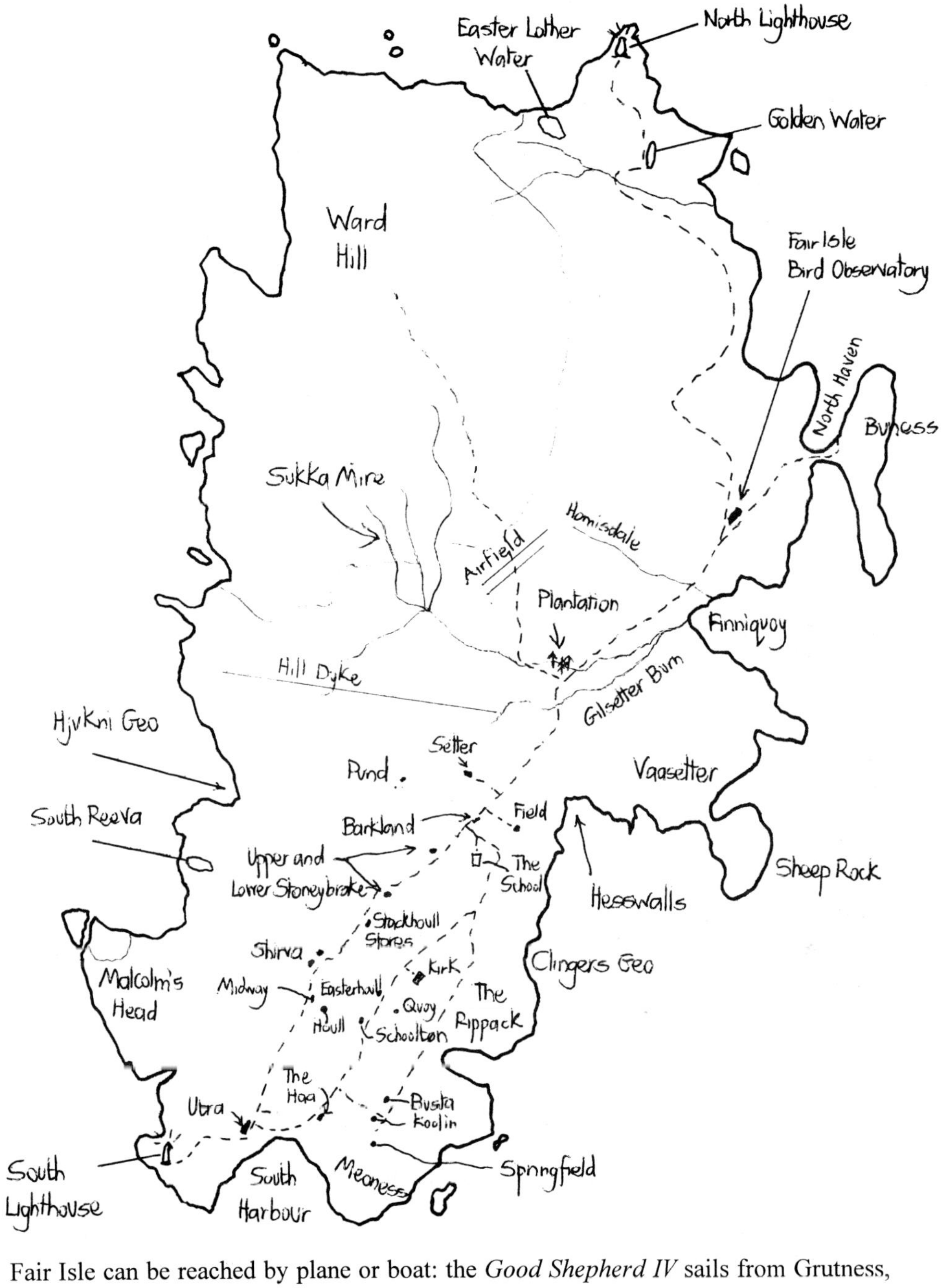

Fair Isle can be reached by plane or boat: the *Good Shepherd IV* sails from Grutness, near Sumburgh (Shetland South Mainland) at 11.30am on Tuesdays, alternate Thursdays and Saturdays between May and September, but note that rough seas and strong winds frequently disrupt sailing schedules. It is essential to book sailings in advance - contact Mr J.W. Stout, Skerryholm, Fair Isle, Shetland. ☎ 013512 222.

Loganair fly from Tingwall (Shetland) on Monday, Wednesday, Friday and Saturdays between May and October, and Mondays and Fridays only during the rest of the year. *British Airways* offer a daily service to Sumburgh Airport, Shetland, from Aberdeen or Edinburgh throughout the year (up to six flights a day). Contact *Loganair* on 0159584 246 (Tingwall) or 0131 344 3341 (Edinburgh) for details of current scheduled flights.

Shetland South Mainland

In the same conditions that create falls of birds on Fair Isle, large numbers of birds are displaced on Shetland Mainland. Light to moderate South-easterly winds with rain or overcast conditions are best and at times, the entire south peninsular of Shetland can appear to be alive with exhausted birds, most of which would have arrived from Scandinavia. The peak times for such falls are from 10th May to 20th June and from 26th August to 16th October.

There are numerous prime locations to consider checking for rare migrants or vagrants in the area and these include:

a) the lighthouse garden and surrounding walls, grassy fields and cliff edges at Sumburgh Head;

b) the quarry to the left of the road as you drive between Grutness and Sumburgh Head and the stone walls bordering the road;

c) the garden at Grutness and the surrounding farmland and hillsides;

d) the Sumburgh Head Hotel garden and surrounding fields;

e) the sheltered tidal basin of the Pool of Virkie, immediately north of the airport, and the gardens at the west end by the A970;

f) Scatness Pool and the nearby rocky coastline and the gardens and crop fields between the pool and the A970;

g) the gardens and crop fields around Toab village;

h) the marsh-fringed, machair Loch of Hillwell, visible from the Quendale road at HU 376 143, and the nearby burn at Quendale Mill (HU 372 133);

i) the Loch of Spiggie and surrounding fields, viewed from the minor road off the B9122;

j) the croft gardens and crop fields in Boddam village (HU 395 160) and Boddam Voe;

k) the well-vegetated burn at Geosetter, east of the B9122 north of Scousburgh (at HU 380 201);

l) the croft gardens and crop fields around Sandwick and at Noness (HU 441 225);

Elsewhere on **Shetland**, sites worth checking include the gardens and plantation at Helendale, just west of Lerwick; Lerwick Harbour, especially around the fish factories at Gremista; the croft gardens around Loch of Tingwall (HU 415 425); the mixed area of woodland at Kergord Plantation (HU 395 542) either side of the B9075 near Setter; the plantations at Voe; the gardens at Eswick and on the Nesting peninsular and the gardens at the head of Swining Voe, Vidlin and sycamore trees near Lunna House near Lunnasting, at the end of the B9071, east of Voe.

The islands of Whalsay and Fetlar are also good for rarities, as well as Baltasound, Haroldswick and Northdale on Unst, and these may all be reached via the inter-island ferries that operate throughout the day. The small group of exposed islets known as the Out Skerries are also excellent and these can be reached on Tuesdays, Fridays and Sundays from Whalsay (bookings essential).

North Ronaldsay - Orkney

North Ronaldsay is the most north-easterly of the islands in Orkney and has an impressive list of rare birds to its credit. It is comprised of predominantly rocky coastline with six main lochs, some of which are surrounded by extensive vegetation, large areas of cultivated and sheep-grazing fields and isolated croft gardens and vegetable plots. The Laird's mansion at Holland House, however, is undoubtedly the best site on the island for migrants, where the sheltered garden full of sycamores and fuschias offers a haven for tired stragglers and is used by the Observatory as a trapping area. The island is served by a Bird Observatory, established by Kevin Woodbridge (the warden and island GP) in 1987, and this accommodates up to 35 people at its site in the SW corner of the island (bookings can be made via North Ronaldsay Bird Observatory, Orkney, KW17 2BE ☎ 018573 267).

The peak period for scarcities on North Ronaldsay are from 16th May to 20th June and from 15th August to 12th October, south-easterly winds and rain being the most productive conditions. Birds to expect include Bluethroat, Wryneck, Red-backed Shrike, Common Rosefinch, Little and Rustic Bunting, Icterine, Marsh, Barred and Yellow-browed Warbler and Richard's Pipit. The site also has an attraction for North American waders and is also renowned for its autumn seawatching productivity. Rare waders are often located amongst the hordes of migrating Golden Plover that seek refuge on the island and inhabit the airfield and nearby fields or on the open moorland at Tor Ness or on Ancum or Bewan Lochs. Seawatching is best from Dennis Head.

Elsewhere in Orkney, Stronsay regularly attracts vagrants in spring, summer and autumn, and many other islands also fare well, particularly in May and June.

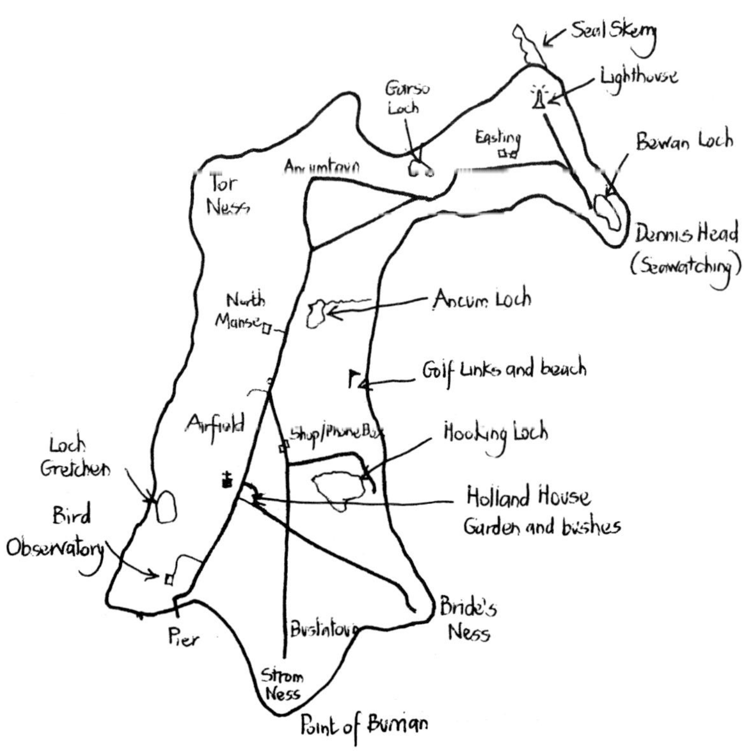

Girdleness - Grampian

The allotments and gardens around Girdle Ness and the golf course bushes and scrub near the lighthouse are also ideal areas in which to find scarce migrants or vagrants, and from mid May to early June and from mid August to late October, this site never fails to turn up surprises. Scandinavian migrants such as Bluethroat, Wryneck, Common Rosefinch, Barred Warbler, Icterine Warbler and Red-backed Shrike are annually recorded at the site, which is best reached from the minor road that runs east along the south side of the River Dee, turning left immediately south of the ferry terminal and dockyard area in Aberdeen Harbour.

The Aberdeenshire coastline offers many other rarity sites, including Rattray Head (the bushes and scrub in the vicinity of the lighthouse), the gardens and coastal scrubland at Cruden Bay; the allotments, gardens and coastal scrub at Collieston, the area of coastal dunes and bushes at Foveran and Drums and the coastal scrub at Balmedie Country Park.

Fife Ness - Fife

Fife Ness lies at the eastern extremity of the Fife peninsula and is renowned for its attraction to vagrants and scarce migrants. The area is characterised by farmland, a golf course, coastal scrub and isolated copses, clumps of gorse and bramble hedgerows. There are several areas of interest at Fife Ness including:

a) the walled garden and large trees at Balcomie Castle and the adjoining shrubbery at Balcomie Farm;

b)the gorse, shrubs and trees surrounding the artificial pools at Fife Ness Muir, immediately inland of the tip;

c) the gardens and trees in the vicinity of Craighead Farm and the quarry below the golf clubhouse;

d) the airfield and quarry immediately below the golf clubhouse at Crail and the coastal footpath from Kilminning car park;

e) the mature wood bordering the churchyard immediately east of Crail (Denburn Wood - NO 614 080).

The entire area is accessed from Crail, Denburn Wood being immediately east of the town and Fife Ness Muir from the car park at Balcomie Golf Course, 1.5 miles to the east. Walk through the gate to the right of the road and then immediately left down to the sea. The Muir is on the hill to the right and this can only be accessed from the cottage on the beach. For Kilminning, turn right off the golf course road to a car park at the far end of the airfield and for Balcomie Farm, turn left at the junction. Do not park anywhere other than at the golf club or Kilminning and always seek permission before entering private property, notably at Balcomie or Craighead Farms. When inspecting areas of cover around habitations, always respect peoples' privacy, especially in the early morning.

St Abb's Head - Borders

St Abb's Head is a rocky headland situated some 13 miles north of Berwick-upon-Tweed and is designated a National Nature Reserve. Apart from its sheer cliffs, that attract over 60,000 nesting seabirds in summer, the habitat largely comprises grazed grassland with a man-made freshwater loch and dense gorse-scrub and birch woodland lining the valley. The Mire Loch, as it is known, and the sheltered valley just inland of the cliffs act as a haven to migrant birds and vagrants, and species such as Bluethroat, Wryneck, Rustic Bunting, Common Rosefinch and Icterine Warbler are regularly recorded in May and June and Barred and Yellow-browed Warblers and Red-breasted Flycatcher in September and October.

Leave the A1 at either Burnmouth or Cockburnspath and continue to Coldingham. Take the B6438 to St Abb's village, turning left at Northfield Farm in Steading and continuing north along the minor road towards the lighthouse. Park either by the lighthouse or at the track to the south end of the loch. The willows and birch scrub at the south end of the loch is good for Bluethroat, Firecrest, warblers and flycatchers and the hillside scrub is good for Wryneck, Red-backed Shrike, Rosefinch and Whinchat. Ideal conditions are light to moderate south-easterlies with rain.

It is also worth checking the scrub at Coldingham Bay, south of St Abb's village.

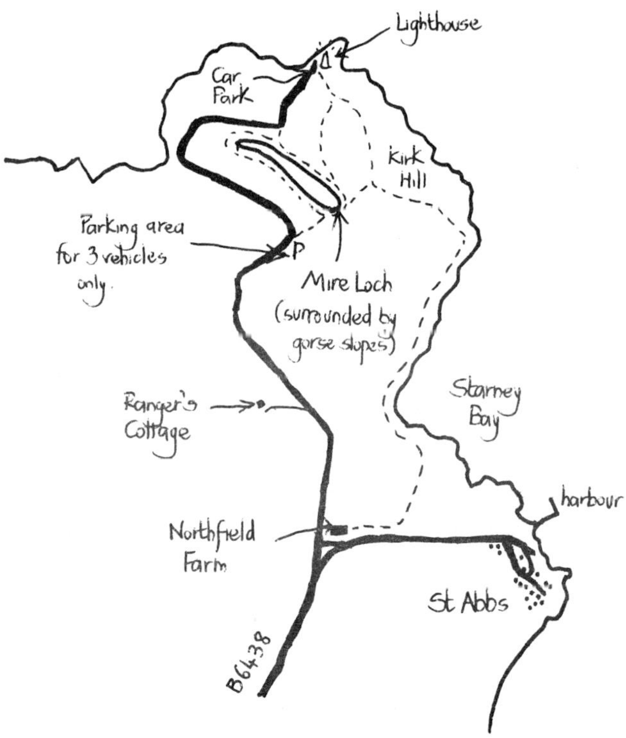

Holy Island - Northumberland

Holy Island is a superb place to be positioned if the wind is in the south-east and it is drizzling or overcast. Huge falls of migrants have been known to occur here, especially in late spring and early autumn, and some may remain for several days following their arrival.

There are several areas that consistently turn up rare birds and scarce migrants:

a) the area of dunes and isolated bushes known as The Snook, at the west end of the island. Wryneck and Icterine Warbler are regularly recorded here;

b) the hedgerows and fields that border the Straight Lonnen, excellent for attracting chats, flycatchers and warblers;

c) the gardens and trees in the village, particularly the churchyard at the south-west end;

d) the areas of scrub and isolated bushes besides the Crooked Lonnen and near Castle Point.

Holy Island is a typical migration watchpoint where birds often drift off during the day after an overnight fall or arrive from mid day onwards following a shift in wind direction.

Leave the A1 at West Mains Inn and continue through Beal to the causeway. Remember that it is impossible to cross to the island two hours before and up to three hours after high tide and it is best to **check the tide tables** before your visit.

Other sites in Northumberland that regularly receive large falls of migrants in suitable conditions include the Farne Islands (access by boat only from Seahouses Harbour), Low Hauxley, south of Amble-by-the-Sea, the coastal dune-scrub north of Newbiggin-by-the-Sea between Newbiggin Point and Beacon Point, St Mary's Island south of Seaton Sluice, Whitley Bay Cemetery and the trees, allotments and tennis court areas around Tynemouth Haven.

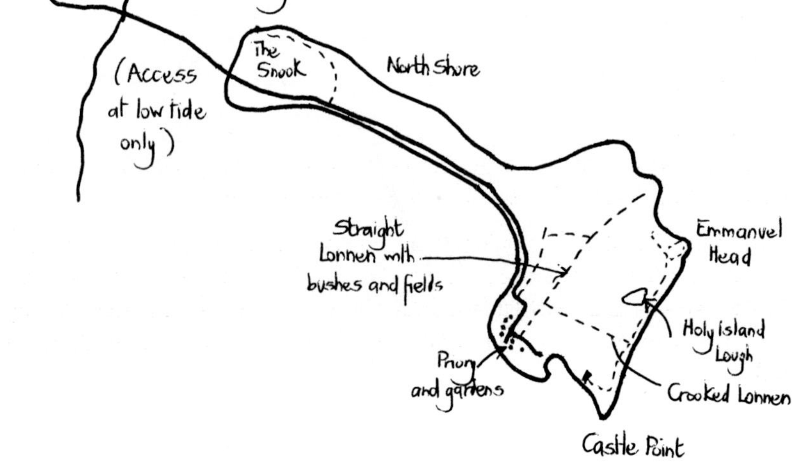

Whitburn Area - Tyne & Wear

The Whitburn area can be exceptionally good for scarce migrants and rarities if weather conditions are conducive. They can be found in the observatory area, in the scrub and bushes around Souter Point; in the gardens of Shearwater Estate bordering the coast road, in the parkland inland of the A183 at the south of the town and in Marsden Quarry, to the north of the town. Specialities of the area include Red-backed Shrike, Bluethroat, Icterine Warbler and Ortolan Bunting.

Teesmouth Area - Cleveland

The complex of habitats on Teeside attracts a wide variety of vagrants, including rare wildfowl, waders, gulls, terns and passerines. It is also an excellent area for seawatching, with Cory's Shearwater and Long-tailed Skua annually recorded.

There are a number of localities that produce rare birds in the Hartlepool area and these are as follows:

a) Hartlepool Headland: in suitable conditions (easterly winds with overcast or wet conditions), large falls of migrants may be found at Hartlepool Headland, with the tennis courts area, bowling green, allotments, doctor's garden and back gardens all offering shelter to tired migrants crossing the North Sea. The North View Cemetery is also worth checking and seawatching can be good from the Head itself.

b) Seaton Carew: the churchyard at the south end of Seaton Carew is regularly visited by rarities in autumn, including Icterine Warbler, Red-breasted Flycatcher, Barred Warbler and Ortolan.

c) Seaton Snook: a large area of intertidal sands regularly used by terns to roost and regularly attracting the Farne Islands Lesser Crested Tern and Roseate Terns. Accessed from the minor road east of the A178, a mile south of Seaton Carew.

d) Greatham Creek: a superb area of creeks and tidal pools regularly attracting rare wildfowl, waders and terns. Park by the bridge on the A178 and follow the footpath that runs along the south side of the creek to view the pools and marshes. Cowpen Marsh lies to the west of the road and is accessed from the same place. Follow the south side of the river westwards to the hide.

e) Saltholme Pools: several roadside pools, south of the A1185 roundabout, which regularly attract Pectoral Sandpiper and Temminck's Stint and quite often much rarer species, such as White-rumped Sandpiper and Long-toed Stint. Saltholme Back Pool can be viewed from a footpath leading down to the pool from the road.

f) Long Drag: an area of willow scrub and freshwater pools which are best accessed from Seal Sands, the extension of the footpath from Greatham Creek. Rare waders and Great Spotted Cuckoo have been recorded here in recent years.

g) Dorman's Pool and Reclamation Pond: a shallow flash and an embanked lagoon which regularly turn up rare wildfowl, waders, gulls and terns and are accessed from the private road that runs east from the A178 towards the oil refineries. A rough track heads off north to a raised bank where both sites can be viewed. A telescope is essential.

h) South Gare: an area of coastal pools and dunes extending north from the golf course to the breakwater. The breakwater itself is an excellent site from which to seawatch and offers unrivalled views over the Tees Mouth, where large numbers of terns and skuas gather from late July to September. The few roadside bushes, Cabin Rocks and Paddy's Hole immediately south of the breakwater regularly produce rarities, including Common Rosefinch, Wryneck, Ortolan, Icterine Warbler and Barred Warbler. Access is from South Gare, along the continuation of Tod Point Road.

i) Coatham Marsh: an area of pools and freshmarsh that regularly attracts wildfowl, migrant raptors and waders, viewable from the reserve car park on Tod Point road, from the A1085 near the fire station or from the footpath along the east boundary.

j) Locke Park, Redcar: a park full of mature trees and bushes that regularly attract migrant passerines, including Greenish Warbler, Yellow-browed Warbler and Red-breasted Flycatcher. Access is off Locke Road and Corporation Road and lies north of the A1085.

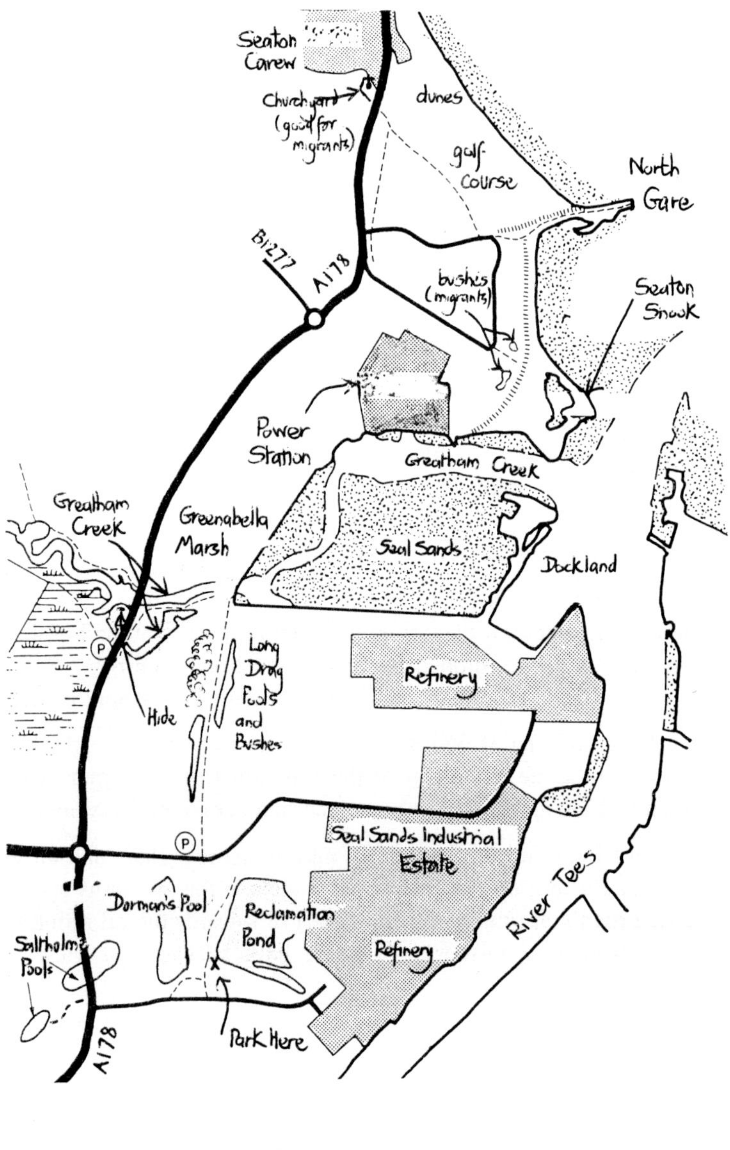

Filey North Cliff Country Park - North Yorkshire

The Filey area in recent years has been an outstanding attraction to scarce migrants and vagrants, with large numbers of birds being found from mid April through to mid November. The area is checked daily by members of the Filey Ornithological Group and a number of sites are particularly attractive. The Brigg itself is excellent for seawatching between late July and mid November and Filey Bay holds wintering grebes, divers, auks and Purple Sandpipers.

Areas to check include the Arndale Ravine, below the cafe, the clifftop vegetation at Coble Landing, Carr Naze, along the top edge of the Brigg (good for Bluethroat and Black Redstart) and the hedgerows that border the car park and the Top Fields. Filey Dams Nature Reserve can also be productive and birds may also be found in Glen Gardens and the ravine that runs south of here.

The Country Park is reached from the A1039 roundabout at the north end of Filey and is well signposted. The Dams reserve is accessed from the A1039 south of the railway crossing.

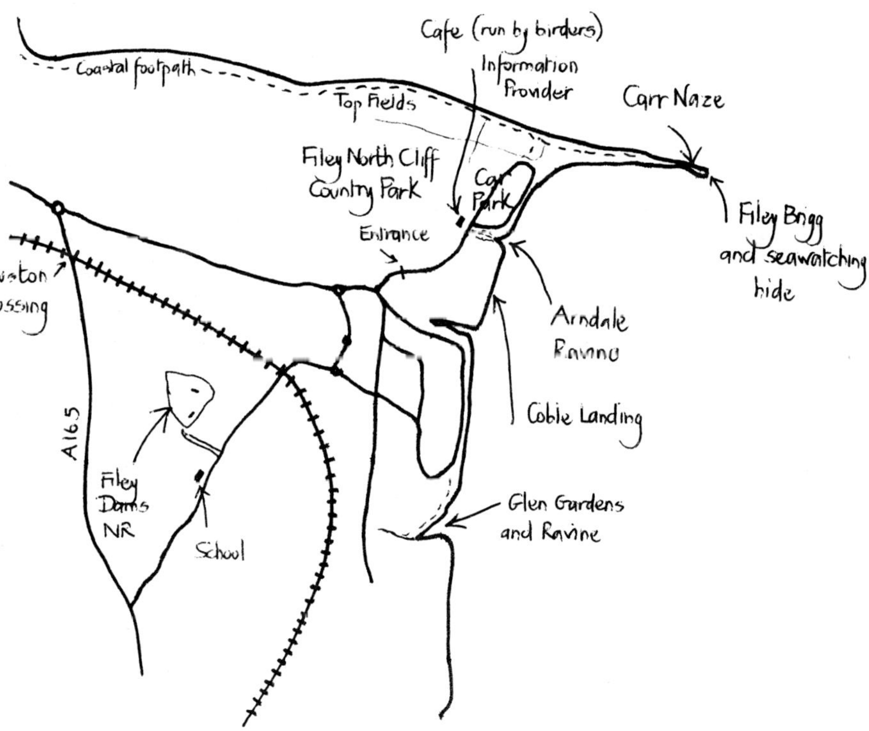

Flamborough Head - East Yorkshire

Flamborough Head is an almost unrivalled site on the Yorkshire coast for rarities and offers birding of a high standard throughout the year. It is undoubtedly the best site on the east coast for seawatching and suitable conditions (winds from the north round to the southeast) may see large numbers of Manx and Sooty Shearwaters, Fulmars, Gannets, Little Gulls, Little Auks, Arctic and Great Skuas streaming past. Rare seabirds such as Cory's and Great Shearwater, Sabine's Gull and Long-tailed and Pomarine Skua are regularly recorded offshore between late July and late September. Migrants also abound on the Head and scarcities such as Firecrest, Black Redstart, Icterine, Barred, Yellow-browed and Pallas's Warbler, Richard's Pipit, Red-breasted Flycatcher, Common Rosefinch, Ortolan, Red-backed Shrike and Bluethroat are regularly recorded in spring and autumn.

The area watched by the Flamborough Ornithological Group is that east of Danes Dyke and areas allowing public access include the Bay Brambles, between the fog signal station and the golf course, the Old Fall hedgerow that leads south from the B1259 and allows a viewpoint over the Old Fall Plantation, the densely vegetated ravine at South Landing and around the Timoneer Hotel and in the wooded valley at Danes Dyke, south of the B1255.

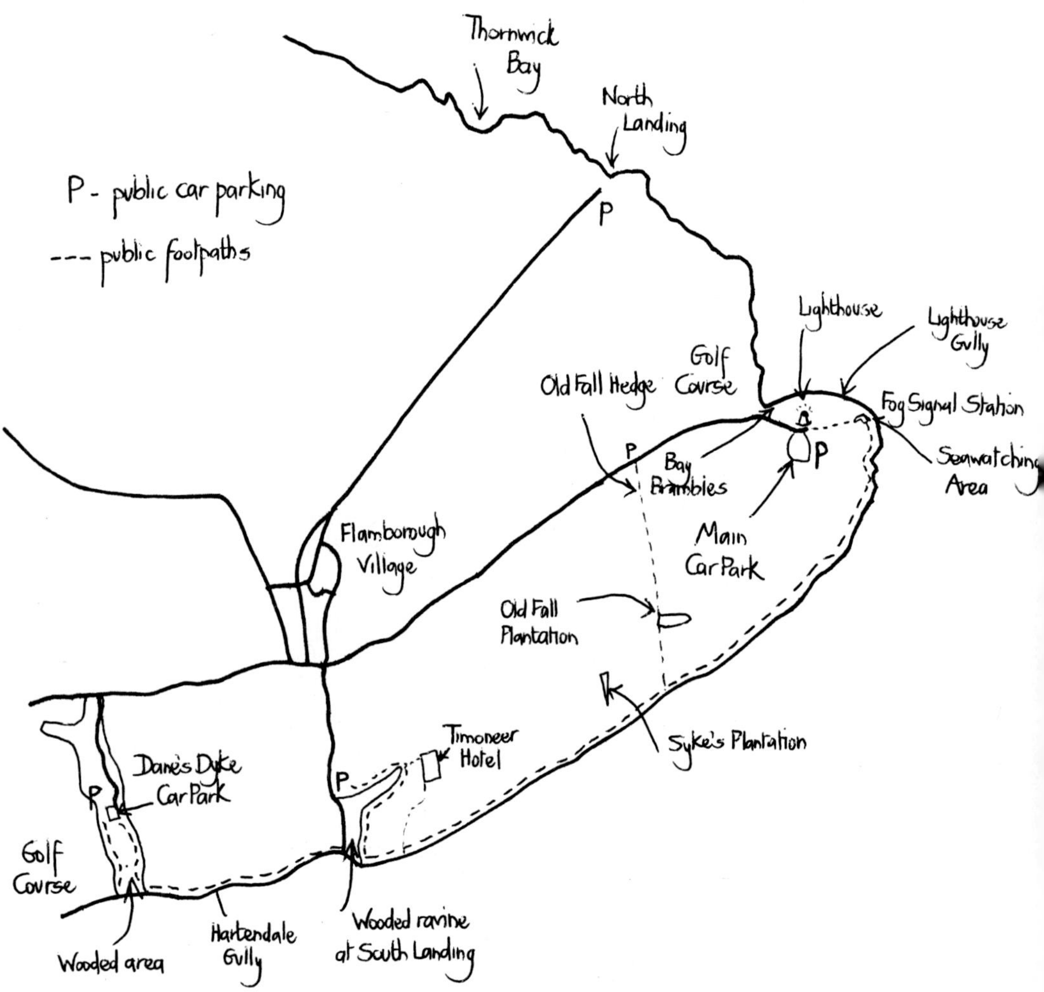

Spurn Point - East Yorkshire

The Spurn peninsula is renowned as being one of the premier migration spots in Britain, with large numbers of common migrants as well as regular scarce migrants and vagrants being recorded. Peak periods are from late March to mid June and from mid August until mid November. Spring specialities include Icterine and Marsh Warblers, Red-backed Shrike, Bluethroat, Wryneck, Common Rosefinch and Subalpine Warbler, whilst in autumn, Richard's Pipit, Wryneck, Icterine, Barred, Greenish, Yellow-browed and Pallas's Warbler and Red-breasted Flycatcher can be expected. Easterly winds are an essential influence on occurrences.

Birds can be found throughout the peninsula but most favoured localities include the Point Camp area and the buckthorn around the lighthouse, the Chalk Bank area and wire dump, the sueda and buckthorn scrub at the Narrow Neck, the Warren north of the Bird Observatory, Clubley's Field to the north of the warren, the Canal Scrape and hide and the bushes around the canal, the bushes and trees around Kilnsea Churchyard and in the gardens opposite the Crown & Anchor pub, the bushes at Sammy's Point, the hedgerow bordering Beacon Lane and opposite the Bluebell Inn and Beacon Ponds and Easington Lagoon to the north. The churchyard at Easington can also be productive.

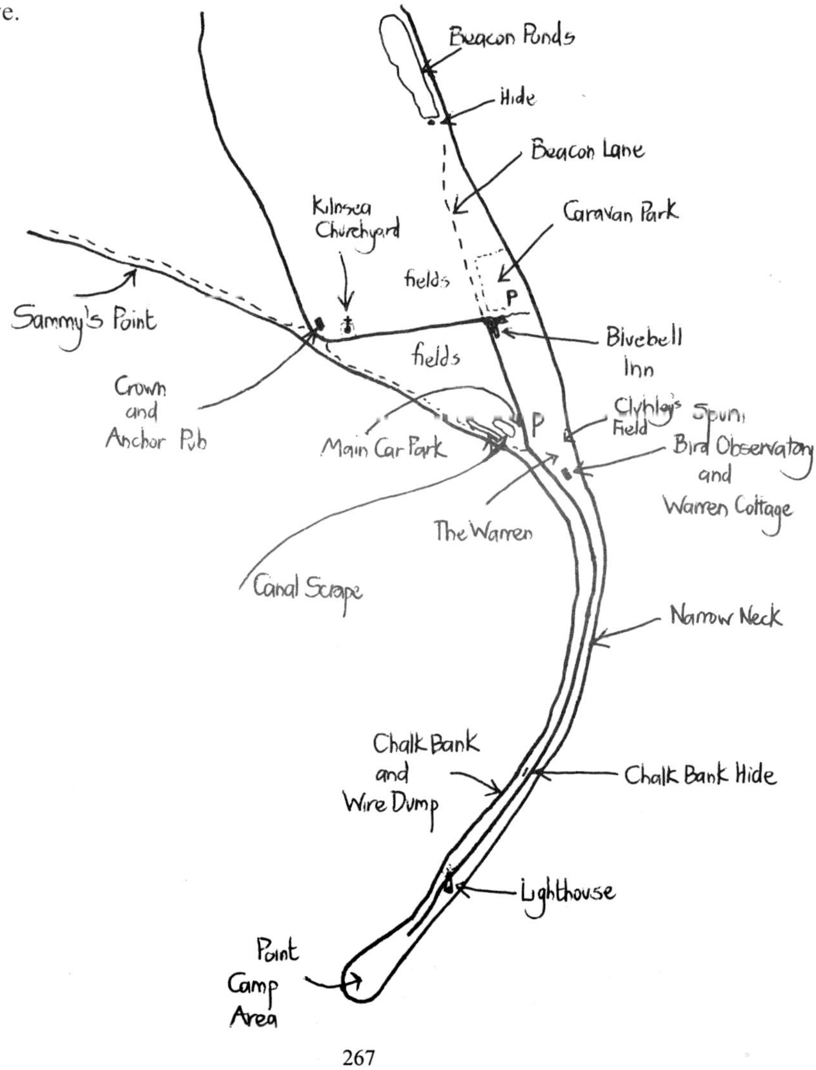

Gibraltar Point - Lincolnshire

Gibraltar Point Bird Observatory is situated south of Skegness and is accessed along the minor road that runs south past Seacroft Golf Course. In suitable conditions (east or south-east winds combined with cloud, rain or poor visibility), moderate falls of birds can be encountered and often these contain a number of scarce migrants amongst them. Regular visitors in May and June include Common Rosefinch, Icterine Warbler, Red-backed Shrike, Serin, Wryneck and Bluethroat, but birds rarely linger and are often extremely difficult to locate in the dense buckthorn scrub. Migrant Spoonbills and rare wildfowl and waders are often seen on The Mere, at the north end of the reserve, and most scarcities are found in the East Dunes area, accessed from the visitor centre.

The Lincolnshire coast north of Skegness has a fine range of habitats which regularly produce rare birds and scarce migrants, including the gardens and dunes at Anderby Creek, the coastline scrub between Theddlethorpe St Helen and Saltfleet Haven (accessed from Saltfleet Haven, where a footpath SE from Saltfleet passes Gowts Farm and crosses the main dyke), Sea View Farm and Rimac) and the Donna Nook area, accessed from the Stonebridge car park.

Holme NOA/NWT Reserve - Norfolk

The reserves at Holme, owned by the NWT and NOA, are an outstanding area of the north Norfolk coast and between mid April and early June and from mid August until mid November regularly attracts many species of scarce migrants and vagrants. Favoured areas include the small pine plantation north of 'The Firs' (the warden's house and information centre) and the elder bushes and blackberry scrub thereabouts, the sea buckthorn scrub to the east of the warden's house (particularly attractive to Bluethroat and rare warblers), the fields to the south of the entrance track (for rare chats, Ring Ouzels and raptors) and the paddocks and gardens immediately north of the chalets at the track entrance (for Wryneck, Red-backed Shrike, Barred Warbler and Common Redstart, *etc.*). The golf course to the west is also worth checking.

At Holme next the Sea on the A149, take the westernmost road north towards the sea and after less than one mile, take the second right along a rough track towards the reserve. The first car parking area on the left gives access to the paddocks behind the chalets, whilst the main car park is after a further 0.75 miles by 'The Firs'. Permits are required for both reserves and available at the information centre or from the access gate in Holme village in summer.

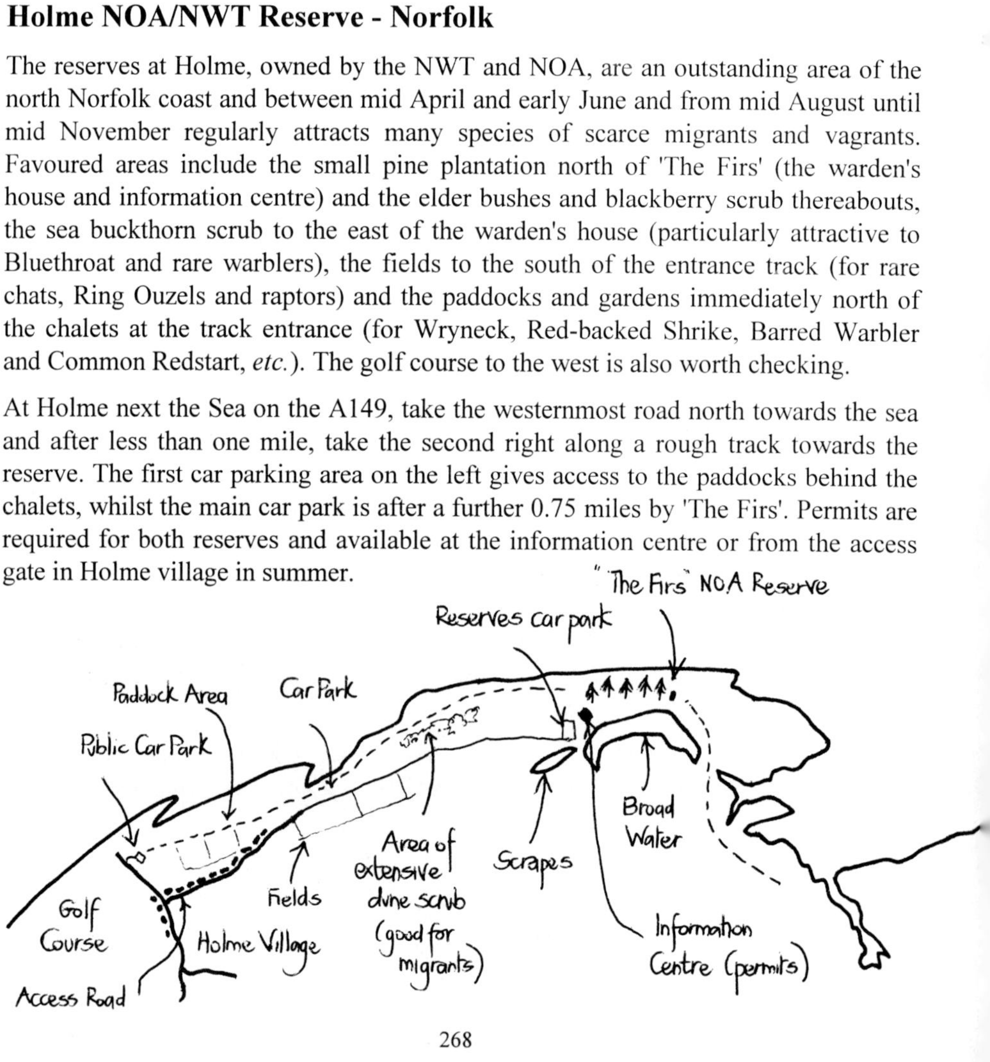

Blakeney Point/Cley NWT Reserve - Norfolk

The stretch of north Norfolk coastline between Blakeney Point and Weybourne is one of the best-known and most rewarding birding areas in Britain with an overwhelming list of rarities to its credit. The area comprises extensive reedbeds, pools, grazing meadows and on the Point, marram grass and Sueda bushes along a shingle spit.

Scarce birds and rarities are almost a daily event at the reserve, with rare herons, wildfowl, waders, gulls, terns and passerines being regularly recorded, and scarce seabirds often being identified offshore of Cley Coastguards and Blakeney Point. There are three NWT car parks on the reserve, the main one being adjacent to the information centre along the A149 east of Cley village and others at the East Bank and at the Coastguards. Up to ten hides overlook Cley Marsh, a permit being required for access. These can be obtained from the reserve centre, open daily except Monday. The North and Maynard's Hides overlook the North Scrape (excellent for passage waders, including regular Red-necked Phalarope, Pectoral Sandpiper and Temminck's Stint), Avocet, Dauke's and Teal Hides overlooking Pat's Pool and Simmond's Scrape (also good for waders, rare wildfowl, Common Bittern, Spoonbill, gulls and terns) and Bittern Hide (often good for Water Rail, Bearded Tit and Water Pipit). The Eye Pool and Field are also worth checking as well as the shingle ridge to North Hide. Arnold's Marsh, immediately east of Cley East Bank, is favoured by waders and terns (Kentish Plover is almost annual) and Walsey Hills and Snipe's Marsh may attract migrant passerines (Firecrest, Yellow-browed Warbler, *etc.*).

Blakeney Point is best visited from mid April to late May and from mid August until late October and can be an arduous walk, the three miles of shingle spit taking at least five hours to cover fully. It is accessed from Cley Coastguards car park (or by boat from Morston or Blakeney Quays) and extends westwards into Blakeney Channel. Ideal conditions involve light to moderate easterly winds with overcast or drizzly conditions, and afternoon visits are recommended as birds often make landfall at this site late in the day. Most rarities are found in the dense Sueda that covers vast stretches of the point, including that at the Marrams (400-800 yards west of the coastguards), Halfway House, The Hood and at the Long Hills and the Yankee Ridge, as well as in the Plantation at the point, in the isolated elder clumps in the dunes and in the lupin beds that lie behind the tea rooms at the point.

Elsewhere in the Cley area, locations worth checking include the isolated elder and bramble bushes on Gramborough Hill, east of the Salthouse Beach car park, the water meadows at Kelling Quags and the adjacent farmland, the seaward end of Weybourne Camp, the scrubby hillside north of the A149 at Muckleburgh Hill (between Weybourne and Kelling), Salthouse and Kelling Heaths and the coastal fields east of Weybourne Coastguard Cottages (accessed from the car park at Weybourne Beach).

Landguard NR - Suffolk

Landguard Point is a shingle spit immediately south of Felixstowe at the mouth of the River Orwell. It is flanked to the north by the Port of Felixstowe, the dockland lights playing a major part in the site's attraction to migrating birds. The nature reserve and observatory are sited at the old Landguard Fort, one of the old gun emplacements being converted into the Observatory's headquarters. Although the ringing compound is inaccessible to visitors, most of the birds recorded there may be seen from the footpath that encircles the perimeter fence, and others may be found on the main common or in the bushes to the north of the car park. Specialities include Wryneck, Barred Warbler, Bluethroat and Red-breasted Flycatcher.

Nearby, the trees and bushes in Adastral Close, to the north of the dockland, are worth checking, and the dense vegetation at Fagbury Reserve can be productive, although the change in intensity of dock lighting at the latter site has had a detrimental effect on bird occurrences.

Sandwich Bay - Kent

The Sandwich Bay Estate regularly attracts scarce migrants and vagrants in spring and autumn, a Bird Observatory there recording daily events at the site. The coast is dominated by a series of sand-dunes, many of which have been converted into a golf course, whilst inland, most of the area is farmland and many waders roost on these fields at high tide. Migrants appear in the gardens and bushes around the estate properties, but birders should take care to respect residents' privacy. The St George's bushes, directly opposite the observatory, are particularly favoured, as well as the small wood at the south end of the estate. Pallas's and Yellow-browed Warblers are regularly recorded here in October, as well as Red-breasted Flycatcher. The fields around New Downs Farm, accessed from the golf clubhouse, regularly attract Dotterel in spring and Quail in summer.

Leave Sandwich east along Sandown Road towards Sandwich Bay Estate, the observatory (at Old Downs Farm) is on the right about 400 yards beyond the toll gate. Please park only by the observatory.

Pegwell Bay, nearby, is also well worth checking, with Kentish Plovers being regularly present in April-May and August-September. Park in the designated car park along the A256 and walk south to the hide to view the bay.

Dungeness - Kent

The Dungeness peninsula is renowned for its attraction to migrating birds and has traditionally recorded scarce migrants and vagrants on a regular basis. The peak period is from mid March to early June and from mid July to mid November, with large falls occurring in April and May and throughout autumn if weather conditions are conducive. South-easterly winds are best, especially if accompanied by rain or overcast conditions, but southerlies may also prove productive. A Bird Observatory borders the power station and an active ringing programme involves the regular trapping of migrants in the dense vegetation to the north of the observatory, as far as the Long Pits. Visiting the trapping area is primarily an early morning operation as many birds move on as the day progresses. Similarly, seawatching at Dungeness is often only very good in the first few hours of daylight with passage often ceasing after 9.00am. However, Black Terns and Little and Mediterranean Gulls can often be seen at 'The Patch' throughout the day. Seawatching can be excellent in spring, with the passage of up to 400 Pomarine Skuas being the annual highlight. These pass through primarily between 24th April and 6th June, with peak numbers moving up channel during the first ten days of May. Peak movements are in south-easterly winds.

The most regularly recorded scarce migrants in spring include April Serins (virtually all brief fly-overs), Woodchat Shrikes, Golden Orioles, Hoopoes and Wrynecks, whilst in autumn, the Observatory Moat or trapping area regularly holds Red-backed Shrike, Barred Warbler, Icterine Warbler, Yellow-browed Warbler and Pallas's Warbler.

From the A259 between Brenzett and New Romney, turn south on the B2075 towards Lydd. Fork left after the railway and then left at the roundabout and follow the road towards Dungeness. Continue past the ARC Pit and New Diggings and take the second right towards the lighthouse. The road eventually meanders around the old lighthouse and continues to the observatory, but it is best to park by the entrance track to the beach, opposite the perimeter fence of the power station.

Dungeness RSPB reserve is also very good and is clearly signposted from the road at Boulderwall Farm. Follow the shingle track to the information centre and hides, from where a well-marked trail gives access to Burrowes Pit and the rest of the gravel pit complex. The reserve is open daily except Tuesdays, 9.00am until dusk, and is free to RSPB members. Spoonbill, Little Egret, Mediterranean Gull, and rarities are regularly recorded.

The nearby ARC pits are always worth a look and these can easily be viewed from the causeway. Further north, Lade Pits may be productive, and these are accessed from Derville Road in Greatstone-on-Sea. View only from the shingle ridge outside the fence.

Dengemarsh Gully, at the seaward end of Dengemarsh Road, is also worth checking for migrants.

Beachy Head - East Sussex

The Beachy Head promontory is undoubtedly one of the best migration watchpoints on the south coast and for this reason regularly attracts a selection of rare and scarce birds in each spring and autumn season. The headland consists of a narrow belt of coastal downland that extends for over four miles between Eastbourne and Birling Gap, with areas of gorse and scrub on the downland and a small wood at Belle Tout.

There are three main areas to check, all connected by the minor road that runs from East Dean to the west end of Eastbourne. The most productive of these is Birling Gap, where the gardens on the western slope regularly attract rare warblers and flycatchers and the fields immediately inland, where Black-headed Bunting, Rose-coloured Starling, Short-toed Lark and Tawny Pipit have appeared in recent years. Birling Gap is also the best site for seawatching at the Head, from the sheltered cove by the lifeboat slipway.

Further east, Belle Tout Wood and the gorse and scrub-covered slope immediately to the east, have recently attracted Sardinian, Melodious, Icterine, Pallas's, Yellow-browed and Dusky Warbler, Common Rosefinch and Golden Oriole. There is a small car park at the west end of the wood.

At the Head itself, the densely-vegetated Whitebread Hollow lies along the coastal footpath north-east of the Beachy Head Hotel. However, due to the nature of the cover, the area is very difficult to work. The area by the hotel is good for aerial migration, with Alpine Swift, Red-rumped Swallow, Bee-eater, Black Kite, Osprey, Red-footed Falcon and Honey Buzzard all being regularly recorded.

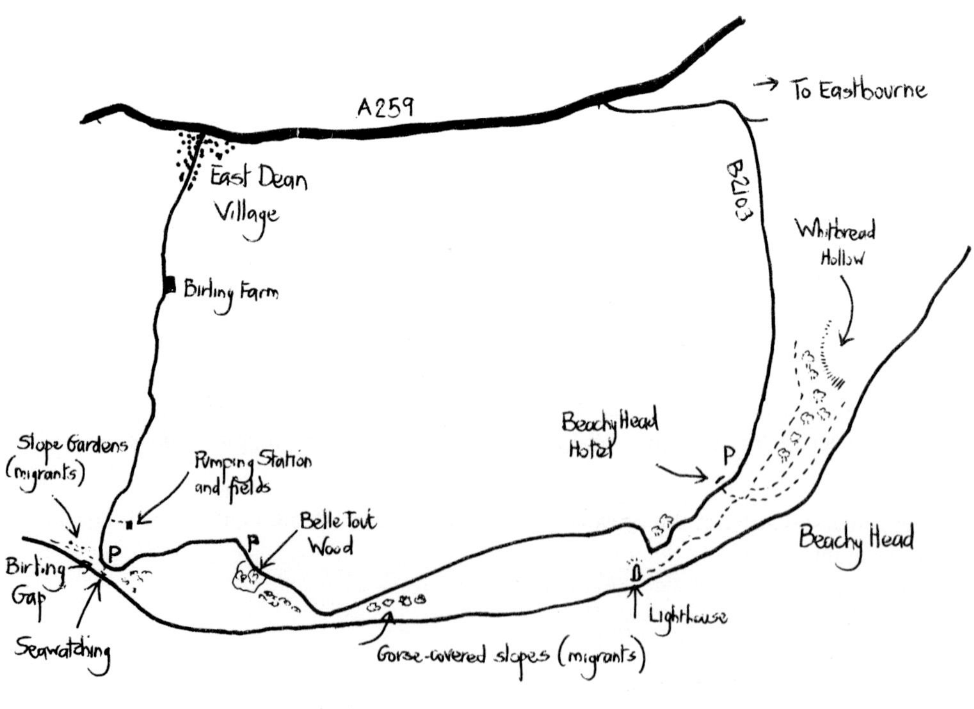

Pagham Harbour/Selsey Bill - West Sussex

The Pagham Harbour area is renowned for attracting rare birds, with wildfowl, herons, waders and passerines being found there on a regular basis. It is a vast area, comprising tidal mudflats intersected by numerous channels and surrounded by farmland and patches of scrub and gorse adorning the harbour banks and seawalls. There are several locations in which vagrants and scarce migrants are continuously found and these include Sidlesham Ferry Pool, adjacent to the B2145 south of Sidlesham, where transatlantic waders and rare herons are annual, the woodland, hedgerows and churchyard at Church Norton and the small wood to the south of the Severalls (accessed from the B2145, one mile south of the Ferry Pool) and Pagham Lagoon and the North Wall, accessed from Pagham village.

Selsey Bill, at the end of the B2145, is relatively attractive to rarities, with the gardens immediately inland of the sea occasionally producing rare warblers or a rare chat and the old Pontins Holiday Camp site, at the east side of the Bill, attracting Woodchat Shrike, Serin and Dartford Warbler. It is also a good site for seawatching, with Pomarine Skuas and Roseate Terns regularly recorded in spring.

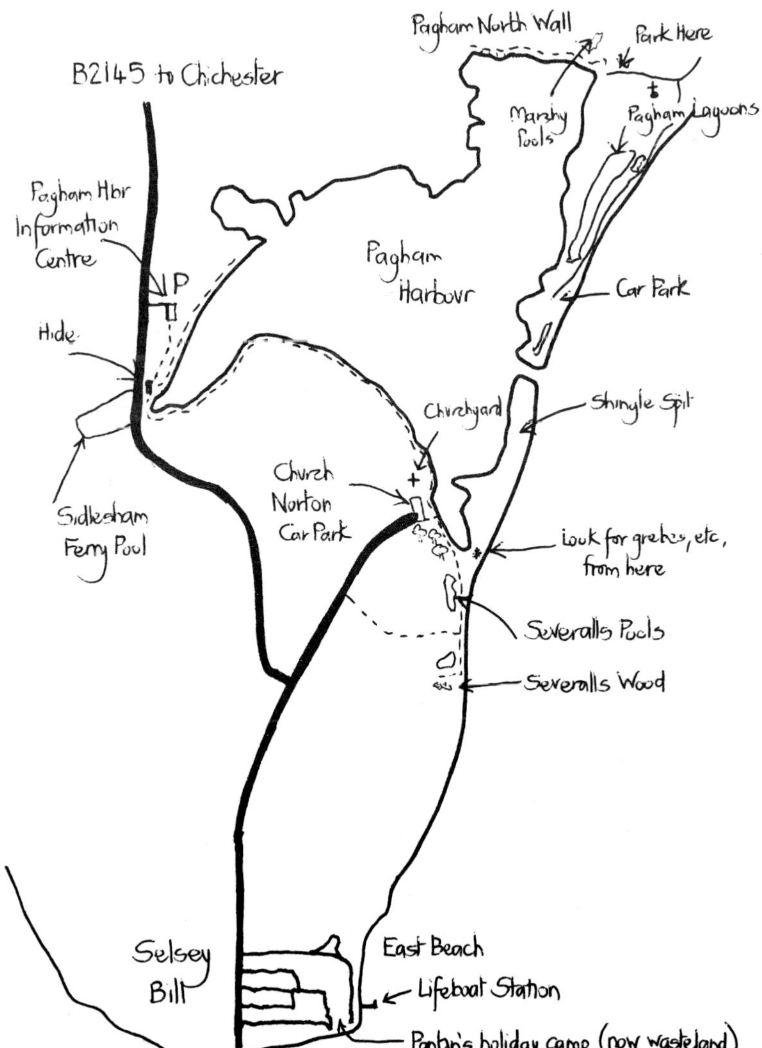

Hengistbury Head/Christchurch Harbour - Dorset

Both Hengistbury Head and Christchurch Harbour are within close proximity of Christchurch and Bournemouth and offer a superb array of birding habitats throughout the year. The harbour itself is excellent for rare wildfowl, waders and crakes, whilst the headland regularly attracts rare passerines, such as Ortolan, Hoopoe, Melodious, Icterine, Yellow-browed and Pallas's Warbler and Red-breasted Flycatcher.

Both sites can be accessed from two main locations: the harbour from the recreation car park along Stanpit Lane, in Christchurch, and Hengistbury Head from the continuation of Broadway, in Bournemouth. (reached from either Belle Vue Road, off the Tuckton roundabout, or Southbourne coast road, and is well signposted).

In Christchurch Harbour, favoured localities include Stanpit Marsh, to the south of the cricket pitch (for wildfowl, waders, gulls and terns), the reed-fringed tidal dyke on the Central Marsh (for Spotted Crake and Aquatic Warbler) and Priory Marsh, to the south of the golf course (for Spotted Crake, phalaropes, Red-rumped Swallow and Aquatic Warbler), whilst at Hengistbury Head, scarcities are most often found in the sallows at Double Dykes (immediately east of the car park), the Common and the extensive woodland at Batters Wood.

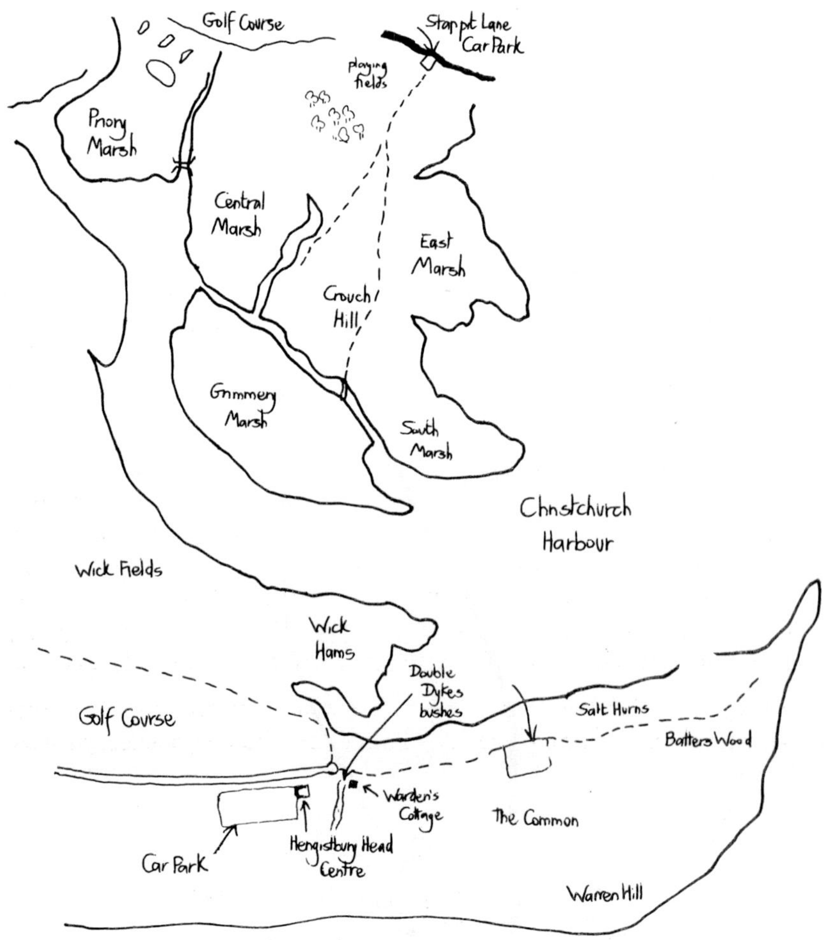

Portland Bill - Dorset

Portland Bill is acknowledged as being one of the premier sites in Britain for bird migration, as well as for attracting regular scarce migrants and extreme vagrants. The main periods of passage are between 20th March and 7th June and from 21st July to 28th October, with large falls often occurring in April or early May and in late August and September.

The Bill itself is good for seawatching, the obelisk offering observers shelter from the elements. Pomarine Skuas pass eastwards from 26th April to 26th May, as well as good numbers of terns, Arctic and Great Skuas, divers and wildfowl, whilst 2-3 pairs of Puffin nest on the West Cliffs nearby and are seen in early morning or dusk. The Bill Quarry and car park area is good for Black Redstart and wheatears, and once held an Alpine Accentor for three weeks, and the Admiralty fence nearby regularly attracts Whinchat and Common Redstart. The Bird Observatory Garden, the overgrown quarry adjacent and the hut fields are the only extensive areas of vegetation near the point and consequently receive some of the better birds, with Wryneck, Bluethroat, Subalpine Warbler, Icterine and Melodious Warbler and Red-breasted Flycatcher being almost annual, whilst the bramble scrub and willow clumps by the Devenish Arms car park can produce rare warblers. The pool and bushes at Culverwell can be excellent, and regularly produce Melodious Warbler, and the Top Fields between the ARE buildings perimeter fence and the West Cliffs are good for Ring Ouzel, Woodchat and Red-backed Shrike and Ortolan Bunting. Eight Kings Quarry and the quarry workings opposite may produce migrants, and the fields to the south bordering the East Cliffs can be good for pipits, larks and buntings, and have attracted Bobolink and Yellow-billed Cuckoo in the past. Early morning visits are essential as many migrants leave Portland as the day progresses.

Portland Bill is situated immediately south of Weymouth and is reached via the A354 at Ferrybridge. Continue south through Fortuneswell, Easton and Southwell to the Bill, where there is a large parking area. The observatory is sited in the Old Lower Lighthouse to the left of the road about half a mile north of the car park and is open March to November and offers comfortable accommodation for up to 25 people - details from The Warden, Old Lower Light, Portland Bill, Dorset, DT5 2JT ☎ 01305 820553.

Outside the Observatory Recording Area, there are many other areas on Portland worth checking, including the garden and coastal footpath around Pennsylvania Castle (on the east side of Portland south of Easton), the fields at Reap Lane and surrounding gardens in Southwell, the quarries and roadside fields between Southwell and Weston, the Tout Quarry Sculpture Park to the north of Weston, the Avalanche Road Hump and surrounding hedgerows immediately south of Weston, Easton Quarry and the undercliff habitat at Grove Cliff and the dense vegetation on the slopes of the Verne.

Close by, in Weymouth, Radipole Lake and Lodmoor NR both offer superb opportunities for birding and regularly attract rare wildfowl, waders, gulls and terns.

Portland Bill - Dorset

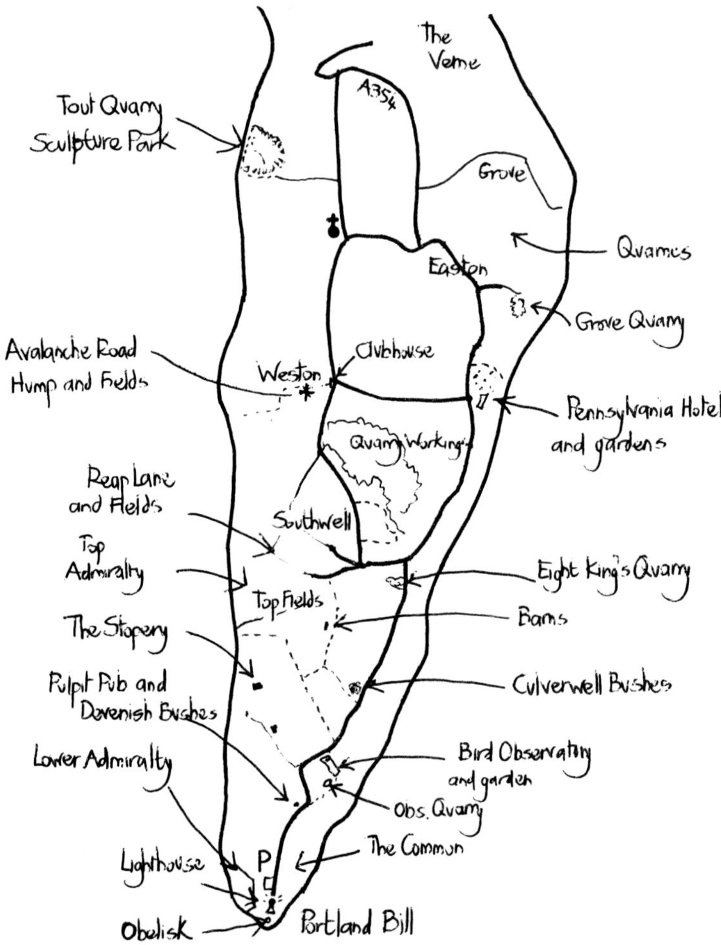

Radipole Lake and Lodmoor - Dorset

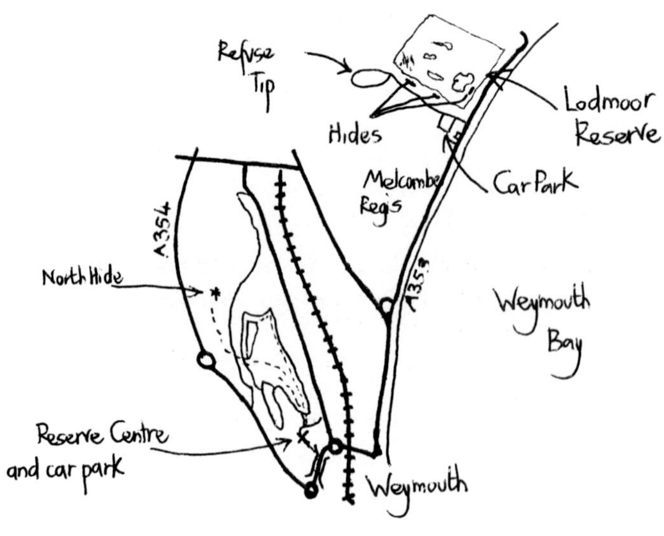

Prawle Point - Devon

Prawle Point is the most southerly point on the south Devon coast and consequently attracts good numbers of migrants and rare birds in spring and autumn. It is particularly good for Hoopoe and Serin in April and Melodious Warbler in August/September and is also one of the better places for the resident Cirl Bunting.

The habitat at Prawle Point is quite diverse, with a small wood at East Prawle and sheltered hollows and valleys containing dense vegetation and bushes along the coast. Most migrants appear in the area around the car park or in the bushes near the Coastguard Cottages and often filter further inland or to the wood as the day progresses.

Leave Kingsbridge east on the A379 and turn right in Frogmore or Chillington along a number of minor roads to East Prawle village. Continue through the village towards the sea and follow the narrow and winding lane to the National Trust car park in the sheltered hollow surrounded by trees. Seawatching is best from the memorial seat below the west side of the point and a full migrant check of the area would entail walking east along the coastal footpath from the point past Langerstone Point to East Prawle Wood and back through the village to the car park.

Lundy Island - Devon

Lundy Island lies in the Bristol Channel twelve miles north of Hartland Point and is a haven for rare migrants and vagrants in both spring and autumn. The gardens and bushes at Millcombe Valley are perhaps the best areas for migrants, with the isolated bushes and sheltered valleys around the Halfway Wall also being productive. The various pools are also attractive to migrants, especially that at Millcombe Pond. The island is also famed for its nesting seabirds, with impressive colonies of auks and Kittiwakes at the North Light. Storm Petrels are regularly seen from the ferry crossing in May and June.

Lundy Island can only reached by sea, with the *MV Oldenburg* making regular sailings in summer from Ilfracombe or Bideford Quay, allowing 2-4 hours on the island. A variety of accommodation is available on Lundy but space is limited, information being obtained from the Island Administrator on ☎ 01271 870870.

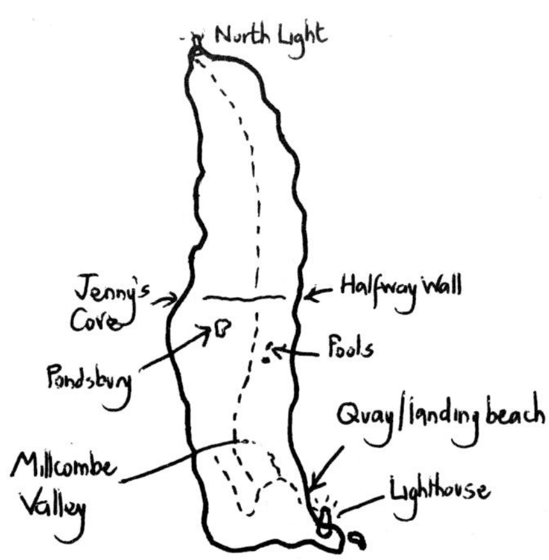

Porthgwarra - Cornwall

The Land's End area is renowned for its attraction to scarce migrants and vagrants and these may be found in the number of sheltered valleys that lie between St Just and Porthgwarra. The optimum periods are between 12th April and 5th June and from 27th August to 12th November and Nearctic passerines are very often the speciality.

Porthgwarra stands out in that it is not only good for passerine migration but also good for seawatching, with Gwennap Head often producing large numbers of Cory's and Great Shearwaters passing in July-September. Many other species of seabird are also recorded. Migrants may be found in the sallows and bushes that border the stream, especially in the area around the car park and that known as 60' foot cover, as well as in the gardens at the seaward end of the cove and along the stone walls to the north. Hoopoes, Wrynecks, Melodious, Icterine, Yellow-browed and Pallas's Warblers and Red-breasted Flycatchers are regularly recorded in autumn, as well as Woodchat and Red-backed Shrikes. Dotterels, Richard's and Tawny Pipits and Ortolan and Lapland Buntings are regularly found in the open moorland above the valley.

Leave Penzance westwards on the A30, turning left at Catchall onto the B3283. Continue through St Buryan and along the B3315 to Polgigga and turn left along the very narrow lane to Porthgwarra. There is a car park at the bottom of the valley and from here, a track leads up past the coastguard cottages to the moorland and the small pool at the head of the valley. For seawatching, follow the coastal footpath west from the car park to the promontory at Gwennap Head, a distance of about 600 yards.

It is also worth checking the heavily vegetated valleys at Porthcurno, Treen and Penberth Cove (immediately east of Porthgwarra), as well as the fields and stone walls at St Levan, and Nanquidno Valley and St Just Aerodrome to the north and Cot Valley and Kenidjack Valley at St Just.

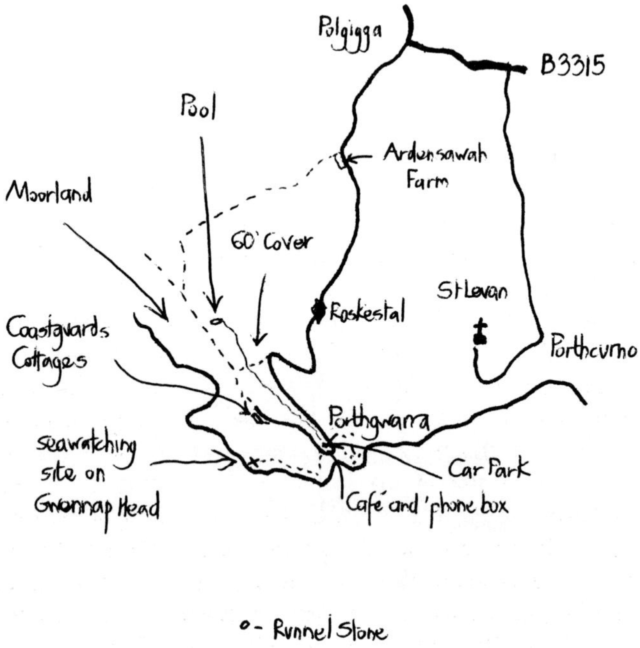

Isles of Scilly

The Isles of Scilly are the Mecca of British ornithology and are legendary for their attraction to rare birds and extreme vagrants. Although Nearctic species predominate and are virtually an annual feature, almost anything can and does turn up. The peak period is from late August through to mid November, with most birders concentrating a minimum two-week stay between late September and October. In addition to the vagrants, which turn with increasing regularity and frequency, Scilly is renowned for the scarce migrants that it attracts, with Red-breasted Flycatcher, Richard's, Tawny and Red-throated Pipit, Short-toed Lark, Wryneck, Red-backed Shrike, Ortolan and Lapland Bunting and Barred, Yellow-browed, Pallas's, Melodious and Icterine Warbler all being virtually guaranteed. Obviously, weather has a major influence on the birds that appear on the islands, but other than north-westerly winds, rarities arrive on virtually every other wind, with south-easterlies or strong south-westerlies being most productive.

The islands can also be productive in spring, with warm southerlies in April and May regularly producing rare herons, Black Kite, Alpine Swift, Bee-eater, Woodchat Shrike, Common Rosefinch, Golden Oriole, Serin, Red-rumped Swallow and Subalpine Warbler.

There are five inhabited islands, St Mary's, St Agnes, Tresco, Bryher and St Martin's, with most birders preferring to stay on the largest, St Mary's, where daily boat trips allow access to each of the others. Although rare and scarce birds can be found virtually anywhere on the islands, many decades of study have shown that certain places attract the majority.

On St Mary's, prime sites include the tall pines and Battery walls of the Garrison (rare warblers, Firecrest, Red-breasted Flycatcher and Wryneck), the fields and bushes along the top walk to Peninnis Head and the rocky coastline between the lighthouse and the allotments in Porthcressa Bay (Ring Ouzel, Ortolan and Lapland Bunting, Short-toed Lark and Richard's Pipit), the elm trees and bushes in Old Town Cemetery (Yellow-browed and Pallas's Warbler, Red-breasted Flycatcher and Red-eyed Vireo), the Lower Moors nature trail (Jack Snipe, Corncrake, Golden Oriole, rare warblers), especially the pool areas and the sallows to the left of the tin shed), the airfield and Giant's Castle (American Golden Plover, Dotterel, Buff-breasted Sandpiper, Upland Sandpiper, Richard's, Tawny and Red-throated Pipit, Short-toed Lark, Wryneck and Lapland Bunting), the track that leads from the airport to Salakee Farm and back towards the road (Little and Rustic Bunting, Rose-coloured Starling, chats and rare warblers), Porthellick Pool and Higher Moors (rare wildfowl and waders, Spotted Crake, Bluethroat and Yellow-browed and Pallas's Warbler), the trail from the road at Higher Moors Pumping Station to the Holy Vale Cottages (Red-breasted Flycatcher, Radde's, Yellow-browed and Pallas's Warbler and rare thrushes, etc.), the elms alongside Watermill Lane and the sallows adjacent to Newford Duckpond (Red-breasted Flycatcher and Yellow-browed and Pallas's Warbler), the gardens and sheltered valley at Watermill Cove (Wryneck, Firecrest, Yellow-browed Warbler), the bulb fields along the Silver Trail and around Maypole (Little and Rustic Bunting), Carn Friars Farm and the fields thereabouts (buntings and finches), Normandy Farm and Kitty Down (larks, pipits, chats, buntings and finches), the fields around Trewince and upper Porthloo Lane (Richard's Pipit, Woodchat and Red-backed Shrike), the fields at Four Lane Ends (rare thrushes and warblers), the golf course (rare waders, Dotterel, rare pipits, Short-toed Lark and Lapland Bunting) and the fields bordering Porthloo Lane and Duckpond and Rocky Hills (Red-backed Shrike and rare warblers, etc.).

St Agnes is the most south-westerly of the inhabited islands, facing directly into the Atlantic and therefore likely the first landfall for transatlantic vagrants. The islet of Gugh is connected to St Agnes by a sandbar, only accessible at low tide. The best sites for migrants on St Agnes include the secluded orchard garden surrounded by elms and hedgerows in the Parsonage, the bulb fields surrounded by hedgerows and bushes above the Chapel and in Troy Town, the Big Pool and the cricket pitch adjacent, Periglis and Porth Killier beaches, the tree-lined Barnaby Lane and adjacent bulb fields, Wingletang Down and Horse Point and the sheltered cove at Covean.

Tresco, with its two large expanses of water, the Great and Abbey Pools, is the most attractive island to wildfowl and regularly hosts North American species such as Black Duck and Ring-necked Duck, whilst the bulb fields and sallows bordering the north side of the pool regularly attracts rare warblers, buntings, larks and outstanding vagrants. Spotted Crakes and waders tend to prefer the shallower Abbey Pool and can be seen feeding along the muddy fringes and rare pipits can regularly be seen on the helipad nearby. The elm hedges and bulb fields at Borough Farm are renowned for their attraction to rarities, but the fields themselves are strictly inaccessible. The fields and hedgerows behind Island Hotel, Old Grimsby, regularly attract rare buntings and finches, as do those near the quay at New Grimsby.

Bryher is one of the least visited islands by birders but the maze of hedgerows immediately inland of Green Bay regularly attract rare warblers and flycatchers and the fields in the middle of the island, rare pipits and larks. The pool on the west side of the island and the gardens around Hell Bay Hotel may also be productive.

St Martin's is a large island and the most north-easterly of the inhabited ones. Much of it is cultivated and the northern side rather barren and heather-clad. The birds tend to congregate in the more sheltered fields and hedges along the south side, the area of trees north of the cricket pitch at Lower Town being the productive on the island.

The Isles of Scilly can be reached by air or sea. British International Helicopters (☎ 01736 63871) operate a daily service (except Sundays) from Penzance to St Mary's and Tresco, taking just twenty minutes, and a similar service (weather allowing) is offered by Skybus (☎ 01736 788771/788601), between St Just Aerodrome and St Mary's. The Isles of Scilly Steamship Company (☎ 01736 62009/64013) also operates a daily service (except Sundays) between Penzance Harbour and St Mary's Quay on the MV *Scillonian*, taking two and a half hours.

Isles of Scilly

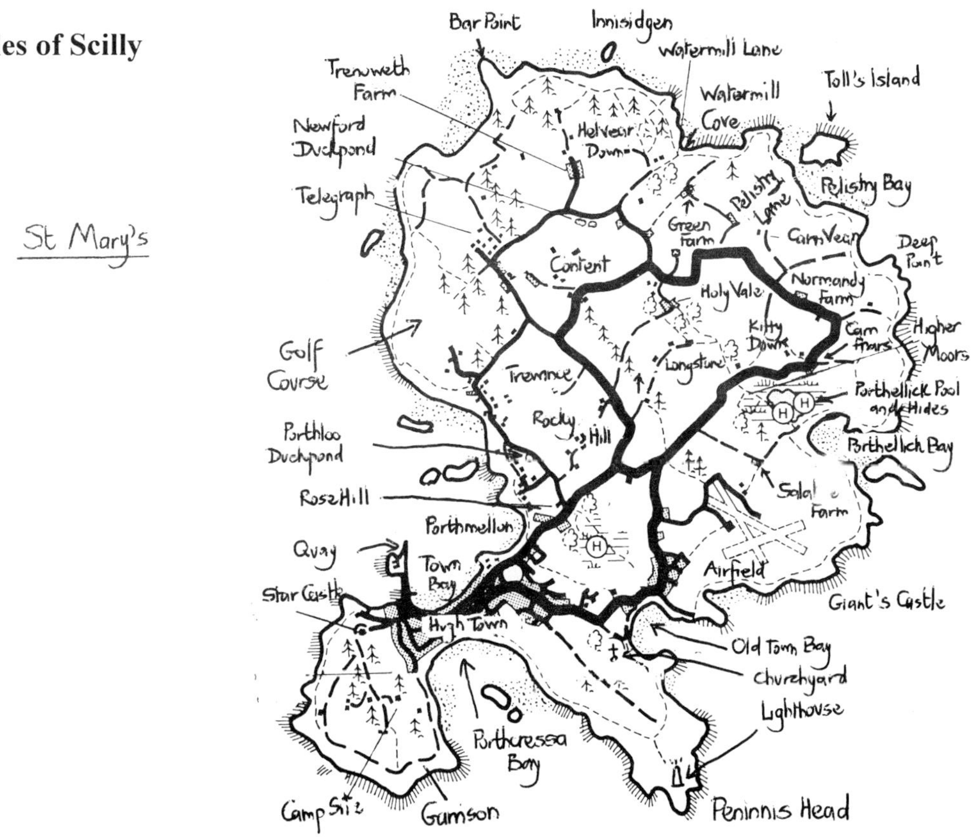

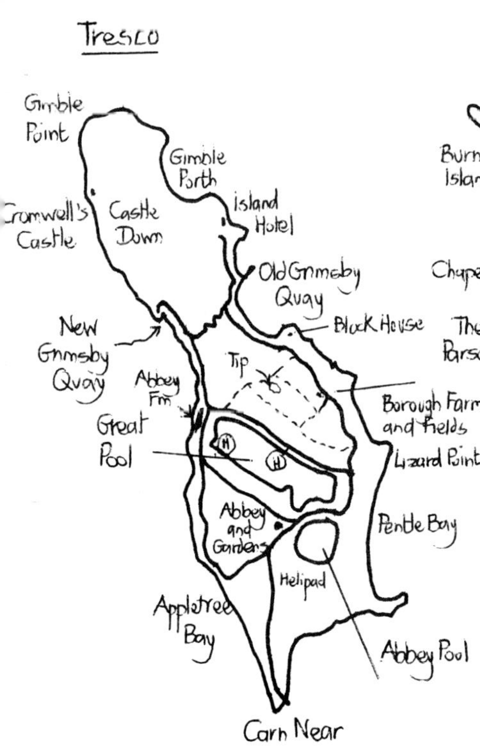

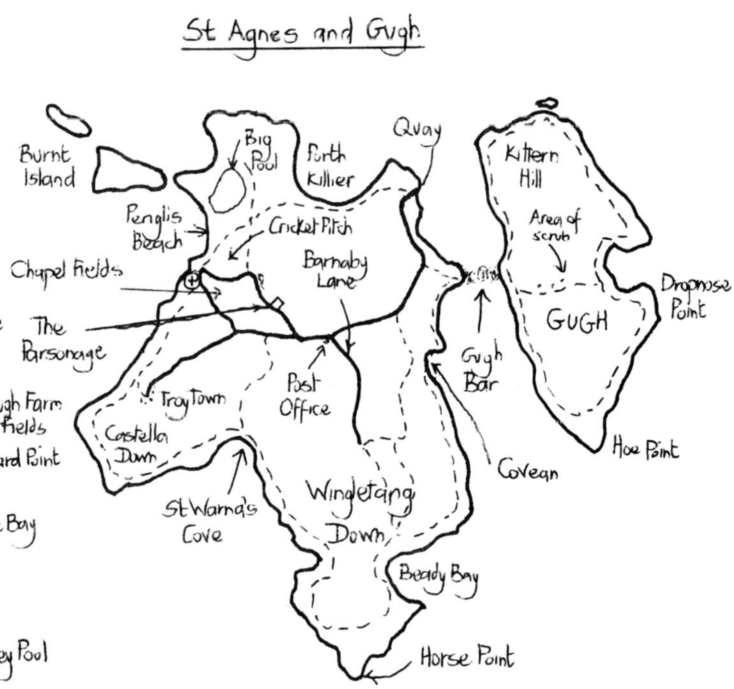

Checklist of the Birds of Britain

This list includes all 575 species that have been recorded in Britain for which there is adequate documentation and is that which is currently recognised by the *UK400 Club*. The order follows that of Voous, and the chosen English names are those preferred by the author.

The status codes used are as defined below, a few species appearing as a result of past introductions and now accepted into Category C (maintain self-supporting populations).

KEY TO CHECKLIST

Category A	*Species which have been recorded in an apparently wild state in Britain since 1900.*
Category B	*Species which have been recorded in an apparently wild state in Britain, but not since the nineteenth century.*
Category C	*Species which, although originally introduced by man, have now established a regular feral breeding stock which apparently maintains itself without necessary recourse to further reintroduction.*
Category D	*Species which have occurred in an apparently wild state in Britain but for which the possibility of escape cannot be fully excluded.*
Category M	*Species only recorded in Britain on the basis of birds found dead.*

The task of separating escapes from genuinely wild vagrants is a particularly difficult one, especially now as there are so many species in captivity or being imported into Western Europe, and there is no satisfactory solution. Given the variety and magnitude of the bird trade in SE Asia and the problems of transit to Europe (customs, quarantine, etc.), it is inevitable that many escapes will occur. The problem of escapes in relation to listing birders is one of great debate, and for the purpose of comparison and to be consistent, the *UK400 Club* considers some species admitted to Category D as being acceptable.

Entries for the commoner species are followed by a classification code of their current British status. These are as follows:-

RB Resident breeder, with at least 10,000 pairs nesting regularly in Britain.

SB Scarce breeder, with only five to 300 pairs nesting annually in Britain.

SV Summer visitor, breeding in large numbers

IB Introduced breeder, mostly resident and sustaining self-supporting populations in the v

OB Occasional breeder, with 1-10 pairs present during some years.

WV Winter visitor in reasonable numbers.

PM Passage migrant in fluctuating numbers, mainly in spring and autumn.

A numeric entry in the status column indicates the number of occurrences in Britain up to 1990.

A	Red-throated Diver	SB	
A	Black-throated Diver	SB	
A	Great Northern Diver	WV	
A	White-billed Diver	124	
A	Pied-billed Grebe	11	
A	Little Grebe	RB	
A	Great Crested Grebe	RB	
A	Red-necked Grebe	WV	
A	Slavonian Grebe	SB	
A	Black-necked Grebe	SB	
A	Black-browed Albatross	12	
A	Fulmar	RB	
A	Soft-plumaged Petrel	1	
B	Black-capped Petrel	2	
A	Bulwer's Petrel	2	
A	Cory's Shearwater	PM	
A	Great Shearwater	PM	
A	Sooty Shearwater	PM	
A	Manx Shearwater	SV	
A	Mediterranean Shearwater	PM	
A	Little Shearwater	59	
A	Wilson's Petrel	197	
B	White-faced Petrel	1	
A	Storm Petrel	SV	
A	Leach's Petrel	SB	
A	Swinhoe's Petrel	3	
A	Madeiran Petrel	1	
A	Gannet	RB	
A	Cormorant	RB	
A	Double-crested Cormorant	2	
A	Shag	RB	
D	White Pelican	15	
A	Magnificent Frigatebird	1	
A	Eurasian Bittern	SB	
A	American Bittern	37	
B	Least Bittern	1	
D	Yellow Bittern	1	
A	Little Bittern	417	
A	Black-crowned Night Heron	474	
A	American Green Heron	3	
A	Squacco Heron	98	
A	Cattle Egret	44	
A	Little Egret	SV	
A	Great White Egret	47	
A	Grey Heron	RB	
A	Purple Heron	560	
A	Black Stork	87	
A	White Stork	432	
A	Glossy Ibis	366	
A	Spoonbill	PM	
D	Greater Flamingo	19	

A	Mute Swan	RB	
A	Bewick's Swan	WV	
A	Whooper Swan	WV	
A	Western Bean Goose	WV	
A	Pink-footed Goose	WV	
A	White-fronted Goose	WV	
A	Lesser White-fronted Goose	99	
A	Greylag Goose	WV	
A	Snow Goose	300+	
A	Canada Goose	IB	
A	Barnacle Goose	WV	
A	Brent Goose	WV	
A	Red-breasted Goose	40	
C	Egyptian Goose	IB	
A	Ruddy Shelduck	70	
A	Shelduck	RB	
C	Mandarin Duck	IB	
A	Eurasian Wigeon	WV	
A	American Wigeon	167	
D	Falcated Duck	7	
A	Gadwall	WV	
A	European Teal	WV	
A	Mallard	RB	
A	Black Duck	13	
A	Pintail	SB	
A	Garganey	SV	
A	Blue-winged Teal	132	
A	Shoveler	RB	
A	Red-crested Pochard	WV	
A	Northern Pochard	WV	
A	Redhead	0	
A	Ring-necked Duck	191	
A	Ferruginous Duck	PM	
A	Tufted Duck	RB	
A	Greater Scaup	WV	
A	Lesser Scaup	3	
A	Eider	RB	
A	King Eider	57	
A	Steller's Eider	13	
A	Harlequin Duck	12	
A	Long-tailed Duck	WV	
A	Common Scoter	SB	
A	Black Scoter	4	
A	Surf Scoter	163	
A	Velvet Scoter	WV	
A	Bufflehead	6	
A	Barrow's Goldeneye	1	
A	Goldeneye	RB	
A	Hooded Merganser	1	
A	Smew	WV	
A	Red-breasted Merganser	RB	

A	Goosander	RB	
C	Ruddy Duck	IB	
A	Honey Buzzard	SB	
A	Black Kite	148	
A	Red Kite	SB	
A	White-tailed Eagle	15	
A	Bald Eagle	1	
A	Egyptian Vulture	3	
A	Griffon Vulture	2	
A	Black Vulture	1	
A	Marsh Harrier	RB	
A	Hen Harrier	SB	
A	Pallid Harrier	3	
A	Montagu's Harrier	SB	
A	Goshawk	SB	
A	Sparrowhawk	RB	
A	Red-shouldered Hawk	2	
A	Buzzard	RB	
A	Rough-legged Buzzard	WV	
A	Spotted Eagle	9	
A	Golden Eagle	RB	
A	Osprey	SB	
A	Lesser Kestrel	22	
A	Kestrel	RB	
A	American Kestrel	2	
A	Red-footed Falcon	464	
A	Merlin	SB	
A	Hobby	SV	
A	Eleonora's Falcon	3	
A	Gyrfalcon	238	
A	Peregrine	RB	
A	Red Grouse	RB	
A	Ptarmigan	RB	
A	Black Grouse	RB	
A	Capercaillie	RB	
C	Red-legged Partridge	IB	
A	Grey Partridge	RB	
A	Quail	SB	
C	Pheasant	RB	
C	Golden Pheasant	IB	
C	Lady Amherst's Pheasant	IB	
A	Water Rail	RB	
A	Spotted Crake	PM	
A	Sora Rail	10	
A	Little Crake	97	
A	Baillon's Crake	70	
A	Corncrake	SB	
A	Moorhen	RB	
A	Allen's Gallinule	1	
A	American Purple Gallinule	1	
A	Coot	RB	

A	American Coot	0	
A	Common Crane	PM	
A	Sandhill Crane	1	
A	Little Bustard	185	
A	Houbara Bustard	5	
A	Great Bustard	30	
A	Oystercatcher	RB	
A	Black-winged Stilt	283	
A	Avocet	SV	
A	Stone Curlew	SB	
A	Cream-coloured Courser	37	
A	Collared Pratincole	83	
A	Oriental Pratincole	2	
A	Black-winged Pratincole	23	
A	Little Ringed Plover	SB	
A	Ringed Plover	RB	
A	Semipalmated Plover	1	
A	Killdeer	35	
A	Kentish Plover	PM	
A	Greater Sand Plover	9	
A	Caspian Plover	4	
A	Dotterel	SB	
A	American Golden Plover	116	
A	Pacific Golden Plover	16	
A	Golden Plover	RB	
A	Grey Plover	WV	
A	White-tailed Plover	3	
A	Sociable Plover	32	
A	Lapwing	RB	
A	Great Knot	1	
A	Knot	WV	
A	Sanderling	WV	
A	Semipalmated Sandpiper	30	
A	Western Sandpiper	5	
A	Red-necked Stint	1	
A	Little Stint	PM	
A	Temminck's Stint	PM	
A	Long-toed Stint	3	
A	Least Sandpiper	25	
A	White-rumped Sandpiper	226	
A	Baird's Sandpiper	105	
A	Pectoral Sandpiper	PM	
A	Sharp-tailed Sandpiper	20	
A	Curlew Sandpiper	PM	
A	Purple Sandpiper	WV	
A	Dunlin	RB	
A	Broad-billed Sandpiper	150	
A	Stilt Sandpiper	18	
A	Buff-breasted Sandpiper	PM	
A	Ruff	SB	
A	Jack Snipe	WV	

A	Common Snipe	RB	
A	Great Snipe	78	
A	Long-billed Dowitcher	124	
A	Woodcock	RB	
A	Black-tailed Godwit	SB	
A	Hudsonian Godwit	2/3	
A	Bar-tailed Godwit	WV	
A	Little Whimbrel	2	
B	Eskimo Curlew	6/7	
A	Whimbrel	PM	
A	Curlew	RB	
A	Upland Sandpiper	30	
A	Spotted Redshank	PM	
A	Redshank	RB	
A	Marsh Sandpiper	70	
A	Greenshank	SB	
A	Greater Yellowlegs	18	
A	Lesser Yellowlegs	154	
A	Solitary Sandpiper	25	
A	Green Sandpiper	PM	
A	Wood Sandpiper	SB	
A	Terek Sandpiper	37	
A	Common Sandpiper	SV	
A	Spotted Sandpiper	90	
A	Grey-tailed Tattler	1	
A	Turnstone	PM	
A	Wilson's Phalarope	164	
A	Red-necked Phalarope	SB	
A	Grey Phalarope	PM	
A	Pomarine Skua	PM	
A	Arctic Skua	SV	
A	Long-tailed Skua	PM	
A	Great Skua	SV	
B	Great Black-headed Gull	1	
A	Mediterranean Gull	SB	
A	Laughing Gull	47	
A	Franklin's Gull	17	
A	Little Gull	PM	
A	Sabine's Gull	PM	
A	Bonaparte's Gull	61	
A	Black-headed Gull	RB	
A	Slender-billed Gull	5	
A	Ring-billed Gull	WV	
A	Common Gull	RB	
A	Herring Gull	RB	
A	Yellow-legged Gull	PM	
A	Iceland Gull	WV	
A	Glaucous Gull	WV	
A	Lesser Black-backed Gull	RB	
A	Great Black-backed Gull	RB	
A	Ross's Gull	42	

A	Kittiwake	RB	
A	Ivory Gull	30	
A	Gull-billed Tern	251	
A	Caspian Tern	183	
A	Royal Tern	4	
A	Lesser Crested Tern	6	
A	Sandwich Tern	SV	
A	Roseate Tern	SB	
A	Common Tern	SV	
A	Arctic Tern	SV	
A	Aleutian Tern	1	
A	Forster's Tern	9	
A	Bridled Tern	13	
A	Sooty Tern	24	
A	Little Tern	SV	
A	American Least Tern	1	
A	Whiskered Tern	96	
A	Black Tern	PM	
A	White-winged Black Tern	675	
A	Guillemot	RB	
A	Brunnich's Guillemot	24	
A	Razorbill	RB	
A	Black Guillemot	RB	
A	Ancient Murrelet	1	
A	Little Auk	WV	
A	Puffin	SV	
A	Pallas's Sandgrouse	7	
A	Rock Dove	RB	
A	Stock Dove	RB	
A	Woodpigeon	RB	
A	Collared Dove	RB	
A	Turtle Dove	SV	
A	Rufous Turtle Dove	5	
A	Mourning Dove	1	
C	Ring-necked Parakeet	IB	
A	Great Spotted Cuckoo	31	
A	Cuckoo	SV	
A	Black-billed Cuckoo	12	
A	Yellow-billed Cuckoo	49	
A	Barn Owl	RB	
A	Scops Owl	71	
B	Eagle Owl	4	
A	Snowy Owl	122	
A	Hawk Owl	11	
C	Little Owl	IB	
A	Tawny Owl	RB	
A	Long-eared Owl	RB	
A	Short-eared Owl	RB	
A	Tengmalm's Owl	47	
A	Nightjar	SV	
B	Red-necked Nightjar	1	

A	Egyptian Nightjar	2	
A	Common Nighthawk	14	
A	White-throated Needletail	5-7	
A	Chimney Swift	5	
A	Swift	SV	
A	Pallid Swift	6	
A	Pacific Swift	(1)	
A	Alpine Swift	271	
A	Little Swift	8	
A	Kingfisher	RB	
A	Belted Kingfisher	2	
A	Blue-cheeked Bee-eater	7	
A	European Bee-eater	PM	
A	Roller	241	
A	Hoopoe	PM	
A	Wryneck	PM	
A	Green Woodpecker	RB	
A	Yellow-bellied Sapsucker	1	
A	Great Spotted Woodpecker	RB	
A	Middle Spotted Woodpecker	0	
A	Lesser Spotted Woodpecker	RB	
A	Eastern Phoebe	1	
A	Calandra Lark	3	
A	Bimaculated Lark	3	
A	White-winged Lark	2	
A	Short-toed Lark	361	
A	Lesser Short-toed Lark	1	
A	Crested Lark	18	
A	Woodlark	RB	
A	Skylark	RB	
A	Shore Lark	WV	
A	Sand Martin	SV	
A	Tree Swallow	1	
A	Crag Martin	3	
A	Swallow	SV	
A	Red-rumped Swallow	206	
A	Cliff Swallow	2	
A	House Martin	SV	
A	Richard's Pipit	PM	
A	Blyth's Pipit	3	
A	Tawny Pipit	PM	
A	Olive-backed Pipit	120	
A	Tree Pipit	SV	
A	Pechora Pipit	37	
A	Meadow Pipit	RB	
A	Red-throated Pipit	217	
A	Rock Pipit	RB	
A	Water Pipit	PM	
A	Buff-bellied Pipit	3	
A	Yellow Wagtail	SV	
A	Citrine Wagtail	53	

A	Grey Wagtail	RB	
A	Pied Wagtail	RB	
A	Waxwing	WV	
A	Cedar Waxwing	1	
A	Dipper	RB	
A	Wren	RB	
A	Northern Mockingbird	3	
A	Brown Thrasher	1	
A	Dunnock	RB	
A	Alpine Accentor	32	
A	Rufous Bushchat	8	
A	Robin	RB	
A	Thrush Nightingale	91	
A	Nightingale	SV	
A	Siberian Rubythroat	1	
A	Bluethroat	PM	
A	Red-flanked Bluetail	10	
A	White-throated Robin	2	
A	Black Redstart	SB	
A	Common Redstart	SV	
A	Moussier's Redstart	1	
A	Whinchat	SV	
A	Stonechat	RB	
A	Isabelline Wheatear	7	
A	Northern Wheatear	SV	
A	Pied Wheatear	19	
A	Black-eared Wheatear	47	
A	Desert Wheatear	36	
A	White-crowned Bl Wheatear	1	
A	Rock Thrush	23	
D	Blue Rock Thrush	3	
A	White's Thrush	41	
A	Siberian Thrush	3	
D	Varied Thrush	1	
A	Wood Thrush	1	
A	Hermit Thrush	3	
A	Swainson's Thrush	14	
A	Grey-cheeked Thrush	36	
A	Veery	2	
A	Ring Ouzel	SV	
A	Blackbird	RB	
A	Eye-browed Thrush	12	
A	Dusky Thrush	8	
A	Naumann's Thrush	1	
A	Black-throated Thrush	22	
A	Red-throated Thrush	0	
A	Fieldfare	WV	
A	Song Thrush	RB	
A	Redwing	WV	
A	Mistle Thrush	RB	
A	American Robin	17	

A	Cetti's Warbler	SV	
A	Fan-tailed Warbler	2	
A	Pallas's Grasshopper Wblr	11	
A	Lanceolated Warbler	49	
A	Grasshopper Warbler	SV	
A	River Warbler	12	
A	Savi's Warbler	PM	
A	Moustached Warbler	7-10	
A	Aquatic Warbler	PM	
A	Sedge Warbler	SV	
A	Paddyfield Warbler	17	
A	Blyth's Reed Warbler	17	
A	Marsh Warbler	SB	
A	Reed Warbler	SV	
A	Great Reed Warbler	153	
A	Thick-billed Warbler	2	
A	Olivaceous Warbler	6	
A	Booted Warbler	32	
A	Icterine Warbler	PM	
A	Melodious Warbler	PM	
A	Marmora's Warbler	1	
A	Dartford Warbler	RB	
A	Spectacled Warbler	0	
A	Subalpine Warbler	243	
A	Sardinian Warbler	25	
A	Ruppell's Warbler	3	
A	Desert Warbler	6	
A	Orphean Warbler	5	
A	Barred Warbler	PM	
A	Lesser Whitethroat	SV	
A	Common Whitethroat	SV	
A	Garden Warbler	SV	
A	Blackcap	SV	
A	Greenish Warbler	199	
A	Two-barred Greenish Wblr	1	
A	Green Warbler	1	
A	Arctic Warbler	156	
A	Pallas's Leaf Warbler	540	
A	Yellow-browed Warbler	PM	
A	Hume's Yellow-browed Wb	17	
A	Radde's Warbler	104	
A	Dusky Warbler	96	
A	Bonelli's Warbler	100	
A	Wood Warbler	SV	
A	Chiffchaff	SV	
A	Willow Warbler	SV	
A	Goldcrest	RB	
A	Firecrest	SB	
A	Brown Flycatcher	1	
A	Spotted Flycatcher	SV	
A	Red-breasted Flycatcher	PM	

A	Collared Flycatcher	17	
A	Pied Flycatcher	SV	
D	Mugimaki Flycatcher	0	
A	Bearded Tit	RB	
A	Long-tailed Tit	RB	
A	Marsh Tit	RB	
A	Willow Tit	RB	
A	Crested Tit	RB	
A	Coal Tit	RB	
A	Blue Tit	RB	
A	Great Tit	RB	
A	Red-breasted Nuthatch	1	
A	Nuthatch	RB	
A	Wallcreeper	10	
A	Treecreeper	RB	
A	Short-toed Treecreeper	16	
A	Penduline Tit	56	
A	Golden Oriole	SB	
A	Brown Shrike	1	
A	Isabelline Shrike	33	
A	Red-backed Shrike	PM	
A	Lesser Grey Shrike	138	
A	Great Grey Shrike	WV	
A	Steppe Shrike	4	
A	Woodchat Shrike	PM	
A	Jay	RB	
A	Magpie	RB	
A	Nutcracker	444	
A	Chough	SB	
A	Jackdaw	RB	
A	Rook	RB	
A	Carrion Crow	RB	
A	Raven	RB	
D	Daurian Starling	1	
A	Starling	RB	
A	Rose-coloured Starling	460	
A	House Sparrow	RB	
A	Spanish Sparrow	3	
A	Tree Sparrow	RB	
A	Rock Sparrow	1	
A	Yellow-throated Vireo	1	
A	Philadelphia Vireo	1	
A	Red-eyed Vireo	55	
A	Chaffinch	RB	
A	Brambling	WV	
A	Serin	PM	
A	Greenfinch	RB	
A	Goldfinch	RB	
A	Siskin	RB	
A	Linnet	RB	
A	Twite	RB	

A	Redpoll	RB	
A	Arctic Redpoll	246	
A	Two-barred Crossbill	130	
A	Common Crossbill	RB	
A	Parrot Crossbill	SB	
A	Trumpeter Finch	6	
A	Common Rosefinch	PM	
A	Pine Grosbeak	13	
A	Bullfinch	RB	
A	Hawfinch	RB	
A	Evening Grosbeak	2	
A	Black-and-white Warbler	9	
A	Golden-winged Warbler	1	
A	Tennessee Warbler	3	
A	Northern Parula	12	
A	Yellow Warbler	3	
A	Cape May Warbler	1	
A	Chestnut-sided Warbler	1	
A	Blackburnian Warbler	2	
A	Magnolia Warbler	1	
A	Yellow-rumped Warbler	10	
M	Palm Warbler	1	
A	Bay-breasted Warbler	0	
A	Blackpoll Warbler	25	
A	American Redstart	5	
A	Ovenbird	3	
A	Northern Waterthrush	5	
A	Common Yellowthroat	4	
A	Wilson's Warbler	1	
A	Hooded Warbler	0	
A	Summer Tanager	1	
A	Scarlet Tanager	4	
A	Rufous-sided Towhee	2	
A	Lark Sparrow	2	
A	Savannah Sparrow	2	
A	Song Sparrow	6	
A	White-crowned Sparrow	2	
A	White-throated Sparrow	16	
A	Dark-eyed Junco	15	
A	Lapland Bunting	WV	
A	Snow Bunting	SB	
D	Black-faced Bunting	0	
A	Pine Bunting	20	
A	Yellowhammer	RB	
A	Cirl Bunting	SB	
A	Rock Bunting	6	
A	Ortolan Bunting	PM	
A	Cretzschmar's Bunting	2	
A	Yellow-browed Bunting	2	
A	Rustic Bunting	232	
A	Little Bunting	550	

A	Yellow-breasted Bunting	141		
D	Chestnut Bunting	4		
A	Reed Bunting	RB		
A	Pallas's Reed Bunting	3		
D	Red-headed Bunting	301		
A	Black-headed Bunting	83		
A	Corn Bunting	RB		
A	Rose-breasted Grosbeak	14		
A	Indigo Bunting	5		
A	Bobolink	13		
A	Brown-headed Cowbird	1		
B	Rusty Blackbird	1		
A	Yellow-headed Blackbird	2		
A	Northern Oriole	17		
	575 Species - 01/07/96			

SCHEDULE 1 SPECIES

Under the provisions of the Wildlife and Countryside Act 1981 the following bird species (listed in Schedule 1 - Part I of the Act) are protected by special penalties at all times.

Avocet
Bee-eater
Bittern
Bittern, Little
Bluethroat
Brambling
Bunting, Cirl
Bunting, Lapland
Bunting, Snow
Buzzard, Honey
Chough
Corncrake
Crake, Spotted
Crossbills (all spp)
Curlew, Stone
Divers (all species)
Dotterel
Duck, Long-tailed
Eagle, Golden
Eagle, White-tailed
Falcon, Gyr
Fieldfare
Firecrest
Garganey
Godwit, Black-tailed
Goshawk
Grebe, Black-necked
Grebe, Slavonian
Greenshank
Gull, Little
Gull, Mediterranean
Harriers (all species)
Heron, Purple
Hobby
Hoopoe
Kingfisher
Kite, Red
Merlin
Oriole, Golden
Osprey
Owl, Barn
Owl, Snowy
Peregrine
Petrel, Leach's
Phalarope, Red-necked
Plover, Kentish
Plover, Little Ringed
Quail, Common
Redstart, Black
Redwing
Rosefinch, Common
Ruff
Sandpiper, Green
Sandpiper, Purple
Sandpiper, Wood
Scaup
Scoter, Common
Scoter, Velvet
Serin
Shorelark
Shrike, Red-backed
Spoonbill
Stilt, Black-winged
Stint, Temminck's
Swan, Bewick's
Swan, Whooper
Tern, Black
Tern, Little
Tern, Roseate
Tit, Bearded
Tit, Crested
Treecreeper, Short-toed
Warbler, Cetti's
Warbler, Dartford
Warbler, Marsh
Warbler, Savi's
Whimbrel
Wryneck

The following species (listed in Schedule 1 - Part II of the Act) are protected by special penalties during the close season, which is 1st February to 31st August, but may be killed outside this period.

Goldeneye, Greylag Goose (in Scotland only) and Pintail.

List of Illustrations

Black-throated Diver - Steven Brown
Red-necked Grebe - Mike Langman
Black-necked Grebe - Steven Brown
Black-browed Albatross - Steven Brown
Balearic Shearwater - Mike Langman
Wilson's Petrel - Steven Brown
Leach's Petrel - Steven Brown
Common Bittern - Steven Brown
Little Egret - Steven Brown
Bewick's Swan - Phil Jones
Whooper Swan - Steven Brown
Lesser White-fronted Goose - Phil Jones
Barnacle Goose - James Hanlon
Ferruginous Duck - Steven Brown
King Eider - Steven Brown
Long-tailed Duck - Phil Jones
Velvet and Surf Scoters - Phil Jones
Honey Buzzard - Steven Brown
Honey Buzzard - Pete Dennis
White-tailed Eagle - Steven Brown
Hen Harrier - Phil Jones
Northern Goshawk - Mike Langman
Rough-legged Buzzard - Steven Brown
Osprey - Phil Jones
Merlin - Steven Brown
Merlin - Mike Langman
Hobby - Steven Brown
Peregrine - Phil Jones
Ptarmigan - Steven Brown
Capercaillie - Phil Jones
Black Grouse - Richard Millington
Golden Pheasant - Richard Millington
Lady Amherst's Pheasant - R. Millington
Common Crane - Phil Jones
Avocet - Steven Brown
Black-winged Stilt - Mike Langman
Stone Curlew - Mike Langman
Kentish Plover - Steven Brown
Dotterel - Steven Brown
Red-necked Phalarope - Steven Brown
Grey Phalarope - Andrew Mackay
Pomarine Skua - Mike Langman
Pomarine Skuas - Steven Brown
Long-tailed Skua - Mike Langman
Mediterranean Gull - Mike Langman
Mediterranean Gull - Phil Jones
Sabine's Gull - Mike Langman
Ring-billed Gull - Steven Brown
Lesser Crested Tern - Phil Jones
Roseate Tern - Mike Langman
Little Auks - Steven Brown
Rock Dove - Richard Millington
Long-eared Owl - Mike Langman
Nightjar - Steven Brown
Hoopoe - Mike Langman
Wryneck - Antony Disley
L S Woodpeckers - R.Millington
Shorelarks - Mike Langman
Richard's Pipit - Steven Brown
Waxwing - Phil Jones
Dipper - Phil Jones
Bluethroat - Paul Semmens
Icterine Warbler - Mike Langman
Barred Warbler - Mike Langman
Pallas's Leaf Warbler - Mike Langman
Yellow-browed Warbler - M. Langman
Firecrest - Mike Langman
Red-breasted Flycatcher - Mike Langman
Bearded Tit - Steven Brown
Willow Tit - Steven Brown
Crested Tit - Steven Brown
Golden Oriole - Phil Jones
Woodchat Shrike - Andrew Mackay
Great Grey Shrike - Phil Jones
Chough - Steven Brown
Serin - Mike Langman
Crossbill - Steven Brown
Hawfinch - Steven Brown
Lapland Bunting - Mike Langman
Cirl Bunting - Paul Semmens
White-tailed Eagle - Phil Jones

Index to Species

American Black Duck, 50
Aquatic Warbler, 203
Avocet, 110

Balearic Shearwater, 24
Barnacle Goose, 45
Barred Warbler, 210
Bearded Tit, 218
Bewick's Swan, 35
Black Grouse, 92
Black Guillemot, 159
Black-browed Albatross, 20
Black-necked Grebe, 18
Black-throated Diver, 10
Black-winged Stilt, 109
Bluethroat, 198
British Storm-Petrel, 26

Capercaillie, 95
Cetti's Warbler, 199
Chough, 231
Cirl Bunting, 254
Common Bittern, 29
Common Crane, 108
Common Crossbill, 240
Common Rosefinch, 243
Corncrake, 107
Cory's Shearwater, 21
Crested Tit, 224

Dartford Warbler, 208
Dipper, 188
Dotterel, 116
Egyptian Goose, 47
European Spoonbill, 33
European White-fronted Goose, 41

Ferruginous Duck, 55
Firecrest, 215

Garganey, 50
Glaucous Gull, 152
Golden Eagle, 79
Golden Oriole, 226
Golden Pheasant, 99
Great Grey Shrike, 229
Great Northern Diver, 12
Great Shearwater, 22
Greater Flamingo, 34
Greenland White-fronted Goose, 42
Grey Phalarope, 128
Greylag Goose, 44

Hawfinch, 244
Hen Harrier, 71
Hobby, 85
Honey Buzzard, 64
Hoopoe, 172

Iceland Gull, 150
Icterine Warbler, 205

Jack Snipe, 125

Kentish Plover, 115
King Eider, 58

Lady Amherst's Pheasant, 102
Lapland Bunting, 250
Leach's Petrel, 27
Lesser Crested Tern, 156
Lesser Spotted Woodpecker, 175
Lesser White-fronted Goose, 40
Little Auk, 161
Little Egret, 31
Little Gull, 141
Long-eared Owl, 165
Long-tailed Duck, 59
Long-tailed Skua, 134

Mandarin Duck, 48
Marsh Warbler, 204
Mediterranean Gull, 136
Melodious Warbler, 207
Merlin, 83
Montagu's Harrier, 74

Nightingale, 194
Nightjar, 166
Northern Goshawk, 75

Osprey, 81

Pale-bellied Brent Goose, 46
Pallas's Leaf Warbler, 212
Peregrine, 87
Pink-footed Goose, 40
Pomarine Skua, 130
Ptarmigan, 91
Purple Sandpiper, 120

Quail, 97

Red Grouse, 90
Red Kite, 68
Red-backed Shrike, 227
Red-breasted Flycatcher, 217
Red-crested Pochard, 52
Red-necked Grebe, 14
Red-necked Phalarope, 127
Richard's Pipit, 184
Ring-billed Gull, 145
Ring-necked Duck, 54
Ring-necked Parakeet, 163
Rock Dove, 162
Roseate Tern, 157
Rough-legged Buzzard, 78

Sabine's Gull, 143
Savi's Warbler, 202
Scaup, 56
Scottish Crossbill, 241
Serin, 237
Shorelark, 183
Slavonian Grebe, 16
Smew, 62
Snow Bunting, 252
Snow Goose, 45
Sooty Shearwater, 23
Spotted Crake, 105
Stone Curlew, 113
Surf Scoter, 60

Tree Sparrow, 233
Twite, 238

Velvet Scoter, 61

Water Pipit, 185
Waxwing, 188
Western Bean Goose, 39
White-tailed Eagle, 70
Whooper Swan, 37
Willow Tit, 221
Wilson's Petrel, 25
Woodchat Shrike, 228
Woodlark, 181
Wryneck, 172

Yellow-browed Warbler, 214
Yellow-legged Gull, 148

Index of Sites for which directions are given in the text.

Abberton Reservoir, 52
Aberlady Bay, 14, 16
Abernethy Forest, 96, **241**
Acres Down, 66, 86
Allenheads, 93
Amberley Wild Brooks, **51**
Amwell GP, 62
Anglesey, 88, 99, 157, 160, 231
Ardivachar Point, 132
Ardnamurchan Point, 12
Arkengarthdale, 94
Arne, 112, 171, 180, 209
Arundel Refuge, 36
Ashdown Forest, 169, 230
Aviemore, 87, 189, 225
Aylesbeare Common, 171, 209
Ayr, 153

Bakethin, 76
Ballaugh Curraghs, 71
Balranald, 107
Balsdean, 119
Bamburgh, 59
Barrowden, 194
Barton-upon-Humber Pits, 218
Beachy Head, **272**
Beaulieu Heath, 182
Beaulieu Road Station, 86, 170, 179, 182, 209
Beddington Sewage Farm, 186, 236
Bedgebury Pinetum, **248**
Bembridge, 124, 138
Benbecula, 107
Berry Head, 131
Bewdley, 246
Bird Rock, 231
Bishop's Dyke,, 170
Black Gutter Bottom, 230
Black Mountain, 119
Black Rock Ledge, 124
Blackpill Beach, 137
Blacktoft Sands, 71, 218
Blackwater Estuary, 17
Blakeney Point, 173, 205, 269
Blean Woods, 196
Blenheim Park, 248
Blithfield Reservoir, 154, 176
Bond's Mill, 191
Bookham Common, 197
Bough Beech Reservoir, 82
Bourne Woods, 194
Bowness-on-Solway, 132
Braco Moor, 93
Brading Marsh, 200
Bradley Wood, 175
Bradwell-on-Sea, 239
Brandon Country Park, 199
Breakheart Hill, 104
Brechfa Woods, 77
Bredon Hill, 118
Bretton Sculpture Park, 245
Breydon Water, 111
Bricket Park, 178
Brickhill, 103
Brighton Marina, 123
Brockhampton Sewage Farm, 187
Brogborough Lake, 155
Broomhead Moor, 167
Broomhill Ings, 153
Broughton Woods, 245
Brownsea, 112
Broxbourne Woods, 248
Brumby Woods, 245
Buckenham, 39
Buddon Wood, 176
Burbage Moor, 117
Burgh Castle, 111
Burghead, 225
Burghead Bay, 12
Burley Wood, 176
Burnham Overy Dunes, 184
Bursdon Moor, 73, 84
Burton Mill Pond, 178
Butterwick Marsh, 239

Caer-Hun, 245
Caerlaverock, 37, 46
Cairngorm, 91, 117, 252
Caithness, 162
Calf of Man, 231
Camel Estuary, 32
Canford Heath, 171, 180
Cannock Chase, 168
Cannock Tip, 154
Cannop Ponds, 49
Capel Fleet, 72, 84
Carn Ban More, 91, 117
Carnsew Basin, 15
Carrbridge, 242
Castletown, 11
Catcott Lows, 36
Cemlyn Bay, 157
Charle Wood, **104**
Chasewater, 154
Cheddar Reservoir, 15
Chelson Meadow Tip, **138**
Chew Valley Lake, 140, **187**
Chobham Common, 169
Christchurch Harbour, **274**
Church Norton, 17, 273
Church Wood, 196
Clangar Wood, 197
Cley, 30, 51, 269
Cliffe, **34**, 112
Clipstone Forest, 168
Clo Mor, 159
Clophill, 103
Clowes Wood, 169
Clumber Park, 66, 246
Coalhouse Fort, **148**
Coatham Marsh, **264**
Combley Great Wood, 197
Coombes Valley, 191
Copperhouse Creek, **145**
Copt Point, **136**
Cotswold Water Park, 53, 86
Cowpen Marsh, 250, 251
Craig-yr-Aderyn, 231
Craigvinian Forest, 76
Crail, 260
Cromarty Firth, 56
Cromford Canal, 246
Crookham Common, 197
Crosby Marina, 27
Cruden Bay, 260
Crychan Forest, 77, 166
Cwm Clydach, 179, 192
Cwmystwyth, 68

Danbury Common, 196
Danbury Country Park, 247
Danby Beacon, 117
Dawlish Warren, 17, **158**
Deadmoor Common, 197
Dearne Valley, 38
Dee Estuary, 185
Denburn Wood, 260
Dengie Flats, 239
Denny Wood, 179
Derwentdale, 76
Devil's Bridge, **68**
Dinas, 68, 179, 192
Dinton Pastures, 197
Dipton Woods, 167
Ditchford Gravel Pits, 185
Dorman's Pool, **263**
Dosthill GP, 63
Dowles Brook, 191
Draycote Water, 53, 154
Druridge Pools, 38

Dulnain Bridge, 189
Dungeness, 14, 52, 62, 130, 271
Dunnet Bay, 12
Dunnet Head, 159, 162
Dunsford Woods, 180
Dunwich Forest, 169
Dyffryn Wood, 179
Dyfi Estuary b, 43

East Tilbury, **148**
Eckington Bridge, 204
Eden Valley, 44
Elmbridge Leisure Centre, **163**
Elmley, 111, 203
Elmley Marshes, **251**
Epping Forest, 196
Estuary Farm, 74
Etherow Country Park, 175
Eye Brook Reservoir, 154
Eye Water, 48

Fagg's Wood, 196
Fair Isle, 256, **257**
Fairburn Ings, 38, 54
Falls of Clyde, 189
Farlington Marsh, 89, **139**, 147
Farne Islands, 156, 157
Fawsley Park, 176
Fedw Fawr, **160**
Fenham Flats, 46
Ferny Ness, 14
Ferries, 26, 42, 56, 70, 107, 127, 132
Ferrybridge, 11
Fetlar, 127
Fife Ness, 260
Filey, 251, **265**
Filey Brigg, 122
Findhorn Valley, 80, 87, 188
Fineshade Wood, 246
Fingringhoe Wick, 196
Fisher's Green, 29, 63
Flackwell Heath, 246
Flamborough Head, 23, 123, 161, 173, **266**
Fleet Pond, 179
Fobbing Marshes, 219
Folkestone, 136
Fordham, 226
Foreland, 124
Foreness Point, 123
Forest Lodge, 96, **224**
Forest of Dean, 168
Formby Point, 27
Foxhole Heath, **114**
Fred Dale, 191
Frensham Common, 182, 208, 229
Fridaywood, 196
Funton Creek, 112

Gardener's Hill, 119
Gerrans Bay, 10
Gibraltar Point, 268
Gigrin Farm, 69
Girdleness, 120, 260
Glapthorn Cow Pasture, **99**
Glapthorn Cow Pastures, 194
Glen Affric, 80
Glen Clova, 80
Glen Tanar, 80, 95, 189, **242**
Glenshee, 90
Glyne Gap, 123
Goathurst Common, 169
Godrevy Point, 124
Goodrington Bay, 19
Goodrington Beach, 17
Gosford Bay, 14, 16
Goswick Sands, 252
Grantown-on-Spey, **96**, 242
Great Haldon, 65,171
Greatham Creek, **263**
Greenabella Marsh, 238
Gringley Carr, 118
Grisdale, 94
Grovely Wood, 170
Gruinard Bay, 12
Gruinard Island, 70
Guisborough Woods, 245
Gwenffrwd, 179
Gwennap Head, 21
Gwydir Forest, 166

Hainault Forest, 178
Haldon watchpoint, 65
Hampstead Heath, 178
Hamstreet, 196
Handa Island, 162
Hanningfield Reservoir, 15
Happisburgh, 118
Harewood Forest, 180
Harpford Commons, 209
Hartlepool Headland, 23, **263**
Hatfield Forest, 247
Hatfield Moors, 167
Haweswater, **79**, 87
Hayle Estuary, 15, 32, **145**
Hayling Island, 19
Heacham, 123
Hengistbury Head, **274**
Hermaness, 20
Heysham Harbour, 121
Hickling, 219
Hickling Broad, 202
Higham Bight, 34
Hilbre Island, 27, 121
Hobhole Drain, 165
Holkham, 47, 177, 247
Holkham Pines, **212**
Holland Haven, 123
Holme, 110, 173, 211, 227, 268
Holy Island, 206, 252, **262**
Holyfield GP, 63
Hope's Nose, 124
Hornsea Mere, 141
Horseshoe Pass, 90
Horsey, 38, 108
Horsey Broad, 72, 83
Hulne Park, 245
Hutton, 245

Inglestone Common, 197
Insh Marshes, 37
Inshriach Forest, 96
Inverdruie Fish Farm, 54, 82
Islay, 42, 231
Islay Marsh, 33
Isle of Man, 231
Isle of Sheppey, 84
Isle of Wight, 124
Isles of Scilly, 279, 281

Jubilee Grove, 215

Kielder Forest, 76
Killington Reservoir, 54
Kingley Vale NR, **101**
Kyle of Lochalshe, 159

Ladybower Reservoir, 76, 88
Lake District, **79**
Landguard NR, 211, 270
Langdon Beck, 94
Langstone Harbour, 19
Langtoft Pits, 63
Largo Bay, 14, 56
Larkeyvalley Wood, 178
Laughton Forest, 167
Launde Park Wood, 176
Lea Marston, **55**
Lea Valley, 29
Lea Valley Park, 63
Leasowe Embankment, 27
Leighton Moss, 30, 87
Lindisfarne, 38, 46
Linwood Warren, 175
Lipson Vale, 146
Little Paxton GP, 63, 85 195
Livermere Park, 47
Loch an Eilean, **225**, 242
Loch Eck, 48
Loch Eye, 44
Loch Garten, 81, 96, **224**, 225, **241**

Loch Gruinart, 42
Loch Indaal, 56
Loch Ken, 42
Loch Morlich, 54
Loch of Funzie, 127
Loch of Kinnordy, 18
Loch of Mey, 42
Loch of Skene, 44
Loch of Strathbeg, 62
Loch of the Lowes, 82
Loch Ruthven, 16, **92**
Loch Ryan, 19, **57**
Locke Park, **264**
Lodmoor, 187, **276**
Loganair, 42, 257
Long Drag, **263**
Lower Largo Bay, 11
Lowes Wood, 103
Lowestoft, 123
Luffenham Heath, 194
Lundy Island, **277**
Lydlinch Common, 197
Lynford Arboretum, 177, 247

Marazion Marsh, **106**, 201, **203**
Marbury Country Park, 175
Marford Quarry, 221
Margam Forest, 167
Margate, **164**
Martin Down, **98**
Martin Mere, 35, 37, 40, 89
Maulden Woods, 103
Mayday Farm, **77**, 169, 182, 229
Mere Sands, 48
Mereworth Woods, 169
Mersea Island, 239
Mhor-mor picnic site, **225**
Minsmere, 51, 110, 195, 219
Moneypot Hill, 104
Mount's Bay, 124
Mull of Galloway, 159
MV Oldenburg, 277
Mynydd Du, 119

Nag's Head, 192
Nagshead Reserve, 177, 246
Nanquidno, 207
Nant Irfon, 192
Nercwys Mountain, 166
Ness Point, 123
New Brighton, 27
New Forest, 66, 86, 170, 179, 182, 208, 215, 230, 249
New Swillington Ings, 185
Newsam Green Tip, 154
North Ronaldsay, **259**
North Shields, 151
North Warren, 195
North Wootton, 74
Northdown Park, **164**

Ogston Reservoir, 154
Old Baths, 125, 185
Old Hall Marshes, 111, 219, 239
Oldenburg, 277
Orkney, 259
Otmoor, 222
Otter Estuary, 187
Ouse Washes, 35, 50
Overleigh Bridge, 244
Oxwich NR, 199

Pagham Harbour, 17, **273**
Par Beach Pool, **146**
Parkgate Marsh, 71
Parkhurst Forest, 170
Pegwell Bay, 270
Pelagic trips, 25
Pendeen, 21, 143
Pendinas, 94
Pendle Hill, 117
Pennington Flash, 165, 221
Penzance, 124
Pig's Nose Valley, 254
Pitlochry, 93
Pitsea Mars, 219
Pitsford Reservoir, 15, 154
Pitts Wood Inclosure, 249
Plym Estuary, **138**, 146
Plymouth, **138**
Polgigga, 21
Poole Harbour, 33, 112
Porthgwarra, 21, 132, **278**
Portland Bill, 124, 174, **207**, 275, **276**
Portland Harbour, 11, 19
Portsmouth, 124
Potter Heigham, 202
Prawle Point, 172, **254**, 277
Puddletown, 180
Pulborough Brooks, **51**

Quantock Hills, 171, 180

Radipole, 220
Radipole Lake, 137, 201, **276**
Rainham Marsh, 186
Raleigh Common, 171
Ramsgate Cemetery, 164
Rattray Head, 260
Reclamation Pond, 263
Redcar, 264
Rhayader, 69
Rhinefield Inclosure, 249
Rhosneigr, 157
Rimac, 238
River Lynher, 32, 33
River Plym, 155
River Taw, 33
Rockland Broad, 199
Rockley, 245
Roseisle Forest, 225
Ross Back Sands, 59
Rothiemurchus, 96
Round Woods, 248
Roydon Common, 72, 83, 168
Royston, **97**
Rubha Beag, 12
Ruddon's Point, 11, 120
Rutland Water, 15, 53, 82
Rye Meads, 186

Saito Outcrop, 20
Saligo Bay, 231
Saltholme Pools, 250, **263**
Salthouse Heath, 168, 195
Sandsfoot Castle, 11, 19
Sandwich Bay, 173, **270**
Santon Downham, 181
Scalby Mills, 122
Scaling Dam Reservoir, 71
Scammonden Reservoir, 238
Scarborough, 122
Scillonian, 25
Scilly, 279, **281**
Scone Palace, 244
Seaford, 131
Seaforth, 27, 137, 142, 153
Seahouses, 156
Seaton Carew, **263**
Seaton Common, 250
Seaton Sluice, 23
Seaton Snook, **263**
Sedgemoor, 36, 197
Selsey, 131, **273**
Selsey Bill, 155, 237
Shapwick Heath, 171, 197
Shell Ness, 239
Sheringham, 184
Shetland, 127, 258
Sidlesham Ferry, 273
Silverdale, 87, 244
Sizergh Castle, 244
Skybus, 280
Slapton Ley, 173, **201**, 254
Slimbridge, 35, 41, 89
Somerset Levels, 36
Souter Point, 263
South Downs, 119
South Gare, **264**
South Stack, 88, 231
Southerness Point, 121
Southsea Castle, 124
Speyside, 80

Sprotborough Flash, 175
Spurn Point, 173, **267**
St Abb's, **261**
St Abb's Head, 121, 172
St Austell Bay, 14
St Bees Head, 160
St David's Peninsula, 73
St Ives Island, 22, **128**, 143
St Margaret's Bay, 204
Stainborough, 245
Staines Reservoirs, 19, 142
Stanney Woods, 175
Stanton Harcourt, 53
Stapleton Wood, 175
Start Point, 173
Steppingly Wood, **104**
Stewartby Lake, 155
Stodmarsh, 73, 84, 85, 186, 202, 220
Stoke Water Meadows, 186
Stour Valley, 202
Stradey Woods, 179
Strines, 76
Strumble Head, 129, 232
Strumpshaw Fen, 199
Stubb Mill, 108
Studland, 209
Studland Bay, 19
Studley Royal Park, 245
Sunbury-on-Thames, **163**
Swale, 239
Swallow Moss, **94**
Swanton Novers, 65
Swell Wood, 197
Swillington Ings, 154
Swithland Reservoir, 176
Symonds Yat, 88

Talybont Reservoir, 192
Taw Estuary, 33
Tayport Harbour, 58
Teesmouth, 238, **263**
Teifi Marshes, 200
Teign Estuary, 147
Temple Wood, 194
Thames Estuary, **148**
The Dell, 206, 212
The Hirsel, 244
The Naze, 239
The Patch, 130
Thetford Forest, 168
Thorne Moors, 167
Thorney Deeps, 31, 200
Thorney Island, 89
Thursley Common, 86, 182, 208, **230**
Tiddenfoot CP, 222
Tilehouse, 62
Timsbury GP, 54
Titchfield Haven, **200**
Titchwell, **109**, 218
Tollesbury, 111, 239
Tomatin, 87
Tor Bay, 17, 19
Torquay, 124
Tortworth Lake, 49
Totley Moor, 118
Treaddur Bay, 122
Tregaron Bog, **69**
Tregonetha Downs, 73, 84
Trimley Marshes, 110
Tring Reservoirs, 85
Tunbury Woods, 248
Turnberry Point, 12
Tyninghame CP, 250

Uists, 107, **132**
Unst, 20
Upper Derwent Valley, 76
Upton Country Park, 180
Uwchmynydd Headland, 232

Venn Ottery Common, 171
Veryan Bay, 10
Vyrnwy Confluence, 190

Wacker Quay, 32
Waddock Farm, 187
Walberswick, 219
Walberswick NNR, 202
Walderslade, 248
Wareham Forest, 171, 180, **182**
Warham Greens, 72
Warleigh Marsh, 187
Wat Tyler Country Park, 219
Wats Ness, 133
Waxham, 38, 108
Wayland Wood, **100**
Weeting Heath, 113, **114**, **181**
Weir Wood Reservoir, 82
Wells Woods, 206
Welney, 35, 38
Wembury, **255**
Wendover Woods, 215
Wentwood Forest, 167
Wepre Wood, 179
West Freugh Airfield, 71
West Sedgemoor, 197
Westdean Woods, 223
Western Approaches, 25
Westhay Reserve, 201
Westleton Heath, 169, 229
Westwood Marshes, 72
Wet Moor, 36
Weybourne, 184
Wharncliffe Crags, 167
Wheldrake, 38
Whisby Pits, 194
Whitburn, 23, 263
Whitcliffe Common, 246
Wicken Fen, 72
Wilson's Triangle, 25
Wilstone Reservoir, 85
Windsor Great Park, 248
Windy Corner, 76, 88
Wintersett Reservoir, 154
Winterton Dunes, 174
Wolves Wood, 195
Wood of Cree, 190
Woodbury Common, 209
Woodwalton Fen, 196
Woodwell, 244
Wooler, 190
Woolmer Forest, 170
Wraysbury, 163
Wraysbury GP, 62
Wykeham Forest, 65

Yare Valley , **39**
Yarner Wood, 180
Ynys-hir, 43, 179
Ythan Estuary, 58, 252